Embedded C Programming and the Atmel AVR, 2e

THOMSON
DELMAR LEARNING

Australia Brazil Canada Mexico Singapore Spain United Kingdom United States

Embedded C Programming and the Atmel AVR, 2e

RICHARD BARNETT
LARRY O'CULL
SARAH COX

Embedded C Programming and the Atmel AVR, 2e

Richard Barnett, Larry O'Cull, and Sarah Cox

Vice President, Technology and Trades ABU:
David Garza

Director of Learning Solutions:
Sandy Clark

Senior Acquisitions Editor:
Stephen Helba

Senior Product Manager:
Michelle Cannistraci

Marketing Director:
Deborah S. Yarnell

Channel Manager:
Dennis Williams

Marketing Coordinator:
Stacey Wiktorek

Production Director:
Mary Ellen Black

Senior Production Manager:
Larry Main

Production Coordinator:
Benj Gleeksman

Art/Design Coordinator:
Francis Hogan

Printed in Canada
2 3 4 5 XX 06

For more information contact Thomson Delmar Learning
Executive Woods
5 Maxwell Drive, PO Box 8007,
Clifton Park, NY 12065-8007
Or find us on the World Wide Web at
www.delmarlearning.com

Library of Congress Cataloging-in-Publication Data:
Barnett, Richard H.
Embedded C programming and the Atmel AVR/Richard Barnett, Larry O'Cull, Sarah Cox.
p. cm.
Includes bibliographical references and index.
ISBN 1-4180-3959-4 (alk. paper)
1. C (Computer program language) 2. Embedded computer systems—Design and construction. 3. RISC microprocessors. I. O'Cull, Larry. II. Cox, Sarah (Sarah Alison) III. Title.

QA76.73.C15B378 2006
005.13'3—dc22
2006007153

NOTICE TO THE READER

Publisher does not warrant or guarantee any of the products described herein or perform any independent analysis in connection with any of the product information contained herein. Publisher does not assume, and expressly disclaims, any obligation to obtain and include information other than that provided to it by the manufacturer.

The reader is expressly warned to consider and adopt all safety precautions that might be indicated by the activities herein and to avoid all potential hazards. By following the instructions contained herein, the reader willingly assumes all risks in connection with such instructions.

The publisher makes no representation or warranties of any kind, including but not limited to, the warranties of fitness for particular purpose or merchantability, nor are any such representations implied with respect to the material set forth herein, and the publisher takes no responsibility with respect to such material. The publisher shall not be liable for any special, consequential, or exemplary damages resulting, in whole or part, from the readers' use of, or reliance upon, this material.

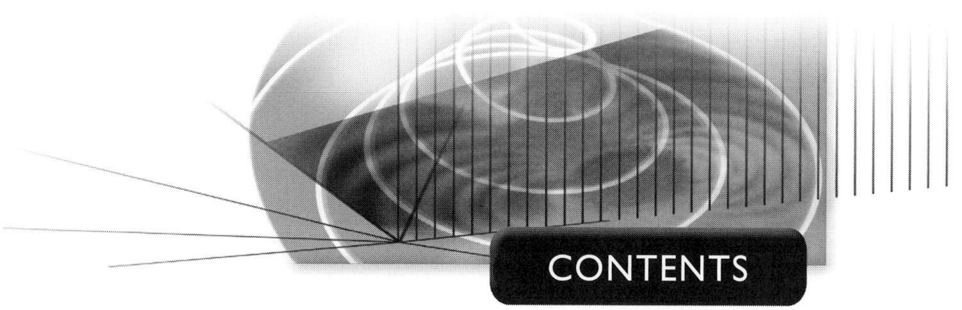

CONTENTS

CHAPTER 5 PROJECT DEVELOPMENT

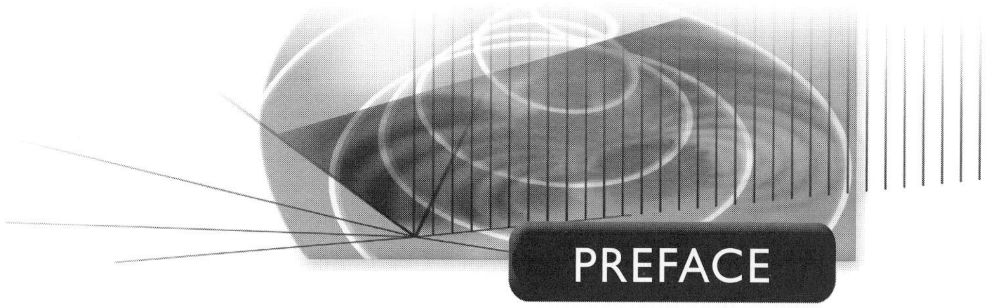

PREFACE

This text is designed both to teach C language programming as it applies to embedded microcontrollers and to provide knowledge in the application of the Atmel® family of AVR® RISC microcontrollers.

INTENDED AUDIENCE

This book is designed to serve two diverse audiences:

1. Students in Electrical and Computer Engineering, Electronic Engineering, Electrical Engineering Technology, and Computer Engineering Technology curricula. Two scenarios for students fit the book very well.

   1.1 Beginning students who have not yet had a C programming course: The book serves a two-semester or four-quarter sequence in which students learn C language programming, learn how to apply C to embedded microcontroller designs, and advance to the more sophisticated embedded applications. In this instance, the programs can all be run on an embedded microcontroller with very little hardware knowledge required. After Chapter 1, "Embedded C Language Tutorial," is completed, it will serve as a programming reference for the balance of the courses.

   1.2 Students who have already taken a C programming course can use the book for a one-semester or two-quarter course in embedded microcontrollers. In this instance, the students study only those portions of Chapter 1 that relate to programming for the embedded environment and move quickly to the advanced hardware concepts. Chapter 1 is organized (as are all the chapters in the book) to provide a usable reference for information needed in other courses.

2. Practicing engineers, technologists, and technicians who want to add a new microcontroller to their areas of expertise: Chapter 1 can be used as needed (depending on the user's level of programming experience) either to learn needed concepts or as a reference. The chapters about the Atmel AVR microcontroller hardware will lead such an individual through the steps of learning a new microcontroller and serve as a reference for future projects.

PREREQUISITES

Some knowledge of digital systems, number systems, and logic design is required. Preliminary versions of Chapter 1, "Embedded C Language Tutorial," have been used successfully in a fundamental microcontrollers course (sophomore-level class—no prerequisite programming) following two semesters of basic digital logic courses. It has also proven to be an excellent text for an advanced microcontrollers elective course. In many cases, the students have elected to keep the book and use it as a reference through their senior project design courses and have taken it with them into industry as a useful reference.

ORGANIZATION

The text is organized in logical topic units so that instructors can either follow the text organization, starting with the C language and progressing through the AVR hardware and into more advanced topics, or can choose the order of the topics to fit their particular needs. Topics are kept separate and identified for easy selection. The chapter exercises and laboratory exercises are also separated by topic to make it easy to select those that apply in any particular instance.

CHAPTER CONTENTS SUMMARY

Chapter 1, Embedded C language Tutorial

The C language is covered in detail in a step-by-step method as it applies to programming embedded microcontrollers. One or more example programs accompany each programming concept to illustrate its use. At the conclusion of the chapter, students are able to create C language programs to solve problems.

Chapter 2, The Atmel RISC Processors

The AVR RISC processors are covered from basic architecture through use of all of the standard peripheral devices included in the microcontrollers. Example programs are used to demonstrate common uses for each of the peripherals. Upon completion of Chapters 1 and 2, students are able to apply AVR RISC processors to solve problems.

Chapter 3, Standard I/O and Preprocessor Functions

Chapter 3 introduces students to the built-in functions available in C and to their use. Again, example programs are used to illustrate how to use the built-in functions. Finishing Chapter 3 prepares students to use the built-in functions to speed their programming and efforts at problem solving.

Chapter 4, The CodeVisionAVR C Compiler and IDE

This chapter is the handbook for using the CodeVisionAVR compiler and its accompanying integrated development environment (IDE). Students can learn to use the CodeVisionAVR and its IDE effectively to create and debug C programs.

Chapter 5, Project Development

This chapter focuses on the orderly development of a project using microcontrollers. A wireless indoor/outdoor weather station is developed in its entirety to illustrate the process. Students learn to efficiently develop projects to maximize their successes.

Appendices:

Appendix A, Library Functions Reference. A complete reference to the built-in library functions available at the time of publication.

Appendix B, Getting Started with CodeVisionAVR and the STK500. This is the quick-start guide to CodeVisionAVR when used with the Atmel STK500.

Appendix C, Programming the AVR Microcontrollers. This is a guide to actually programming the FLASH memory area of the AVR devices, so that students can understand the programming function.

Appendix D, Installing and Using TheCableAVR, This users guide covers the installation and use of TheCableAVR hardware and software.

Appendix E, The MegaAVR-DEV Development Board

Appendix F, ASCII Character Table.

Appendix G, an assembly code instruction summary for use with the assembly code programming examples.

Appendix H, Answers to Selected Exercises

Appendix I, Fast Start Guide. This guide is a quick reference to installing the software, connecting the hardware, and getting a simple programming up and going and a MegaAVR-DEV Development Board.

RATIONALE

The advancing technology surrounding microcontrollers continues to provide amazingly greater amounts of functionality and speed. These increases have led to the almost universal use of high-level languages such as C to program even time-critical tasks that used to require assembly language programs to accomplish. Simultaneously, microcontrollers have become easier and easier to apply, making them excellent vehicles for educational use. Many schools have adopted microcontroller vehicles as target devices for their courses. And the price of microcontroller development boards has dropped to the level where a number of schools have the students buy the board as a part of their "parts kit" so that all students have their own development board. Some of these courses require C programming as a prerequisite, and others teach C language programming and the application of embedded microcontrollers in an integrated manner.

This book is an answer to the need for a text that is usable in courses with and without a C language prerequisite course and that can be a useful reference in later coursework. The CD-ROM included with this book contains the compiler and other software so that students with their own development boards have everything they need to work outside of class as well as in the school labs.

HARDWARE USED

Most of the programming application examples in this text were developed using an AVR evaluation board provided by Progressive Resources, LLC (refer to Appendix E for specifics). This board is particularly well suited for educational use and is a good general-purpose development board. However, the Atmel AVR microcontrollers are very easy to use and can be run perfectly well by simply plugging them into a prototype board, adding the oscillator crystal, along with two capacitors, and connecting four wires for programming. Students have been very successful with either method.

The ATMega8535, ATMega8515, ATMega16, and ATMega48 microcontrollers have been used to work out the examples for the text. One of the major advantages of the AVR microcontrollers is that they are parallel in their architecture and the programming approach for the devices. This means that the examples shown will work on virtually any Atmel AVR microcontroller, provided that it contains the peripherals and other resources to do the job—it is not necessary to make changes to use the code for other members of the AVR family. Consequently, the text is useful with other members of the AVR family as well.

The more common peripherals are covered by this text, and the code can be used as a template to apply to more exotic peripherals in some of the other AVR family members.

CD-ROM CONTENTS AND SOFTWARE USED IN THE TEXT

The software used with the text includes the Atmel AVR Studio (free from http://www.atmel.com/) and the CodeVisionAVR C compiler from HP InfoTech S.R.L. (evaluation version free from http://www.hpinfotech.ro/). All of the programs in the text can be compiled and run using the included evaluation version of CodeVisionVAVR.

The CD-ROM included with this book contains the source code for all of the software examples from the text. These can be used as references or as starting points for specific assignments. Also included on the CD-ROM is the CodeVisionAVR evaluation version compiler that was current at the time of publication. Also refer to the Web site information to obtain the latest version of the compiler. More information about purchasing the full version may be found at http://www.prllc.com/.

ACKNOWLEDGEMENTS

The material contained in this text is not only a compilation of years of experience but of information available from Atmel Corporation, Forest Technology Systems, Inc., Progressive Resources LLC, and HP InfoTech S.R.L.

Pavel Haiduc
HP InfoTech S.R.L
Str. Liviu Rebreanu 13A
Bl.N20 Sc.B Ap.58
Sector 3
Bucharest 746311
ROMANIA
http://www.hpinfotech.ro/

Forest Technology Systems, Inc.
Suite F, 4131 Mitchell Way
Bellingham, WA USA 98226
FTS Forest Technology Systems Ltd.
113 - 2924 Jacklin Road
Victoria, BC Canada V9B 3Y5
Phone: 1-250-478-5561 or
1-800-548-4264
Fax: 1-250-478-8579 or 1-800-905-7004
http://www.ftsinc.com/

Atmel Corporation
2325 Orchard Parkway
San Jose, CA 95131
1-408-441-0311
http://www.atmel.com

Progressive Resources LLC [Priio]
4105 Vincennes Road
Indianapolis, IN 46468
1-317-471-1577
http://www.prllc.com/ and
http://www.priio.com

AUTHOR-SPECIFIC ACKNOWLEDGEMENTS

The support of my family, Gay, Laura, and April, has made this book both a pleasure to work on and a joy to complete. It is also important to acknowledge the motivation supplied by Larry O'Cull and the fantastic pleasure of working with Larry and Sarah on this project.

Richard H. Barnett, PE, Ph.D.
May 2006

It was a great pleasure to work on this project with Dr. Barnett, teacher and mentor, and Sarah Cox, partner and co-author. They kept this project exciting. This work would not have been possible without the patience and support of my wife, Anna, and children, James, Heather, Max, and Alan, who have been willing to give up some things now to build bigger and better futures.

Larry D. O'Cull
May 2006

This book has been a challenging and exciting endeavor. I have a tremendous amount of respect for Larry O'Cull and Dr. Barnett and have considered it a great privilege to work with them. I must specifically thank Larry for having the vision for this project. I also want to thank my husband, Greg, daughter Meredith, and son David, for their support throughout the entire process.

Sarah A. Cox
May 2006

AUTHORS

This text is definitely a highly collaborative work by the three authors. Each section was largely written by one author and then critically reviewed by the other two, who rewrote chunks as needed. It is not possible to delineate who is responsible for any particular part of the book. The authors:

Richard H. Barnett, PE, Ph.D.
Professor of Electrical Engineering Technology
Purdue University

Dr. Barnett has been instructing in the area of embedded microcontrollers for the past eighteen years, starting with the Intel 8085, progressing to several members of the 8051 family of embedded microcontrollers, and now teaching Advanced Embedded Microcontrollers using the Atmel AVR devices. During his summers and two sabbatical periods, he worked extensively with multiple-processor embedded systems, applying them in a variety of control-oriented applications. In addition, he consults actively in the field. Prior to his tenure at Purdue University, he spent ten years as an engineer in the aerospace electronics industry.

In terms of teaching, Dr. Barnett has won a number of teaching awards, including the Charles B. Murphy Award as one of the best teachers at Purdue University. He is also listed on Purdue University's Book of Great Teachers, a list of the 225 most influential teachers over Purdue's entire history. This is his second text.

He may be contacted with suggestions/comments at Purdue University at 765-494-7497 or by email at rbarnett@purdue.edu.

Larry D. O'Cull
Senior Operating Member
Progressive Resources LLC

Mr. O'Cull received a B.S. degree from the School of Electrical Engineering Technology at Purdue University. His career path started in the design of software and control systems for CNC (computer numeric controlled) machine tools. From there he moved to other opportunities in electronics engineering and software development for vision systems, LASER-robotic machine tools, medical diagnostic equipment, and commercial and consumer products, and he has been listed as inventor/co-inventor on numerous patents.

Mr. O'Cull started Progressive Resources in 1995 after several years of working in Electrical and Software Engineering and Engineering Management. Progressive Resources LLC, now Priio (www.priio.com), specializes in innovative commercial, industrial, and consumer product development. Priio is an Atmel AVR consultant member.

He may be contacted with suggestions/comments by email at ldocull@priio.com.

Sarah A. Cox
Director of Software Development
Progressive Resources LLC

Ms. Cox has a Bachelor of Science degree in both Computer and Electrical Engineering from Purdue University, where she focused her studies on software design.

After a short career for a large consulting firm working on database management systems, she was lured away by the fast pace and the infinite possibilities of microprocessor designs. She worked independently on various pieces of medical test equipment before becoming a partner at Progressive Resources LLC.

At Progressive Resources, Ms. Cox has developed software for projects ranging from small consumer products to industrial products and test equipment. These projects have spanned several fields, among them automotive, medical, entertainment, child development, public safety/education, sound and image compression, and construction. Along the way she has been listed as co-inventor on numerous patent applications. She has also written the software for in-system programming and development tools targeting Atmel AVR processors.

She may be contacted with suggestions/comments by email at scox@priio.com.

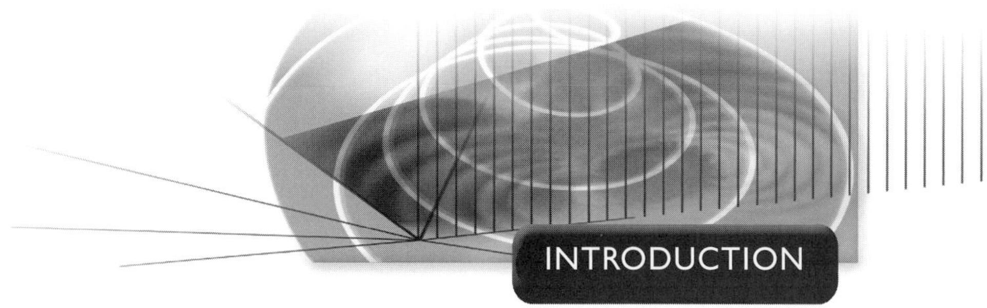

INTRODUCTION

An embedded microcontroller is a microcomputer that contains most of its peripherals and required memory inside a single integrated circuit along with the CPU. It is in actuality "a microcomputer on a chip."

Embedded microcontrollers have been in use for more than three decades. The Intel 8051 series was one of the first microcontrollers to integrate the memory, I/O, arithmetic logic unit (ALU), program ROM, as well as some other peripherals, all into one very neat little package. These processors are still being designed into new products. Other companies that followed Intel's lead into the embedded microcontroller arena were General Instruments, National Semiconductor, Motorola, Philips/Signetics, Zilog, AMD, Hitachi, Toshiba, and Microchip, among others.

In recent years, Atmel has become a world leader in the development of FLASH memory technology. FLASH technology is a non-volatile yet reprogrammable memory often used in products such as digital cameras, portable audio devices, and PC motherboards. This memory technology really pushed Atmel ahead in the microcontroller industry by providing an in-system programmable solution. This, coupled with the development of the AVR RISC (reduced instruction set computing) core architecture, provides for very low-cost yet amazing solutions.

The next great step in this high-tech evolution was the implementation of high-level language compilers that are targeted specifically for use with these new microprocessors. The code generation and optimization of the compilers is quite impressive. The C programming language, with its "make your own rules" structure, lends itself to this application by its ability to be tailored to a particular target system, while still allowing for code to be portable to other systems. The key benefit of a language like this is that it creates pools of intellectual property that can be drawn from again and again. This lowers development costs on an on-going basis by shortening the development cycle with each subsequent design.

One of the finest C language tools developed to date is CodeVisionAVR. Written by Pavel Haiduc of HP InfoTech S.R.L., this completely *integrated development environment* (IDE) allows editing, compiling, part programming, and debugging to be performed from one PC Windows application.

The motivation that has led to the development of this book is the growing popularity of the AVR and other RISC microcontrollers, the ever increasing level of integration (more on a chip and fewer chips on a circuit board), and the need for "tuned thinking" when it comes to developing products utilizing this type of technology. You may have experience writing C for a PC, or assembler for a microcontroller. But when it comes to writing C for an embedded microcontroller, the approach must be modified to get the desired final results: small, efficient, reliable, reusable code. This text is designed to provide a good baseline for the beginner as well as a helpful reference tool for those experienced in embedded microcontroller design.

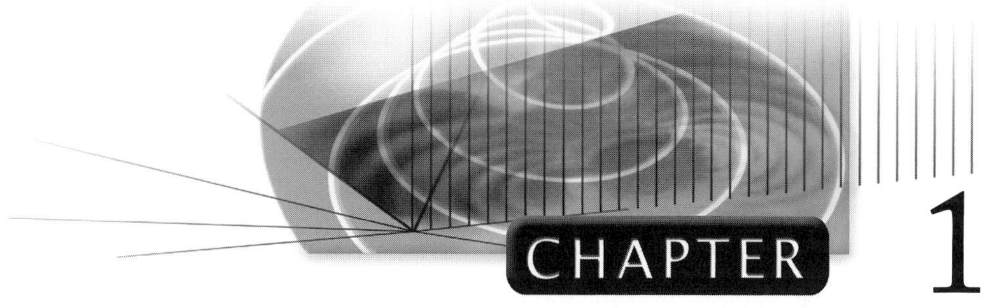

Embedded C Language Tutorial

1.1 OBJECTIVES

At the conclusion of this chapter, you should be able to

- Define, describe, and identify variable and constant types, their scope, and uses
- Construct variable and constant declarations for all sizes of numeric data and for strings
- Apply enumerations to variable declarations
- Assign values to variables and constants by means of the assignment operator
- Evaluate the results of all of the operators used in C
- Explain the results that each of the control statements has on program flow
- Create functions that are composed of variables, operators, and control statements to complete tasks
- Apply pointers, arrays, structures, and unions as function variables
- Create C programs that complete tasks using the concepts in this chapter

1.2 INTRODUCTION

This chapter provides a baseline course in the C programming language as it applies to embedded microcontroller applications. The chapter includes extensions to the C language that are a part of the CodeVisionAVR® C language. You will go from beginning concepts through writing complete programs, with examples that can be implemented on a microcontroller to further reinforce the material.

The information is presented somewhat in the order that it is needed by a programmer:

- Declaring variables and constants
- Simple I/O, so that programs can make use of the parallel ports of the microcontroller

- Assigning values to the variables and constants, and doing arithmetic operations with the variables
- C constructs and control statements to form complete C programs

The final sections cover the more advanced topics, such as pointers, arrays, structures, and unions, and their use in C programs. Advanced concepts such as real-time programming and interrupts complete the chapter.

1.3 BEGINNING CONCEPTS

Writing a C program is, in a sense, like building a brick house: A foundation is laid, sand and cement are used to make bricks, these bricks are arranged in rows to make a course of blocks, and the courses are then stacked to create a building. In an embedded C program, sets of instructions are put together to form functions; these functions are then treated as higher-level operations, which are then combined to form a program.

Every C language program must have at least one function, namely *main()*. The function *main()* is the foundation of a C language program, and it is the starting point when the program code is executed. All functions are invoked by *main()* either directly or indirectly. Although functions can be complete and self-contained, variables and parameters can be used to cement these functions together.

The function *main()* is considered to be the lowest-level task, since it is the first function called from the system starting the program. In many cases, *main()* will contain only a few statements that do nothing more than initialize and steer the operation of the program from one function to another.

An embedded C program in its simplest form appears as follows:

```
void main()
{
        while(1)     // do forever..
            ;
}
```

The program shown above will compile and operate perfectly, but you will not know that for sure, because there is no indication of activity of any sort. We can embellish the program such that you can actually see life, review its functionality, and begin to study the syntactical elements of the language:

```
#include <stdio.h>

void main()
{
        printf("HELLO WORLD"); /* the classic C test program.. */
        while(1)           // do forever..
            ;
}
```

This program will print the words "HELLO WORLD" to the standard output, which is most likely a serial port. The microcontroller will sit and wait, forever or until the microcontroller is reset. This demonstrates one of the major differences between a personal computer program and a program that is designed for an embedded microcontroller: namely, that the embedded microcontroller applications contain an infinite loop. Personal computers have an operating system, and once a program has executed, it returns control to the operating system of the computer. An embedded microcontroller, however, does not have an operating system and cannot be allowed to fall out of the program at any time. Hence, every embedded microcontroller application has an infinite loop built into it somewhere, such as the *while(1)* in the example above. This prevents the program from running out of things to do and doing random things that may be undesirable. The **while** construct will be explained in a later section.

The example program also provides an instance of the first of the common preprocessor compiler directives. **#include** tells the compiler to include a file called stdio.h as a part of this program. The function *printf()* is provided for in an external library, and it is made available to us because its definition is located in the stdio.h file. As we continue, these concepts will come together quickly.

These are some of the elements to take note of in the previous examples:

;	A semicolon is used to indicate the end of an expression. An expression in its simplest form is a semicolon alone.
{ }	Braces "{}" are used to delineate the beginning and the end of the function's contents. Braces are also used to indicate when a series of statements is to be treated as a single block.
"text"	Double quotes are used to mark the beginning and the end of a text string.
// or /* ... */	Slash-slash or slash-star/star-slash are used as comment delimiters.

Comments are just that, a programmer's notes. Comments are critical to the readability of a program. This is true whether the program is to be read by others or by the original programmer at a later time. The comments shown in this text are used to explain the function of each line of the code in the example. The comments should always explain the actual *function* of the line in the program, and not just echo the specific instructions that are used on the line.

The traditional comment delimiters are the slash-star (/*), star-slash (*/) configuration. Slash-star is used to create block comments. Once a slash-star (/*) is encountered, the compiler will ignore the subsequent text, even if it encompasses multiple lines, until a star-slash (*/) is encountered. Refer to the first line of the *main()* function in the previous program example for an example of these delimiters.

The slash-slash (//) delimiter, on the other hand, will cause the compiler to ignore the comment text only until the end of the line is reached. These are used in the second line of the *main()* function of the example program.

As we move into the details, a few syntactical rules and some basic terminology should be remembered:

- An identifier is a variable or function name made up of a letter or underscore (_), followed by a sequence of letters and/or digits and/or underscores.
- Identifiers are case sensitive.
- Identifiers can be any length, but some compilers may recognize only a limited number of characters, such as the first thirty-two. So beware!
- Particular words have special meaning to the compiler and are considered reserved words. These reserved words must be entered in lowercase and should never be used as identifiers. Table 1–1 lists reserved words.

auto	defined	float	long	static	while
break	do	for	register	struct	
bit	double	funcused	return	switch	
case	eeprom	goto	short	typedef	
char	else	if	signed	union	
const	enum	inline	sizeof	unsigned	
continue	extern	int	sfrb	void	
default	flash	interrupt	sfrw	volatile	

Table 1–1 *Reserved Word List*

- Since C is a free-form language, "white space" is ignored unless delineated by quotes. This includes blank (space), tab, and new line (carriage return and/or line feed).

1.4 VARIABLES AND CONSTANTS

It is time to look at data stored in the form of variables and constants. Variables, as in algebra, are values that can be changed. Constants are fixed. Variables and constants come in many forms and sizes; they are stored in the program's memory in a variety of forms that will be examined as we go along.

1.4.1 VARIABLE TYPES

A variable is declared by the reserved word indicating its type and size followed by an identifier:

```
unsigned char Peabody;
int dogs, cats;
long int total_dogs_and_cats;
```

Variables and constants are stored in the limited memory of the microcontroller, and the compiler needs to know how much memory to set aside for each variable without wasting memory space unnecessarily. Consequently, a programmer must declare the variables, specifying both the size of the variable and the type of the variable. Table 1–2 lists variable types and their associated sizes.

Type	Size (Bits)	Range
bit	1	0, 1
char	8	–128 to 127
unsigned char	8	0 to 255
signed char	8	–128 to 127
int	16	–32768 to 32767
short int	16	–32768 to 32767
unsigned int	16	0 to 65535
signed int	16	–32768 to 32767
long int	32	–2147483648 to 2147483647
unsigned long int	32	0 to 4294967295
signed long int	32	–2147483648 to 2147483647
float	32	±1.175e-38 to ±3.402e38
double	32	±1.175e-38 to ±3.402e38

Table 1–2 *Variable Types and Sizes*

1.4.2 VARIABLE SCOPE

As noted above, constants and variables must be declared prior to their use. The *scope* of a variable is its accessibility within the program. A variable can be declared to have either *local* or *global* scope.

Local Variables

Local variables are memory spaces allocated by the function when the function is entered, typically on the program stack or a compiler-created heap space. These variables are not accessible from other functions, which means their scope is limited to the functions in which they are declared. The local variable declaration can be used in multiple functions without conflict, since the compiler sees each of these variables as being part of that function only.

Global Variables

A global or external variable is a memory space that is allocated by the compiler and can be accessed by all the functions in a program (unlimited scope). A global variable can be modified by any function and will retain its value to be used by other functions.

Global variables are typically cleared (set to zero) when *main()* is started. This operation is most often performed by startup code generated by the compiler, invisible to the programmer.

An example piece of code is shown below to demonstrate the scope of variables:

```
unsigned char globey;    //a global variable

void function_z (void)   //this is a function called from main()
{
    unsigned int tween;           //a local variable

    tween = 12;                   //OK because tween is local
    globey = 47;                  //OK because globey is global
    main_loc = 12;        // This line will generate an error
                          // because main_loc is local to main.
}

void main()
{
    unsigned char main_loc;   //a variable local to main()
    globey = 34;                  //Ok because globey is a global
    tween = 12;           //will cause an error - tween is local
                            // to function_z

    while(1)              // do forever..
        ;
}
```

When variables are used within a function, if a local variable has the same name as a global variable, the local will be used by the function. The value of the global variable, in this case, will be inaccessible to the function and will remain untouched.

1.4.3 CONSTANTS

As described earlier in the text, constants are fixed values—they may not be modified as the program executes. Constants in many cases are part of the compiled program itself, located in read-only memory (ROM), rather than an allocated area of changeable random access memory (RAM). In the assignment

```
x = 3 + y;
```

the number *3* is a constant and will be coded directly into the addition operation by the compiler. Constants can also be in the form of characters or a string of text:

```
printf("hello world");
            // The text "hello world" is placed in program memory
            // and never changes.
x = 'B';    // The letter 'B' is permanently set
            // in program memory.
```

You can also declare a constant by using the reserved word **const** and indicating its type and size. An identifier and a value are required to complete the declaration:

```
const char c = 57;
```

Identifying a variable as a constant will cause that variable to be stored in the program code space rather than in the limited variable storage space in RAM. This helps preserve the limited RAM space.

Numeric Constants

Numeric constants can be declared in many ways by indicating their numeric base and making the program more readable. Integer or long integer constants may be written in

- Decimal form without a prefix (such as 1234)
- Binary form with **0b** prefix (such as 0b101001)
- Hexadecimal form with **0x** prefix (such as 0xff)
- Octal form with **0** prefix (such as 0777)

There are also modifiers to better define the intended size and use of the constant:

- Unsigned integer constants can have the suffix **U** (such as 10000U).
- Long integer constants can have the suffix **L** (such as 99L).
- Unsigned long integer constants can have the suffix **UL** (such as 99UL).
- Floating point constants can have the suffix **F** (such as 1.234F).
- Character constants must be enclosed in single quotation marks, **'a'** or **'A'**.

Character Constants

Character constants can be printable (like 0–9 and A–Z) or non-printable characters (such as new line, carriage return, or tab). Printable character constants may be enclosed in single quotation marks (such as 'a'). A backslash followed by the octal or hexadecimal value in single quotes can also represent character constants:

```
't'   can be represented by   '\164'       (octal)
                          or
't'   can be represented by   '\x74'       (hexadecimal)
```

Table 1–3 lists some of the "canned" non-printable characters that are recognized by the C language.

Backslash (\) and single quote (') characters themselves must be preceded by a backslash to avoid confusing the compiler. For instance, '\'' is a single quote character and '\\' is a backslash. BEL is the bell character and will make a sound when it is received by a computer terminal or terminal emulator.

1.4.4 ENUMERATIONS AND DEFINITIONS

Readability in C programs is very important. Enumerations and definitions are provided so that the programmer can replace numbers with names or other more meaningful phrases.

Character	Representation	Equivalent Hex Value
BEL	'\a'	'\x07'
Backspace	'\b'	'\x08'
TAB	'\t'	'\x09'
LF (new line)	'\n'	'\x0a'
VT	'\v'	'\x0b'
FF	'\f'	'\x0c'
CR	'\r'	'\x0d'

Table 1–3 *Non-printable Character Notations*

Enumerations are listed constants. The reserved word **enum** is used to assign sequential integer constant values to a list of identifiers:

```
int num_val;                //declare an integer variable

//declare an enumeration
enum { zero_val, one_val, two_val, three_val );

num_val = two_val;          // the same as:  num_val = 2;
```

The name *zero_val* is assigned a constant value of 0, *one_val* of 1, *two_val* of 2, and so on. An initial value may be forced, as in

```
enum { start=10, next1, next2, end_val };
```

which will cause *start* to have a value of 10, and then each subsequent name will be one greater. *next1* is 11, *next2* is 12, and *end_val* is 13.

Enumerations are used to replace pure numbers, which the programmer would have to look up, with the words or phrases that help describe the number's use.

Definitions are used in a manner somewhat similar to the enumerations in that they will allow substitution of one text string for another. Observe the following example:

```
enum { red_led_on = 1, green_led_on, both_leds_on };
#define leds PORTA
.
.
PORTA = 0x1;    //means turn the red LED on
leds = red_led_on;  //means the same thing
```

The "*#define leds PORTA*" line causes the compiler to substitute the label *PORTA* wherever it encounters the word *leds*. Note that the **#define** line is not ended with a semicolon and may not have comments. The enumeration sets the value of *red_led_on* to 1, *green_led_on* to 2, and *both_leds_on* to 3. This might be used in a program to control the red and green LEDs where outputting 1 turns the red LED on, 2 turns the green LED on, etc. The point is that "*leds = red_led_on*" is much easier to understand in the program's context than "*PORTA = 0x1*".

#define is a preprocessor directive. Preprocessor directives are not actually part of the C language syntax, but they are accepted as such because of their use and familiarity. The preprocessor is a step separate from the actual compilation of a program, which happens before the actual compilation begins. More information on the preprocessor can be found in Chapter 3, "Standard I/O and Preprocessor Functions."

1.4.5 STORAGE CLASSES

Variables can be declared in three storage classes: **auto**, **static**, and **register**. Auto, or automatic, is the default class, meaning the reserved word **auto** is not necessary.

Automatic

An automatic class local variable is uninitialized when it is allocated, so it is up to the programmer to make sure that it contains valid data before it is used. This memory space is released when the function is exited, meaning the values will be lost and will not be valid if the function is reentered. An automatic class variable declaration would appear as follows:

```
auto int value_1;
      or
int value_1;         // This is the common, default form.
```

Static

A static local variable has only the scope of the function in which it is defined (it is not accessible from other functions), but it is allocated in global memory space. The static variable is initialized to zero the first time the function is entered, and it retains its value when the function is exited. This allows the variable to be current and valid each time the function is reentered.

```
static int value_2;
```

Register

A register local variable is similar to an automatic variable in that it is uninitialized and temporary. The difference is that the compiler will try to use an actual machine register in the microprocessor as the variable in order to reduce the number of machine cycles required to access the variable. There are very few registers available compared to the total memory in a typical machine. Therefore, this would be a class used sparingly and with the intent of speeding up a particular process.

```
register char value_3;
```

1.4.6 TYPE CASTING

There are times when a programmer might wish to temporarily force the type and size of the variable. Type casting allows the previously declared type to be overridden for the duration of the operation being performed. The cast, called out in parentheses, applies to the expression it precedes.

Given the following declarations and assignment,

```
int x;        // a signed, 16-bit integer (-32768 to +32767)
char y;       // a signed, 8-bit character (-128 to +127)
x = 12;       // x is an integer, (but its value will fit
              // into a character)
```

type cast operations on these variables could appear as

```
y = (char)x + 3;    // x is converted to a character and then
                    // 3 is added,
                    // and the value is then placed into y.
x = (int)y;         // y is extended up to an integer, and
                    // assigned to x.
```

Type casting is particularly important when arithmetic operations are performed with variables of different sizes. In many cases, the accuracy of the arithmetic will be dictated by the variable of the smallest type. Consider the following:

```
int z;              // declare z
int x = 150;        // declare and initialize x
char y = 63;        // declare and initialize y

z = (y * 10) + x;
```

As the compiler processes the right side of the equation, it will look at the size of *y* and make the assumption that *(y * 10)* is a character (8-bit) multiplication. The result placed on the stack will have exceeded the width of the storage location, one byte or a value of 255. This will truncate the value to 118 (0x76) instead of the correct value of 630 (0x276). In the next phase of the evaluation, the compiler would determine that the size of the operation is integer (16 bits) and 118 would be extended to an integer and then added to *x*. Finally, *z* would be assigned 268 . . . WRONG!!

Type casting should be used to control the assumptions. Writing the same arithmetic as

```
z = ((int)y * 10) + x;
```

will indicate to the compiler that *y* is to be treated as an integer (16 bits) for this operation. This will place the integer value of 630 onto the stack as a result of a 16-bit multiplication. The integer value *x* will then be added to the integer value on the stack to create the integer result of 780 (0x30C). *z* will be assigned the value 780.

C is a very flexible language. It will give you what you ask for. The assumption the compiler will make is that you, the programmer, know what you want. In the example above, if the value of *y* were 6 instead of 63, there would have been no errors. When writing expressions, you should always think in terms of the maximum values that could be given to the expression and what the resulting sums and products may be.

A good rule to follow: "When it doubt—cast it out." Always cast the variables unless you are sure you do not need to.

1.5 I/O OPERATIONS

Embedded microcontrollers must interact directly with other hardware. Therefore, many of their input and output operations are accomplished using the built-in parallel ports of the microcontroller.

Most C compilers provide a convenient method of interacting with the parallel ports through a library or header file that uses the **sfrb** and **sfrw** compiler commands to assign labels to each of the parallel ports and other I/O devices. These commands will be discussed in a later section, but the example below will serve to demonstrate the use of the parallel ports:

```
#include <MEGA8535.h>          // register definition header file for
                               // an Atmel ATMega8535

unsigned char z;               // declare z

void main(void)
{
    DDRB = 0xff;           // set all bits of port B for output

    while(1)
    {
       z = PINA;          // read the binary value on the
                          // port A pins (i.e., input from port A)
       PORTB = z + 1;     // write the binary value read from port A
                          // plus 1 to port B
    }
}
```

The example above shows the methods to both read and write a parallel port. *z* is declared as an unsigned character size variable because the port is an 8-bit port and an unsigned character variable will hold 8-bit data. Notice that the labels for the pins of port A and output port B are all capitalized because these labels must match the way the labels are defined in the included header file MEGA8535.h.

The DDRx register is used to determine which bits of port x (A, B, and so on depending on the processor) are to be used for output. Upon reset, all of the I/O ports default to input by the microcontroller writing a 0 into all the bits of the DDRx registers. The programmer then sets the DDRx bits depending on which bits are to be used for output. For example,

```
DDRB = 0xc3;    // set the upper 2 and lower 2 bits
                // of port B for output
```

This example configures the upper two bits and the lower two bits to be used as output bits.

1.6 OPERATORS AND EXPRESSIONS

An expression is a statement in which an operator links identifiers such that when evaluated, the result may be true, false, or numeric in value. Operators are symbols that indicate to the compiler which type of operation is to be performed using its surrounding identifiers. There are rules that apply to the precedence or order in which operations are performed. When you combine operators in a single expression, those rules of precedence must be applied to obtain the desired result.

1.6.1 ASSIGNMENT AND ARITHMETIC OPERATORS

Once variables have been declared, operations can be performed on them using the assignment operator, an equal sign (=). A value assigned to a variable can be a constant, a variable, or an expression. An expression in the C language is a combination of operands (identifiers) and operators. A typical assignment may appear as follows:

```
dog = 35;
val  = dog;
dog = dog + 0x35;
val  = (2 * (dog + 0172)) + 6;
y = (m * x) + b;
```

The compiled program arithmetically processes expressions just as you would process them by hand. Start from inside the parentheses and work outward from left to right: In the above expression *y = (m * x) + b*, the *m* and *x* would be multiplied together first and then added to *b*; finally, *y* would be assigned the resulting value. In addition to the parentheses, there is an inherent precedence to operators themselves. Multiplication and division are performed first, followed by addition and subtraction. Therefore, the statement

```
y = m * x + b;
```

is evaluated the same as

```
y = (m * x) + b;
```

It should be noted that the use of the parentheses improves the readability of the code.

Table 1–4 shows the typical arithmetic operators in order of precedence.

Operator	Symbol
Multiply	*
Divide	/
Modulo	%
Addition	+
Subtraction or Negation	–

Table 1–4 *Arithmetic Operators*

There are other types of operators besides arithmetic and assignment operators. These include bitwise, logical, relational, increment, decrement, compound assignment, and conditional.

Bitwise Operators

Bitwise operators perform functions that will affect the operand at the bit level.

These operators work on non–floating point operands: **char, int**, and **long.** Table 1–5 lists the bitwise operators in order of precedence.

<table>
<tr><td>Ones Complement</td><td>~</td></tr>
<tr><td>Left Shift</td><td><<</td></tr>
<tr><td>Right Shift</td><td>>></td></tr>
<tr><td>AND</td><td>&</td></tr>
<tr><td>Exclusive OR</td><td>^</td></tr>
<tr><td>OR (Inclusive OR)</td><td>|</td></tr>
</table>

Table 1–5 *Bitwise Operators*

The following is a description of each bitwise operator:

- The ones complement operator converts the bits within an operand to 1 if they were 0, and to 0 if they were 1.
- The left shift operator will shift the left operand to the left, in a binary fashion, the number of times specified by the right operand. In a left shift operation, zero is always shifted in replacing the lower bit positions that would have been "empty." Each shift left effectively multiplies the operand by 2.
- The right shift operator will shift the left operand to the right, in a binary fashion, the number of times specified by the right operand. Each right shift effectively performs a division by 2. When a right shift is performed, signed and unsigned variables are treated differently. The sign bit (the left or most significant bit) in a signed integer will be replicated. This sign extension allows a positive or negative number to be shifted right while maintaining the sign. When an unsigned variable is shifted right, zeros will always be shifted in from the left.
- The AND operator will result in a 1 at each bit position where both operands were 1.
- The exclusive OR operator will result in a 1 at each bit position where the operands differ (a 0 in one operand and a 1 in the other).
- The OR (inclusive OR) operator will result in a 1 at each bit position where either of the operands contained a 1.

Table 1–6 gives some examples of bitwise operations.

Assume that an unsigned character y = 0xC9.

Operation	Result
x = ~y;	x = 0x36
x = y << 3;	x = 0x48
x = y >> 4;	x = 0x0C
x = y & 0x3F;	x = 0x09
x = y ^ 1;	x = 0xC8
x = y | 0x10;	x = 0xD9

Table 1–6 *Examples of Bitwise Operations*

The AND and OR bitwise operators are useful in dealing with parallel ports. Observe the following example:

```
#include <MEGA8535.h>      // register definition file for
                           // an Atmel ATMEGA8535
unsigned char z;           // declare z

void main(void)
{
   DDRB = 0xff;                // set all bits of port B for output

   while(1)
   {
      z = PINA & 0x6;          // Read the binary value on the
                               // port A pins ANDed with 00000110.
      PORTB = PORTB | 0x60;    // Write the binary value from port B
                               // ORed with 01100000 back to port B.
      PORTB = PORTB & 0xfe;    // Write the binary value from port B
                               // ANDed with 0xfe to PORTB.
   }
}
```

The example above demonstrates *masking* and *bitwise port control.* Masking is the technique used to determine the value of certain bits of a binary value. In this case, the masking is accomplished by ANDing the unwanted bits with 0 ('*x*' AND 0 always results in 0) and the bits of interest with 1 ('*x*' AND 1 always results in '*x*'). In this way, all the bits except for the center two bits of the lower nibble are eliminated (set to 0).

Bitwise port control is a method of changing one or more bits of a parallel port without affecting the other bits of the port. The first *PORTB* line demonstrates using the OR operator to force two bits of the port high without affecting the other bits of the port. The second *PORTB* line shows how to force a bit of the port low by ANDing the bit with a 0.

1.6.2 LOGICAL AND RELATIONAL OPERATORS

Logical and relational operators are all binary operators but yield a result that is either TRUE or FALSE. TRUE is represented by a nonzero value, and FALSE by a zero value.

These operations are usually used in control statements to guide the flow of program execution.

Logical Operators

Table 1–7 shows the logical operators in order of precedence.

AND	&&
OR	||

Table 1–7 *Logical Operators*

These differ greatly from the bitwise operators in that they deal with the operands in a TRUE and FALSE sense. The AND logical operator yields a TRUE if both operands are TRUE, otherwise, a FALSE is the result. The OR logical operator yields a TRUE if either of the operators is TRUE. In the case of an OR, both operators must be FALSE in order for the result to be FALSE. To illustrate the difference:

```
Assume x = 5 and y = 2;

(x && y)     is TRUE, because both are non-zero.
(x & y)      is FALSE, because the pattern 101b and 010b
             ANDed bitwise are zero.
(x || y)     is TRUE, because either value is non-zero.
(x | y)      is TRUE, because the pattern 101b and 010b
             Inclusive-ORed bitwise is 111b (non-zero).
```

Relational Operators

Relational operators use comparison operations. As in the logical operators, the operands are evaluated left to right and a TRUE or FALSE result is generated. They effectively "ask" about the relationship of two expressions in order to gain a TRUE or FALSE reply.

```
"Is the left greater than the right?"
"Is the left less than or equal to the right?"
```

Table 1–8 shows the Relational operators.

Is Equal to	==
Is Not Equal to	!=
Less Than	<
Less Than or Equal to	<=
Greater Than	>
Greater Than or Equal to	>=

Table 1–8 *Relational Operators*

Examples are presented in Table 1–9.

Assume that x = 3 and y = 5.

Operation	Result
(x == y)	FALSE
(x != y)	TRUE
(x < y)	TRUE
(x <= y)	TRUE
(x > y)	FALSE
(x >= y)	FALSE

Table 1–9 *Examples of Relational Operations*

1.6.3 INCREMENT, DECREMENT, AND COMPOUND ASSIGNMENT

When the C language was developed, a great effort was made to keep things concise but clear. Some "shorthand" operators were built into the language to simplify the generation of statements and shorten the keystroke count during program development. These operations include the increment and decrement operators, as well as compound assignment operators.

Increment Operators

Increment operators allow for an identifier to be modified, in place, in a pre-increment or post-increment manner. For example,

```
x = x + 1;
```

is the same as

```
++x;          // pre-increment operation
```

and as

```
x++;          // post-increment operation
```

In this example, the value is incremented by 1. *++x* is a pre-increment operation, whereas *x++* is a post-increment operation. This means that during the evaluation of the expression by the compiled code, the value is changed pre-evaluation or post-evaluation. For example,

```
i = 1;
k = 2 * i++;        // at completion, k = 2  and  i = 2
i = 1;
k = 2 * ++i;        // at completion, k = 4  and  i = 2
```

In the first case, *i* is incremented after the expression has been resolved. In the second case, *i* was incremented before the expression was resolved.

Decrement Operators

Decrement operators function in a similar manner, causing a subtraction-of-one operation to be performed in a pre-decrement or post-decrement fashion:

```
j--;            // j = j-1
--j;            // j = j-1
```

Compound Assignment Operators

Compound assignment operators are another method of reducing the amount of syntax required during the construction of a program. A compound assignment is really just the combining of an assignment operator (=) with an arithmetic or logical operator. The expression is processed right to left, and syntactically it is constructed somewhat like the increment and decrement operators. Here are some examples:

```
a += 3;         // a = a + 3
b -= 2;         // b = b - 2
c *= 5;         // c = c * 5
d /= a;         // d = d / a
```

This combining of an assignment with another operator works with modulo and bitwise operators (%, >>, <<, &, |, and ^) as well as the arithmetic operators (+, –, *, and /), as shown below:

```
a |= 3;         // a = a OR 3
b &= 2;         // b = b AND 2
c ^= 5;         // c = c exclusively ORed with 5

PORTC &= 3; // Write the current value on PORTC
            // ANDed with 3 back to port C. Forcing
            // all of the bits except the lower 2 to 0
            // and leaving the lower 2 bits unaffected.
```

1.6.4 THE CONDITIONAL EXPRESSION

The conditional expression is probably the most cryptic and infrequently used of the operators. It was definitely invented to save typing time, but in general, it is not easy for beginning programmers to follow. The conditional expression is covered here in the "Operators and Expressions" section due to its physical construction, but it is really more closely associated with control statements covered later in this chapter (Section 1.7, "Control Statements"). If the control sequence

```
if(expression_A)
      expression_B;
else
      expression_C;
```

is reduced to a conditional expression, it reads as follows:

```
expression_A ? expression_B : expression_C;
```

In either of the forms shown above, the logical expression *expression_A* is evaluated: If the result is TRUE, then *expression_B* is executed; otherwise, *expression_C* is executed.

In a program, a conditional expression might be written as follows:

```
(x < 5) ? y = 1 : y = 0;          // if x is less than 5, then
                                  // y = 1, else y = 0
```

1.6.5 OPERATOR PRECEDENCE

When multiple expressions are in a single statement, the operator precedence establishes the order in which expressions are evaluated by the compiler. In all cases of assignments and expressions, the precedence, or order of priority, must be remembered. When in doubt, you should either nest the expressions with parentheses—to guarantee the order of process—or look up the precedence for the operator in question. Some of the operators listed previously actually share an equal level of precedence. Table 1–10 shows the operators, their precedence, and the order in which they are processed (left to right or right to left). This processing order is called grouping or association.

<table>
<tr><th>Name</th><th>Level</th><th>Operators</th><th>Grouping</th></tr>
<tr><td>Primary</td><td>1 (High)</td><td>() . [] -></td><td>Left to Right</td></tr>
<tr><td>Unary</td><td>2</td><td>! ~ - (type) * & ++ -- sizeof</td><td>Right to Left</td></tr>
<tr><td>Binary</td><td>3</td><td>* / %</td><td>Left to Right</td></tr>
<tr><td>Arithmetic</td><td>4</td><td>+ -</td><td>Left to Right</td></tr>
<tr><td>Shift</td><td>5</td><td><< >></td><td>Left to Right</td></tr>
<tr><td>Relational</td><td>6</td><td>< <= > >=</td><td>Left to Right</td></tr>
<tr><td>Equality</td><td>7</td><td>== !=</td><td>Left to Right</td></tr>
<tr><td>Bitwise</td><td>8</td><td>&</td><td>Left to Right</td></tr>
<tr><td>Bitwise</td><td>9</td><td>^</td><td>Left to Right</td></tr>
<tr><td>Bitwise</td><td>10</td><td>|</td><td>Left to Right</td></tr>
<tr><td>Logical</td><td>11</td><td>&&</td><td>Left to Right</td></tr>
<tr><td>Logical</td><td>12</td><td>||</td><td>Left to Right</td></tr>
<tr><td>Conditional</td><td>13</td><td>? :</td><td>Right to Left</td></tr>
<tr><td>Assignment</td><td>14 (low)</td><td>= += -= /= *= %= <<= >>= &= ^= |=</td><td>Right to Left</td></tr>
</table>

Table 1–10 *Operator Precedence*

Some operators in the table are covered later in this chapter in the "Pointers and Arrays" and "Structures and Unions" sections (Section 1.9 and Section 1.10) . These would include the primary operators like dot (.), bracket ([]), and indirection (->), which are used to indicate the specifics of an identifier, such as an array or structure element, as well as pointer and indirection unary operators, such as contents-of (*) and address-of (&). Some examples will help to make this clear:

```
y = 3 + 2 * 4; //would yield y = 11 because the * is higher
               //in precedence and would be done before the +

y = (3 + 2) * 4; //would yield y = 20 because the () are the
               //highest priority and force the + to be done
               //first

y = 4 >> 2 * 3; //would yield y = 4 >> 6 because the * is
               //higher in priority than the shift right
               //operation

y = 2 * 3 >> 4; //would yield y = 6 >> 4 due to the precedence of
               //the * operator
```

The precedence of operators forces the use of a rule similar to the one for casting: If in doubt, use many parentheses to ensure that the math will be accomplished in the desired order. Extra parentheses do not increase the code size and do increase its readability and the likelihood that it will correctly accomplish its mission.

1.7 CONTROL STATEMENTS

Control statements are used to control the flow of execution of a program. **if/else** statements are used to steer or branch the operation in one of two directions. **while, do/while,** and **for** statements are used to control the repetition of a block of instructions. **switch/case** statements are used to allow a single decision to direct the flow of the program to one of many possible blocks of instructions in a clean and concise fashion.

1.7.1 WHILE LOOP

The **while** loop appears early in the descriptions of C language programming. It is one of the most basic control elements. The format of the **while** statement is as follows:

```
while (expression)       or    while(expression)
{                                    statement;
      statement1;
      statement2;
      ...
}
```

When the execution of the program enters the top of the **while** loop, the expression is evaluated. If the result of the expression is TRUE (non-zero), then the statements within the **while** loop are executed. The "loop" associated with the **while** statement are those lines of code contained within the braces *{}* following the **while** statement, or, in the case of a single statement **while** loop, the statement following the **while** statement. When execution reaches the bottom of the loop, the program flow is returned to the top of the **while** loop, where the expression is tested again. Whenever the expression is TRUE, the loop is executed. Whenever the expression is FALSE, the loop is completely bypassed and execution continues at the first statement following the **while** loop. Consider the following:

```
#include <stdio.h>

void main(void)
{
        char c;

        c = 0;
        printf(" Start of program \n");
        while(c < 100)              // if c less than 100 then ..
        {
                printf("c = %d\n",(int)c);   // print c's value each
                                            // time through the loop
                c++;                        // increment c
        }
        printf(" End of program \n");  // indicate that the
                                       //program is finished
        while(1)       // since 1 is always TRUE, then just sit here..
          ;
}
```

In this example, *c* is initialized to 0 and the text string "Start of program" is printed. The **while** loop will then be executed, printing the value of *c* each pass as *c* is incremented from 0 to 100. When *c* reaches 100, it is no longer less than 100 and the **while** loop is bypassed. The "End of program" text string is printed and the program then sits forever in the *while(1)* statement.

The functioning of the *while(1)* statement should now be apparent. Since the "1" is the expression to be evaluated and is a constant (1 is always non-zero and therefore considered to be TRUE), the **while** loop, even a loop with no instructions as in the example above, is entered and is never left because the 1 always evaluates to TRUE. In this case, the *while(1)* is used to stop execution by infinitely executing the loop so that the processor does not keep executing non-existent code beyond the program.

Also note the cast of *c* to an integer inside the *printf()* function, inside the **while** loop. This is necessary because the *printf()* function in most embedded C compilers will handle only integer-size variables correctly.

The **while** loop can also be used to wait for an event to occur on a parallel port.

```
void main(void)
{
        while (PINA & 0x02) //hangs here waiting while the
          ;                 //second bit of port A is high

        while(1)       // since 1 is always TRUE, then just sit here..
          ;
}
```

In this example, the **while** loop is used to await a bit going low. The expression being evaluated is "*PINA & 0x02*", which will mask all but the second bit (bit 1) of the data read from port A. While the bit is at logic 1, the result will be nonzero (TRUE), and so the program will remain in the **while** loop until the value on the second bit drops to zero. At a later point, it should become clear that in real-time programming this construction is not appropriate, but it is a correct use of a **while** statement.

1.7.2 DO/WHILE LOOP

The **do/while** loop is very much like the **while** loop, except that the expression is tested after the loop has been executed one time. This means that the instructions in a **do/while** loop are always executed *once* before the test is made to determine whether or not to remain in the loop. In the **while** construct, the test is made *before* the instructions in the loop are executed even once. The format of the **do/while** statement is as follows:

```
do                                    or          do
{                                                        statement;
        statement1;                               while (expression);
        statement2;
        . . .
} while (expression);
```

When execution reaches the bottom of the **do/while** construct, the expression is evaluated. If the result of the expression is TRUE (non-zero), then the program flow is returned to the top of the **do/while** loop. Each time execution reaches the bottom of the construct, the expression is tested again. Whenever the expression is TRUE, the loop is executed, but if the expression is FALSE, the program continues on with the instructions that follow the construct. The previous example, coded using the **do/while** construct, would appear as follows:

```
#include <stdio.h>

void main(void)
{
        char c;

        c = 0;
        printf(" Start of program \n");

        do
        {
                printf("c = %d\n",(int)c);  // print c's value each
                                            // time through the loop
                c++;                        // increment c
        } while(c < 100);                   // if c less than 100 then
                                            //repeat the operation

        printf(" End of program \n"); // indicate that the
                                      //program is finished
```

```
        while(1)     // since 1 is always TRUE, then just sit here..
            ;
    }
```

In this example, *c* is initialized to 0 and the text string "Start of program" is printed. The **do/while** loop will then be executed, printing the value of *c* each pass as *c* is incremented from 0 to 100. When *c* reaches 100, it is no longer less than 100 and the **do/while** loop is bypassed. The "End of program" text string is printed, and the program then sits forever in the *while(1)* statement.

1.7.3 FOR LOOP

A **for** loop construct is typically used to execute a statement or a statement block a specific number of times. A **for** loop can be described as an initialization, a test, and an action that leads to the satisfaction of that test. The format of the **for** loop statement is as follows:

```
for (expr1; expr2; expr3)        or    for(expr1; expr2; expr3)
{                                              statement;
        statement1;
        statement2;
        ...
}
```

expr1 will be executed only one time at the entry of the **for** loop. *expr1* is typically an assignment statement that can be used to initialize the conditions for *expr2*. *expr2* is a conditional control statement that is used to determine when to remain in the **for** loop. *expr3* is another assignment that can be used to satisfy the *expr2* condition.

When the execution of the program enters the top of the **for** loop, *expr1* is executed. *expr2* is evaluated and if the result of *expr2* is TRUE (non-zero), then the statements within the **for** loop are executed—the program stays in the loop. When execution reaches the bottom of the construct, *expr3* is executed, and the program flow is returned to the top of the **for** loop, where the *expr2* expression is tested again. Whenever *expr2* is TRUE, the loop is executed. Whenever *expr2* is FALSE, the loop is completely bypassed. The **for** loop structure could be represented with a **while** loop in this fashion:

```
expr1;
while(expr2)
{
        statement1;
        statement2;
        ...
        expr3;
}
```

Here is an example:

```
#include <stdio.h>

void main(void)
```

```
{
    char c;

    printf(" Start of program \n");
    for(c = 0; c < 100; c++)     // if c less than 100 then ..
    {
        printf("c = %d\n",(int)c);  // print c's value each time
                                    // through the loop
    }                               // c++ is executed before the
                                    // loop returns to the top

    printf(" End of program \n");   // indicate that the program is
                                    // finished

    while(1)     // since 1 is always TRUE, then just sit here..
        ;
}
```

In this example, the text string "Start of program" is printed. *c* is then initialized to 0 within the **for** loop construct. The **for** loop will then be executed, printing the value of *c* each pass as *c* is incremented from 0 to 100, also within the **for** loop construct. When *c* reaches 100, it is no longer less than 100 and the **for** loop is bypassed. The "End of program" text string is printed, and the program then sits forever in the *while(1)* statement.

1.7.4 IF/ELSE

if/else statements are used to steer or branch the operation of the program based on the evaluation of a conditional statement.

If Statement

An **if** statement has the following form:

```
if (expression)           or     if(expression)
{                                        statement;
        statement1;
        statement2;
        ...
}
```

If the expression result is TRUE (nonzero), then the statement or block of statements is executed. Otherwise, if the result of the expression is FALSE, then the statement or block of statements is skipped.

If/Else Statement

An **if/else** statement has the following form:

```
if(expression)            or     if(expression)
{                                        statement1;
        statement1;              else
```

```
            statement2;                     statement2;
            ...
      }
      else
      {
            statement3;
            statement4;
            ...
      }
```

The **else** statement adds the specific feature to the program flow that the statement or block of statements associated with the **else** will be executed only if the expression result is FALSE. The block of statements will be skipped if the expression result is TRUE.

An **else** statement must always follow the **if** statement it is associated with.

A common programming technique is to cascade **if/else** statements to create a selection tree:

```
      if(expr1)
            statement1;
      else if (expr2)
            statement2;
      else if(expr3)
            statement3;
      else
            statement4;
```

This sequence of **if/else** statements will select and execute only one statement. If the first expression, *expr1*, is TRUE, then *statement1* will be executed and the remainder of the statements will be bypassed. If *expr1* is FALSE, then the next statement, *if (expr2),* will be executed. If *expr2* is TRUE, then *statement2* will be executed and the remainder bypassed, and so on. If *expr1, expr2,* and *expr3* are all FALSE, then *statement4* will be executed.

Here is an example of **if/else** operation:

```
#include <stdio.h>

void main(void)
{
  char c;

  printf(" Start of program \n");

  for(c = 0; c < 100; c++)    // while c is less than 100 then ..
  {
     if(c < 33)
          printf("0<c<33    "); // use if/else to show the range of
     else if((c >32) && (c < 66)) // numbers that c is in
          printf("33<c<66   ");
```

```
    else
          printf("66<c<100 ");
          printf("c= %d\n",(int)c);    // print c's value each time
                                       //through the loop
  }

  printf(" End of program \n"); // indicate that the program is finished
  while(1)       // since 1 is always TRUE, then just sit here..
      ;
}
```

In this example, the text string "Start of program" is printed. *c* is then initialized to 0 within the **for** loop construct. The **for** loop will then be executed, printing the value of *c* each pass as *c* is incremented from 0 to 100, also within the **for** loop construct.

If the value of *c* is less than 33, then the text string "0<c<33" is printed. If the value of *c* is between 32 and 65, the text string "33<c<66" is printed. If the value of *c* is not within either of the preceding cases, the text string "66<c<100" is printed.

When *c* reaches 100, it is no longer less than 100 and the **for** loop is bypassed. The "End of program" text string is printed, and the program then sits forever in the *while(1)* statement.

Using the constructs and other techniques covered so far, it is possible to create a program that efficiently tests each bit of an input port and prints a message to tell the state of the bit.

```
#include <MEGA8535.h>    // register definition file for
                      // an Atmel ATMEGA8535
#define test_port PORTA

void main(void)
{
      unsigned char cnt, bit_mask;    //variables
      bit_mask = 1;       //start with lowest bit

      for (cnt=0;cnt<8;cnt++)  //for loop to test 8 bits
       {
                 // the instructions below test port bits
                 // and print result
             if (test_port & bit_mask)
                 printf("Bit  %d is high.\n",(int)cnt);
             else
                 printf("Bit  %d is low.\n",(int)cnt);

             bit_mask <<= 1;   //shift bit to be tested
       }

       while(1) // since 1 is always TRUE, then just sit here..
         ;
}
```

The example above uses a **for** loop that is set to loop eight times, once for each bit to be tested. The variable *bit_mask* starts at a value of 1 and is used to mask all of the bits except the one being tested. After being used as a mask, it is shifted one bit to the left, using the compound notation, to test the next bit during the next pass through the **for** loop. During each loop, the **if/else** construct is used to print the correct statement for each bit. The conditional statement in the **if** construct is a bitwise AND using *bit_mask* to mask unwanted bits during the test.

Conditional Expression

Another version of the **if/else** is the conditional expression, designed to save the programmer time and steps through a simplified syntax. The **if/else** sequence

```
if(expression_A)
        statement1;
else
        statement2;
```

can be reduced to a conditional expression that would read as follows:

```
expression_A ? statement1  : statement2;
```

In both of the forms shown above, the logical expression *expression_A* is evaluated and if the result is TRUE, then *statement1* is executed; otherwise, *statement2* is executed.

In a program, a conditional expression might be written as follows:

```
(x < 5) ? y = 1 : y = 0;      // if x is less than 5, then
                              // y = 1, else y = 0
```

1.7.5 SWITCH/CASE

The **switch/case** statement is used to execute a statement, or a group of statements, selected by the value of an expression. The form of this statement is as follows:

```
switch (expression)
{
        case const1:
                statement1;
                statement2;
        case const2:
                statement3;
                    ...
                statement4;
        case constx:
                statement5;
                statement6;
        default:
                statement7;
                statement8;
}
```

The *expression* is evaluated and its value is then compared against the constants (*const1*, *const2*, . . . *constx*). Execution begins at the statement following the constant that matches the value of the expression. The constants must be integer or character values. All of the statements following the matching constant will be executed, to the end of the **switch** construct. Since this is not normally the desired operation, **break** statements can be used at the end of each block of statements to stop the "fall-through" of execution and allow the program flow to resume after the **switch** construct.

The default case is optional, but it allows for statements that need to be executed when there are no matching constants.

```
#include <stdio.h>
#include <MEGA8535.h>    // register definition header file for
                                    // an Atmel ATMEGA8535


void main(void)
{
   unsigned char c;
   while(1)
   {
       c = PINA & 0xf; // read the lower nibble of port A
       switch(c )
       {
              case '0':
              case '1':       // you can have multiple cases
              case '2':       // for a set of statements..
              case '3':
                    printf(" c is a number less than 4 \n");
                    break;   // break to skip out of the loop
              case '5':      // or just one is ok..
                    printf(" c is a 5 \n");
                    break;
              default:
                    printf(" c is 4 or is > 5 \n");
              }
       }
    }
}
```

This program reads the value on port A and masks the upper four bits. It compares the value of the lower nibble from port A against the constants in the **case** statement. If the character is a 0, 1, 2, or 3, the text string "c is a number less than 4" will be printed to the standard output. If the character is a 5, the text string "c is a 5" will be printed to the standard output. If the character is none of these (a 4 or a number greater than 5), the **default** statements will be executed and the text string "c is 4 or is > 5" will be printed. Once the appropriate **case** statements have been executed, the program will return to the top of the **while** loop and repeat.

1.7.6 BREAK, CONTINUE, AND GOTO

The **break**, **continue**, and **goto** statements are used to modify the execution of the **for**, **while**, **do/while**, and **switch** statements.

Break

The **break** statement is used to exit from a **for, while, do/while,** or **switch** statement. If the statements are nested one inside the other, the **break** statement will exit only from the immediate block of statements.

The following program will print the value of *c* to the standard output as it is continuously incremented from 0 to 100, then reinitialized to 0. In the inner **while** loop, *c* is incremented until it reaches 100, and then the **break** statement is executed. The **break** statement causes the program execution to exit the inner **while** loop and continue execution of the outer **while** loop. In the outer **while** loop, *c* is set to 0, and control is returned to the inner **while** loop. This process continues to repeat itself forever.

```
#include <stdio.h>
void main(void)
{
        int c;

        while(1)
        {
             while(1)
             {
                   if (c > 100)
                   break;   // this will take us out of this while
                   ++c;     // block, clearing c..
                   printf("c = %d\n",c);
             }
             c = 0;    // clear c and then things will begin again
        }              // printing the values 0-100, 0-100, etc..
}
```

Continue

The **continue** statement will allow the program to start the next iteration of a **while, do/while,** or **for** loop. The **continue** statement is like the **break** statement in that both stop the execution of the loop statements at that point. The difference is that the **continue** statement starts the loop again, from the top, where **break** exits the loop entirely.

```
#include <stdio.h>

void main(void)
{
        int c;

        while(1)
```

```
    {
        c = 0;
        while(1)
        {
            if (c > 100)
                continue;   // This will cause the printing to
                            // stop when c>100, because the
                            // continue will cause the rest of
                            // this loop to be skipped!!
            ++c;
            printf("c = %d\n",c);   // no code after the continue
                                    //will be executed
        }
    }
}
```

In this example, the value of *c* will be displayed until it reaches 100. The program will appear at this point as if it has stopped, when in fact, it is still running. It is simply skipping the increment and *printf()* statements.

Goto

The **goto** statement is used to literally "jump" execution of the program to a label marking the next statement to be executed. This statement is very unstructured and is aggravating to the "purist," but in an embedded system it can be a very good way to save some coding and the memory usage that goes with it. The label can be before or after the **goto** statement in a function. The form of the **goto** statement is as follows:

```
goto identifier;            or          identifier:
        ...                                     statement;
                                                ...
identifier:
        statement;                      goto identifier;
```

The label is a valid C name or identifier, followed by a colon (:).

```
#include <stdio.h>

void main(void)
{
        int c,d;

        while(1)
        {

start_again:

        c = 0;
        d = -1;
```

```
            while(1)
            {
                    if (d == c)
                            goto start_again;       // (stuck? bail out!)
                    d = c;              // d will remember where we were
                    if (c > 100)
                            continue;    // this will reinitiate this
                                         // while loop
                    ++c;       // d will be checked to see if c is stuck
                    printf("c = %d\n",c);
                                  // because c won't change ( > 100!!)
            }
    }
}
```

In this example, *c* and *d* are initialized to different values. The *if (d == c)* statement checks the values and, as long as they are not equal, execution falls through to the assignment *d = c*. If *c* is less than or equal to 100, execution falls through to the increment *c* statement. The value of *c* is printed and the **while** loop begins again. The values of *c* and *d* will continue to differ by 1 until *c* becomes greater than 100. With *c* at a value of 101, the **continue** statement will cause the increment *c* and *printf()* statements to be skipped. The *if (d==c)* will become TRUE, since they are now equal, and the **goto** will cause the program execution to jump to the *start_again* label. The result would be the value of *c* printing from 0 to 100, over and over again.

Chapter 1 Example Project: Part A

This is the first portion of a project that will be used to demonstrate some of the concepts presented here in Chapter 1. This example project will be based on a simple game according to the following situation:

You have been asked to build a Pocket Slot Machine as a mid-semester project by your favorite instructor, who gave you some basic specifications that include the following:

1. The press and release of a button will be the "pulling the arm of the One-Armed Bandit."
2. The duration of the press and the release can be used to generate a pseudo-random number to create three columns of figures.
3. A flashing light is used to indicate the machine is "moving."
4. There are four kinds of figures that can appear in a column:
    a. Bar
    b. Bell
    c. Lemon
    d. Cherry

5. The payout of the machine has the following rules:
   a. Three of a kind pays a Nickel.
   b. A Cherry anywhere pays a Dime.
   c. Three Cherries are a Jackpot.
   d. Everything else loses.

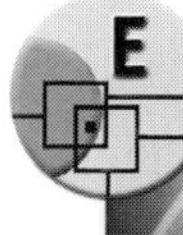

In this first step we will develop a brief program that touches on several of the concepts. We will develop a *main()* that contains various types of control statements to create the desired operation. This first pass will get the press and release, work up a pseudo random number, and print the results in a numeric form using standard I/O functions. **while, do/while,** and **for** loops will be demonstrated as well as **if/else** control statements.

```
//
// "Slot Machine" -- The Mini-Exercise
//
//  Phase 1..

#include <MEGA8515.h>  /* processor specific information */

#define xtal 4000000L  /* quartz crystal frequency [Hz] */
#define baud 9600      /* Baud rate */

#include <stdio.h>     /* this gets the printf() definition */

// Global Variable Declarations..
char    first,second,third; // columns in the slot machine
char    seed;               // number to form a seed for random #s
char    count;              // general purpose counter
int     delay;              // a variable for delay time

void main(void)
{
     PORTA = 0xFF;          // Initialize the I/O ports
     DDRA = 0x00;           // port A all inputs
     PORTB = 0x00;          // port B.1 is an output (light)
     DDRB = 0x02;

  //  Initialize the UART control register
  //  RX & TX enabled, 8 data bits
     UCSRB=0x18;
  //  initialize the UART's baud rate
     UBRRL=xtal/16/baud-1;

     while(1)          // do forever..
```

```
    {
       while(PINA.0)   // Wait for a button and
           seed++;     // let the counter roll over and
                       // over to form a seed.

       first = second = third = seed; // preload the columns

    do                       // Mix up the numbers
    {                        // while waiting for button release.
       first ^= seed>>1;     // Exclusive ORing in the moving seed
       second^= seed>>2;     // can really stir up the numbers.
       third ^= seed>>3;
       seed++;               // keep rolling over the seed pattern
    } while(PINA.0 == 0);    // while the button is pressed

    for(count = 0; count < 5; count++) // flash light when done..
    {
           for(delay = 0; delay < 10000; delay++)
               ;                  // just count up and wait
           PORTB.1 = 0;           // turn the LED on..
           for(delay = 0; delay < 10000; delay++)
               ;
           PORTB.1 = 1;           // turn the LED off..
    }

    first  &= 3;      // limit number to values from 0 to 3
    second &= 3;
    third  &= 3;
                      // show the values..
    printf("--> %d, %d ,%d <--\n", first, second, third);

                      // determine a payout..
    if((first == 3) && (second == 3) && (third == 3))
        printf("Paid out: JACKPOT!!\n");// Three "Cherries"
    else if((first == 3) || (second == 3) || (third == 3))
        printf("Paid out: One Dime\n"); // One "Cherry"
    else if((first == second) && (second == third))
        printf("Paid out: One Nickel\n");     // Three of a kind
    else
        printf("Paid out: ZERO\n");           // Loser..
    }
}
```

1.8 FUNCTIONS

A function is an encapsulation of a block of statements that can be used more than once in a program. Some languages refer to functions as subroutines or procedures. Functions are generally written to break up the operational elements of a program into its fundamental parts. This allows a programmer to debug each element and then use that element again and again.

One of the prime advantages in using functions is the development of a "library." As functions are developed that perform a certain task, they can be saved and used later by a different application or even a different programmer. This saves time and maintains stability, since the functions developed get used and reused, tested and retested.

A function may perform an isolated task requiring no parameters whatsoever. A function may accept parameters in order to have guidance in performing its designed task. A function may not only accept parameters but return a value as well. Even though a function may accept multiple parameters, it can only return one. Some examples are as follows:

```
sleep();        // this function performs a 1 second delay,
                // with no parameters, in or out

printf("this is a parameter %d",x);
                // printf will print its parameters

c = getchar();
                // getchar will return a value from the
                // standard input
```

The standard form of a function is

```
type function_name ( type param1, type param2, ... )
{
      statement1;
      statement2;
            ...
      statementx;
}
```

which is a type, followed by a name, followed by the primary operator *()*. The parentheses, or function operator, indicate to the compiler that the name is to be executed as a function.

The type of a function or its parameters can be any valid variable type such as **int, char,** or **float.** The type can also be empty or void. The type **void** is a valid type and is used indicate a parameter or returned value of *zero size*. The default type of a function is **int**.

So a typical function declaration might be as follows:

```
unsigned char getchar(void)
{
      while((UCSRA & 0x80) == 0)
```

```
                ;          // wait for a character to arrive
            return UDR;    // return its unsigned char value to the
                           // caller
}
```

In this example, *getchar()* is a function that requires no parameters and returns an unsigned character value when it has completed execution. The *getchar()* function is one of the many library functions provided in the C compiler. These functions are available for the programmer's use and will be discussed in more depth later.

1.8.1 PROTOTYPING AND FUNCTION ORGANIZATION

Just as in the use of variables and constants, the function type and the types of its parameters must be declared before it is called. This can be accomplished in a couple of ways: One is the order of the declarations of the functions, and the other is the use of function prototypes.

Ordering the functions is always a good idea. It allows the compiler to have all the information about the function in place before it is used. This would also mean that all programs would have the following format:

```
// declaration of global variables would be first
int var1, var2, var3;

// declarations of functions would come next
int function1(int x, int y)
{
}

void function2(void)
{
}

// main would be built last
void main(void)
{
        var3 = function1(var1, var2);
        function2();
}
```

This is all nice and orderly, but it is also sometimes impossible. There are many occasions when functions use one another to accomplish a task and you simply cannot declare them in such a way that they are all in a top-down order.

Function prototypes are used to allow the compiler to know the requirements and the type of the function before it is actually declared, reducing the need for a top-down order.

The previous example can be organized differently:

```
int var1, var2, var3;     // declaration of global variables
```

```
void main(void)          // main
{
      // these functions are not yet known to the
      // compiler
      var3 = function1(var1, var2);

      function2();
}
// declarations of functions here now generate a
// "Function Redefined Error"
int function1(int x, int y)
{
}

void function2(void)
{
}
```

This organization would typically lead to an error message from the compiler. The compiler simply would not have enough information about the functions that are called in *main()*, or their format, and would be unable to generate the proper code. The prototypes of the functions can be added to the top of the code like this:

```
// the prototype of a function tells the compiler what to expect
int function1(int, int);

void function2(void);

int var1, var2, var3;    // declaration of global variables

void main(void)          // main
{
      var3 = function1(var1, var2);
      function2();
}
// the declaration of the functions here is perfectly OK, since
// the format of the functions are presented in the prototypes
int function1(int x, int y)
{
}

void function2(void)
{
}
```

The compiler now has all the required information about each function as it processes the function name. As long as the actual declared function matches the prototype, everything is fine.

The C compiler provides many library functions. They are made available to the program by the "***#include** <filename.h>*" statement, where *filename.h* will be the file containing the function prototypes for the library functions. In previous examples, we used the function *printf()*. The *printf()* function itself is declared elsewhere, but its prototype exists in the stdio.h header file. This allows the programmer to simply include the header file at the top of the program and start writing code. The library functions are each defined in Appendix A, "Library Functions Reference."

1.8.2 FUNCTIONS THAT RETURN VALUES

In many cases, a function is designed to perform a task and return a value or status from the task performed. The control word **return** is used to indicate an exit point in a function or, in the case of a non-void type function, to select the value that is to be returned to the caller.

In a function of type **void,**

```
int z;          // global variable z
void sum(int x, int y)
{
        z = x + y;   // z is modified by the function, sum()
}
```

the **return** is implied and is at the end of the function. If a **return** were placed in the **void** function, like this,

```
void sum(int x, int y)
{
        return;      // this would return nothing..
        z = x + y;   // and this would be skipped
}
```

the return statement would send the program execution back to the caller, and the statements following the return would not be executed.

In a function whose type is not **void,** the **return** control word will also send the execution back to the caller. In addition, the **return** will place the value of the expression, to the right of the **return,** on the stack in the form of the type specified in the declaration of the function. The following function is of type **float.** So a **float** will be placed on the stack when the return is executed:

```
float cube(float v)
{
        return (v*v*v);     // this returns a type float value
}
```

The ability to return a value allows a function to be used as part of an expression, such as an assignment. For example, in the program below,

```
void main(void)
```

```
{
        float a;

        b = 3.14159;        // put PI into b

        a = cube(b);        // pass b to the cubed function, and
                            // assign its return value to a

        printf("a = %f, b = %f \n",a ,b);     // print the result

        while(1)              // done
                ;
}
```

a would be assigned the value of the function *cube(b)*. The result would appear as

```
a = 31.006198, b = 3.14159
```

1.8.3 RECURSION

A recursive function is one that calls itself. The ability to generate recursive code is one of the more powerful aspects of the C language. When a function is called, its local variables are placed on a stack or heap-space, along with the address of the caller, so that it knows how to get back. Each time the function is called, these allocations are made once again. This makes the function "reentrant" since the values from the previous call would be left within the previous allocations, untouched.

The most common example of a recursive operation is calculating factorials. A factorial of a number is the product of that number and all of the numbers that lead up to it. For example,

```
5! = 5 * 4 * 3 * 2 * 1 = 120
```

This can be stated algebraically as

```
n! = n * (n-1) * (n-2) * ... * 2 * 1     or     n! = n * (n-1) !
```

So a program to demonstrate this operation might look like this:

```
#include <stdio.h>

int fact(int n)
{
if(n == 0)
    return 1;                // if n is zero, return a one, by
                             // definition of factorial
else
    return (n * fact(n-1));      // otherwise, call myself with
                                 // n - 1 until n = 0, then
                                 // return n * the result the
                                 // call to myself
}
```

```
void main(void)
{
        int n;

        n = fact(5);       // compute 5!, recursively

        printf(" 5! = %d \n", n);       // print the result

        while(1)                        // done
            ;
}
```

When the function *fact()* is called with argument *n*, it calls itself $n-1$ times. Each time it calls itself it reduces *n* by 1. When *n* equals 0, the function returns instead of calling itself again. This causes a "chain reaction" or "domino effect" of returns, which leads back to the call made in *main()*, where the result is printed.

As powerful as recursion can be to perform factorials, quick sorts, and linked-list searches, it is a memory-consuming operation. Each time the function calls itself, it allocates memory for the local variables of the function, the return value, the return address for the function, and any parameters that are passed during the call. In the previous example, this would include the following:

```
int n(passed parameter)     2 bytes
Return value                2 bytes
Return address              2 bytes
        Total               6 bytes, per recursion, minimum.
```

In the case of "5!", at ***least*** a total of 30 bytes of memory would be allocated during the factorial operation and deallocated upon its return. This makes this type of operation dangerous and impractical for a small microcontroller. If the depth of the allocation due to recursion becomes too great, the stack or heap-space will overflow, and the program's operation will become unpredictable.

Chapter 1 Example Project: Part B

In this phase of the development of the Pocket Slot Machine, we are going to create a couple of functions to break up the main loop into smaller chunks of code and make the code slightly more readable. The parts of code that flash the lights and compute payout will be moved into functions, shortening the *main()* function and making it a bit easier to follow. Also, a **switch/case** statement will be used in the indication of payout.

```
//
// "Slot Machine" -- The Mini-Exercise
//
//     Phase 2..

#include <MEGA8515.h>  /* processor specific information */
```

```
#define xtal 4000000L  /* quartz crystal frequency [Hz] */
#define baud 9600      /* Baud rate */

#include <stdio.h>     /* this gets the printf() definition */

// Global Variable Declarations..
char    first,second,third; // columns in the slot machine
char    seed;               // number to form a seed for random #s
char    count;              // general purpose counter
int     delay;              // a variable for delay time

#define JACKPOT  3     /* This defines give different payouts */
#define DIME     2     /* names in order to make the code more */
#define NICKEL   1     /* readable to humans. */
#define ZERO     0

void flash_lights(char n)   // flash the lights a number of times
{
     for(count = 0; count < n; count++) // flash light when done..
     {
         for(delay = 0; delay < 10000; delay++)
             ;                    // just count up and wait
         PORTB.1 = 0;             // turn the LED on..
         for(delay = 0; delay < 10000; delay++)
             ;
         PORTB.1 = 1;             // turn the LED off..
     }
}

int get_payout(void)
{
    if((first == 3) && (second == 3) && (third == 3))
        return JACKPOT;    // if all "cherries"..
    else if((first == 3) || (second == 3) || (third == 3))
        return DIME;  // if any are "cherries"..
    else if((first == second) && (second == third))
        return NICKEL;  // if three are alike, of any kind..
    else
        return ZERO;  // otherwise -- you lose..
}

void main(void)
{
  int i;                  // declare a local variable for
                          // temporary use..
```

```
  PORTA = 0xFF;          // Initialize the I/O ports
  DDRA = 0x00;           // port A all inputs
  DDRB = 0x00;           // port B.1 is an output (light)
  DDRB = 0x02;

//  Initialize the UART control register
//  RX & TX enabled, 8 data bits
    UCSRB=0x18;
//  initialize the UART's baud rate
    UBRRL=xtal/16/baud-1;

while(1)                 // do forever..
{
  while(PINA.0)          // Wait for a button and
      seed++;            // let the counter roll over and
                         // over to form a seed.

  first = second = third = seed; // preload the columns

  do                     // Mix up the numbers
  {                      // while waiting for button release.
      first ^= seed>>1; // Exclusive ORing in the moving seed
      second^= seed>>2; // can really stir up the numbers.
      third ^= seed>>3;
      seed++;            // Keep rolling over the seed pattern
  } while(PINA.0 == 0); // while the button is pressed.

  flash_lights(5);       // flash the lights 5 times..

  first  &= 3;      // limit number to values from 0 to 3
  second &= 3;
  third  &= 3;
                         // show the values..
  printf("--> %d, %d, %d <--\n", first, second, third);

                         // determine a payout..
  i = get_payout();

  switch(i)       // now print the payout results
  {
      case    ZERO:
          printf("Paid out: ZERO\n");
          break;
      case    NICKEL:
          printf("Paid out: One Nickel\n");
          break;
```

```
            case     DIME:
                printf("Paid out: One Dime\n");
                break;

            case     JACKPOT:
                printf("Paid out: JACKPOT!!\n");
                break;
        }
    }
}
```

EXAMPLE

1.9 POINTERS AND ARRAYS

Pointers and arrays are widely used in the C language because they allow programs to perform more generalized and more efficient operations. Operations that require gathering data may use these methods to easily access and manipulate the data without moving the data around in memory. They also allow for the grouping of associated variables such as communications buffers and character strings.

1.9.1 POINTERS

Pointers are variables that contain the address or location of a variable, constant, function, or data object. A variable is declared to be a pointer with the indirection or dereferencing operator (*):

```
char *p;     // p is a pointer to a character
int *fp;     // fp is a pointer to an integer
```

The pointer data type allocates an area in memory large enough to hold the machine address of the variable. For example, the address of a memory location in a typical microcontroller will be described in 16 bits. So in a typical microcontroller, a pointer to a character will be a 16-bit value, even though the character itself is only an 8-bit value.

Once a pointer is declared, you are now dealing with the address of the variable it is pointing to, not the value of the variable itself. You must think in terms of locations and contents of locations. The address operator (&) is used to gain access to the address of a variable. This address may be assigned to the pointer and is the pointer's value. The indirection or dereferencing operator (*) is used to gain access to the data located at the address contained in the pointer. For example,

```
char *p;     // p is a pointer to a character
char a, b;   // a and b are characters

p = &a;      // p is now pointing to a
```

In this example, *p* is assigned the address of *a*, so *p* is "pointing to" *a*.

To get to the value of the variable that is pointed to by *p*, the indirection operator (*) is used. When executed, the indirection operator causes the value of *p*, an address, to be used to look up a location in memory. The value at this location is then read from or written to according to the expression containing the indirection operator. So, in the following code, **p* would cause the value located at the address contained in *p* to be read and assigned to *b*.

```
b = *p;        // b equals the contents pointed to by p
```

Therefore, the combined code of the previous two examples would produce the same result as

```
b = a;
```

The indirection operator can also appear on the right side of an assignment.

In this example,

```
char *p;       // p is a pointer to a character
char a, b;     // a and b are characters

p = &a;

*p = b;        // the location pointed to by p is
               // assigned the value of b
```

the memory location, at the address stored in *p* is assigned the value of *b*. This would produce the same result as

```
a = b;
```

Whenever you read these operations, try to read them as "*b* is assigned the value pointed to by *p*" and "*p* is assigned the address of *a*." This helps to avoid making the most common mistake with pointers:

```
b = p;   // b will be assigned a value of p, an address,
         // not what p points to

p = a;   // p will be assigned the value of a, not its address
```

These two assignments are allowed because they are syntactically correct. Semantically speaking, they are most likely not what was intended.

With power and simplicity comes the opportunity to make simple and powerful mistakes. Pointer manipulation is one of the leading causes of programming misfortune. But exercising a little care, and reading the syntax aloud, can greatly reduce the risk of changing memory in an unintended fashion.

Pointers are also an excellent method of accessing a peripheral in a system, such as an I/O port. For instance, if we had an 8-bit parallel output port located at 0x1010 in memory, that port could be accessed through indirection as follows:

```
unsigned char *out_port;   //declare out_port as a pointer
out_port = 0x1010;         //assign out_port with the address value
*out_port = 0xaa;          //now assign out_port's address a value
```

In this code, the location pointed to by *out_port* would be assigned the value 0xAA. It can also be described as "any value assigned to **out_port* will be written to memory address 0x1010."

Because of the structure of the C language, it is possible to have pointers that point to pointers. In fact, there really is no limit to the depth of this type of indirection, except for the confusion that it may cause. For example,

```
int *p1;     // p1 is a pointer to an integer
int **p2;    // p2 is a pointer, to a pointer to an integer
int ***p3;   // p3 is a pointer, to a pointer, to a pointer
             // to an integer
int i, j;

p1 = &i;     // p1 is assigned that address of i
p2 = &p1;    // p2 is now pointing to the pointer to i
p3 = &p2;    // p3 is pointing to the pointer that is pointing
             // to i

             // Therefore,
j = ***p3;   // j is assigned the value pointed to by the value
             // pointed to by the value pointed to by p3.
```

yields the same result as

```
j = i;       // any questions??
```

Since pointers are effectively addresses, they offer the ability to move, copy, and change memory in all sorts of ways with very little instruction. When it comes to performing address arithmetic, the C compiler makes sure that the proper address is computed based on the type of the variable or constant being addressed. For example,

```
int *ptr;
long *lptr;

ptr = ptr + 1;     // moves the pointer to the next integer
                   //    (2 bytes away)

lptr = lptr + 1;   //moves the pointer to the next long integer
                   //    (4 bytes away)
```

ptr and *lptr* are incremented by one location, which in reality is 2 bytes for *ptr* and 4 bytes for *lptr*, because of their subsequent types. This is also true when using increment and decrement operations:

```
ptr++;       // moves the pointer to the next integer location
             //     (2 bytes)

--lptr;      // moves the pointer back to the preceding long
             // integer location
             //     (-4 bytes)
```

Since the indirection (*) and address (&) operators are unary operators and are at the highest precedence, they will always have priority over the other operations in an expression.

Since increment and decrement are also unary operators and share the same priority, expressions containing these operators will be evaluated left to right. For example, listed below are pre-increment and post-increment operations that are part of an assignment. Please take note of the comments in each line as to how the same precedence level affects the outcome of the operation:

```
char c;
char *p;

c = *p++;    // assign c the value pointed to by p, and then
             // increment the address p

c = *++p;    // increment the address p, then assign c the
             // value pointed to by p

c = ++*p;    // increment the value pointed to by p, then
             // assign it to c, leaving the value of p untouched

c = (*p)++; // assign c the value pointed to by p, and then
             // increment the value pointed to by p, leaving
             // the value of p untouched
```

Pointers can be used to extend the amount of information returned from a function. Since a function inherently can return only one item using the **return** control, passing pointers as parameters to a function allows the function an avenue for returning additional values. Consider the following function, *swap2()*:

```
void swap2 (int *a, int *b)
{
    int temp;

    temp = *b; // Place value pointed to by b into temp
    *b = *a;   // move the value pointed to by a into location
               // pointed to by b.

    *a = temp; // Now set the value of location a to the
               // value of temp.
}
```

This sample function swaps the values of *a* and *b*. The caller provides the pointers to the variables it wishes to transpose like this:

```
int v1,v2;
swap2( &v1, &v2);  // pass the addresses of v1 and v2
     ...
```

Since the *swap2()* function is using the addresses that were passed to it, it swaps the values in variables *v1* and *v2* directly. This process of passing pointers is frequently used and can be found in standard library functions like *scanf()*. The *scanf()* function (defined in stdio.h) allows multiple parameters to be gathered from the standard input in a formatted manner and stores them in the specified locations of memory. A typical call to *scanf()* looks like this:

```
int x,y,z;
scanf("%d %d %d", &x, &y, &z);
```

This *scanf()* call will retrieve three decimal integer values from the standard input and place these values in *x*, *y*, and *z*. Details about the *scanf()* function are available in Chapter 3, "Standard I/O and Preprocessor Functions," as well as in Appendix A, "Library Functions Reference."

1.9.2 ARRAYS

An array is another system of indirection. An array is a data set of a declared type, arranged in order. An array is declared like any other variable or constant, except for the number of required array elements:

```
int digits[10];     // this declares an array of 10 integers
char str[20];       // this declares an array of 20 characters
```

The referencing of an array element is handled by an index or subscript. The index can range from 0 to the length of the declared array less 1.

```
str[0], str[1], str[2], . . . . . str[19]
```

Array declarations can contain initializers. In a variable array, the initialization values will be placed in the program memory area and copied into the actual array before *main()* is executed. A constant array differs in that the values will be allocated in program memory, saving RAM memory, which is usually in short supply on a microcontroller. A typical initializer would appear as

```
int array[5] = {12, 15, 27 56, 94 };
```

In this case, *array[0]* = 12, *array[1]* = 15, *array[2]* = 27, *array[3]* = 56, and *array[4]* = 94.

The C language has no provision for checking the boundaries of an array. If the index were to be assigned a value that exceeds the boundaries of an array, memory could be altered in an unexpected way, leading to unpredictable results. For example,

```
char digits[10]={0,1,2,3,4,5,6,7,8,9};//an array of characters

numb = digits[12]; // this reads outside the array
```

Arrays are stored in sequential locations in memory. Reading *digits[5]* in the example above will cause the processor to go to the location of the first index of the array and then read the data 5 locations above that. Therefore, the second line of code above will cause the processor to read the data 12 spaces above the starting point for the array. Whatever data exists at that location will be assigned to *numb* and may cause many strange results. So, as in many other programming areas, some caution and forethought should be exercised.

A primary difference in the use of an array versus that of a pointer is that in an array, an actual memory area has been allocated for the data. With a pointer, only an address reference location is allocated, and it is up to the programmer to declare and define the actual memory areas (variables) to be accessed.

The most common array type is the character array. It is typically referred to as a string or character string. A string variable is defined as an array of characters, while a constant string is typically declared by placing a line of text in quotes. In the case of a constant string, the C compiler will null-terminate, or add a zero to the end of the string. When you declare character strings, constant or variable, the declared array size should be one more than what is needed for the contents, in order to allow for the null terminator:

```
char str[12];               // variable string

printf("Hello World!");     // constant string in program memory

const cstr[16] = "Constant String";
                            // constant string in program memory
```

cstr in the example above is set to contain sixteen values because the string itself contains fifteen characters and one more space must be allowed for the terminator.

An array name followed by an index may reference the individual elements of an array of any type. It is also possible to reference the first element of any array by its name alone. When no index is specified, the array name is treated as the address of the first element in the array. Given the following declarations,

```
char stng[20];
char *p;
```

the assignment

```
p = stng;          // p is pointing to stng[0]
```

is the same as

```
p = &stng[0];      // p is pointing to stng[0]
```

Character strings often need to be handled on a character-by-character basis. Sending a message to a serial device or an LCD (Liquid Crystal Display) are examples of this requirement. The example below shows how array indexing and pointer indirection function nearly interchangeably. This example uses the library function *putchar()* to send one character at a time to the standard output device, most likely, a serial port:

```
#include <stdio.h>

const char s[15] = {"This is a test"};

char i;
char *p;
```

```
void main(void)
{
   for(i=0; i<15; i++)         // Print each character of the array
         putchar(s[i]);        // by using i as an index.

   p = s;                      // point to string as a whole
   for(i=0; i<15; i++)         // Print each character of the array
         putchar(*p++);        // and move to the next element
                               // by incrementing the pointer p.
   while(1)
      ;
}
```

The first portion on this program uses a **for** loop to output each character of the array individually. The **for** loop counter is used as the index to retrieve each character of the array so that it can be passed to the *putchar()* function. The second portion uses a pointer to access each element of the array. The line "*p = s*" sets the pointer to the address of the first character in the array. The **for** loop then uses the pointer (post-incrementing it each time) to retrieve the array elements and pass them to *putchar()*.

1.9.3 MULTIDIMENSIONAL ARRAYS

The C language supports multidimensional arrays. When a multidimensional array is declared, it should be thought of as arrays of arrays. Multidimensional arrays can be constructed to have two, three, four, or more dimensions. The adjacent memory locations are always referenced by the right-most index of the declaration.

A typical two-dimensional array of integers would be declared as

```
int two_d[5][10];
```

In memory, the elements of the array would be stored in sequential rows like this:

```
two_d[0][0], two_d[0][1], two_d[0][2], ... two_d[0][9],
two_d[1][0], two_d[1][1], two_d[1][2], ... two_d[1][9],
two_d[2][0], two_d[2][1], two_d[2][2], ... two_d[2][9],
two_d[3][0], two_d[3][1], two_d[3][2], ... two_d[3][9],
two_d[4][0], two_d[4][1], two_d[4][2], ... two_d[4][9]
```

When a two-dimensional array is initialized, the layout is the same, sequential rows:

```
int matrix[3][4] = {    0, 1,  2,  3,
                        4, 5,  6,  7,
                        8, 9, 10, 11 };
```

Multidimensional arrays are useful in operations such as matrix arithmetic, filters, and look-up-tables (LUTs).

For instance, assume we have a telephone keypad that generates a row and column indication, or scan-code, whenever a key is pressed. A two-dimensional array could be used as a look-up-table to convert the scan-code to the actual ASCII character for the key. We will assume for this example that the routine *getkeycode()* scans the keys and sets the *row* and *col* values to indicate the position of the key that is pressed. The code to perform the conversion operation may look something like this:

```
#include <stdio.h>

#define TRUE 1
#define FALSE 0

char getkeycode(char *row, char *col);
            // the getkeycode routine gets the key press and
            // returns TRUE if a key is pressed, FALSE if not

/* look up table for ASCII values */
const char keys[4][3] = {       '1','2','3',
                                '4','5','6',
                                '7','8','9',
                                '*','0','#' };

void main(void)
{
      char row, col;
      char i;

      while(1)     // while forever....
      {
            i = getkeycode(&row, &col);
                  // TRUE, if there was a key pressed
                  // and row and col contain that key
            if(i == TRUE)
                  // only print the key value that is pressed
                  putchar(keys[row][col]);
      }
}
```

Another, more common, form of a two-dimensional array is an array of strings. This example declares an array of strings initialized to the days of the week:

```
char day_of_the_week[7][10] = {
            "Sunday",
            "Monday",
            "Tuesday",
            "Wednesday",
            "Thursday",
            "Friday",
            "Saturday" };
```

In the array shown above, the strings are of different lengths. The compiler places the null terminator after the last character in the string no matter how long the string may be. Any wasted locations in memory are left uninitialized. Functions such as *printf()* print the string until they encounter the null-terminator character, thus being able to print strings of various lengths correctly.

To access the fourth day of the week and print the associated string, we will use *printf()*. *printf()* requires the address of the first character of a string:

```
printf("%s", &day_of_the_week[3][0]); // prints Wednesday
```

The name of a string is considered to be the address of the first character of that string. Stating only the first dimension is effectively referencing the entire string as the second dimension. So the same string within the array shown above can be accessed as

```
printf("%s", day_of_the_week[3]);    // prints Wednesday
```

1.9.4 POINTERS TO FUNCTIONS

One of the more esoteric and powerful aspects of pointers and indirection is that they can also be applied to functions. Using a pointer to a function allows for functions to be called from the result of look-up-table operation. Pointers to functions also allow functions to be passed by reference to other functions. This can be used to create a dynamic flow of execution, which is sometimes called "self-modifying" code.

Consider an example that calls a function from a table of pointers to functions. In the following example, the *scanf()* function gets a value from the standard input. The value is checked to make sure it is in range (1–6). If it is an appropriate value, *func_number* is used as an index into an array of function pointers. The array value at the *func_number* index is assigned to *fp*, which is a pointer to a function of type **void.**

Remember from the "Functions" section (Section 1.8) that the standard form of a function is

```
type function_name ( type param1, type param2, ... )
{
      statement1;
      statement2;
             ...
}
```

which is a type, followed by a name, followed by the primary operator *()*. The parentheses, or function operator, indicate to the compiler that the identifier is to be treated as a function. In this case, *fp* contains the address of a function. The indirection operator is added to obtain the address of the function from the pointer *fp*. Now the function can be called by simply adding the function operator *()*, like this:

```
(*fp)();     // execute the function pointed to by fp
```

Here is the entire program:

```
#include <stdio.h>
```

```
void do_start_task(void)
{
    printf("start selected\n");
}

void do_stop_task(void)
{
    printf("stop selected\n");
}

void do_up_task(void)
{
    printf("up selected\n");
}

void do_down_task(void)
{
    printf("down selected\n");
}

void do_left_task(void)
{
    printf("left selected\n");
}

void do_right_task(void)
{
    printf("right selected\n");
}

void (*task_list[6])(void) = {
   do_start_task,
   do_stop_task,        /* an array of pointers to functions */
   do_up_task,
   do_down_task,
   do_left_task,
   do_right_task,
};

void main(void)
{
      int func_number;
      void (*fp) (void);     /* fp is a pointer to a function */

      while(1)
```

```
        {
                printf("\nSelect a function 1-6 :");
                scanf("%d",&func_number);

                if((func_number > 0) && (func_number < 7))
                {
                        fp = task_list[func_number-1];
                        (*fp)();
                        /* assign the function address to fp */
                }       /* and call the selected function */
        }
    }
```

Chapter 1 Example Project: Part C

In this phase of the development of the Pocket Slot Machine, we are going to use enumerations and multidimensional arrays to "polish" the project's appearance. Instead of printing numbers for the column values and referencing numbers in our payout calculations, we will give these values names.

```
//
// "Slot Machine" -- The Mini-Exercise
//
// Phase 3..

#include <MEGA8515.h>       /* processor specific information */

#define xtal 4000000L       /* quartz crystal frequency [Hz] */
#define baud 9600           /* Baud rate */

#include <stdio.h>          /* this gets the printf() definition */

// Global Variable Declarations..
char    first,second,third; // columns in the slot machine
char    seed;               // number to form a seed for random #s
char    count;              // general purpose counter
int     delay;              // a variable for delay time

#define JACKPOT  3      /* This defines give different payouts */
#define DIME     2      /* names in order to make the code more */
#define NICKEL   1      /* readable to humans. */
#define ZERO     0

enum { BAR, BELL, LEMON, CHERRY };    // the values for each kind


char kind[4][8] = {            // for kinds of names..
```

```
        "BAR",
        "BELL",
        "LEMON",
        "CHERRY"
};

void flash_lights(char n)   // flash the lights a number of times
{
    for(count = 0; count < n; count++)  // flash light when done..
    {
        for(delay = 0; delay < 10000; delay++)
            ;                    // just count up and wait
        PORTB.1 = 0;             // turn the LED on..
        for(delay = 0; delay < 10000; delay++)
            ;
        PORTB.1 = 1;             // turn the LED off..
    }
}

int get_payout(void)
{
    if((first == CHERRY) && (second == CHERRY) && (third == CHERRY))
        return JACKPOT;   // if all "cherries"..
    else if((first == CHERRY) || (second == CHERRY) ||
                   (third == CHERRY))
        return DIME;  // if any are "cherries"..
    else if((first == second) && (second == third))
        return NICKEL;  // if three are alike, of any kind..
    else
        return ZERO;  // otherwise -- you lose..
}

void main(void)
{
     int i;               // declare a local variable for
                          // temporary use..

     PORTA = 0xFF;        // Initialize the I/O ports
     DDRA = 0x00;         // port A all inputs
     DDRB = 0x00;         // port B.1 is an output (light)
     DDRB = 0x02;

  // Initialize the UART control register
  // RX & TX enabled, 8 data bits
     UCSRB=0x18;
  // initialize the UART's baud rate
```

```
UBRRL=xtal/16/baud-1;

while(1)            // do forever..
{
  while(PINA.0)     // Wait for a button and
      seed++;       // let the counter roll over and
                    // over to form a seed.

  first = second = third = seed; // preload the columns

  do                          // Mix up the numbers
  {                           // while waiting for button release.
     first ^= seed>>1;        // Exclusive ORing in the moving seed
     second^= seed>>2;        // can really stir up the numbers.
     third ^= seed>>3;
     seed++;                  // Keep rolling over the seed pattern
  } while(PINA.0 == 0);       // while the button is pressed.

  flash_lights(5);            // flash the lights 5 times..

  first  &= 3;      // limit number to values from 0 to 3
  second &= 3;
  third  &= 3;
                    // show the values.. BY NAME!!
                    // simply change the %d to %s
                    // and pass the pointer to the string from
                    // the 2D array kind[] to printf()
  printf("--> %s, %s, %s <--\n",
           kind[first], kind[second], kind[third]);

                    // determine a payout..
  i = get_payout();

  switch(i)       // now print the payout results
  {
       case    ZERO:
           printf("Paid out: ZERO\n");
           break;

       case    NICKEL:
           printf("Paid out: One Nickel\n");
           break;

       case    DIME:
           printf("Paid out: One Dime\n");
           break;
```

EXAMPLE

```
            case    JACKPOT:
                printf("Paid out: JACKPOT!!\n");
                break;
        }
      }
    }
```

EXAMPLE

1.10 STRUCTURES AND UNIONS

Structures and unions are used to group variables under one heading or name. Since the word "object" in C programming generally refers to a group or association of data members, structures and unions are the fundamental elements in *object-oriented programming*. Object-oriented programming (OOP) refers to the method in which a program deals with data on a relational basis. A structure or union can be thought of as an object. The members of the structure or union are the properties (variables) of that object. The object name, then, provides a means to identify the association of the properties to the object throughout the program.

1.10.1 STRUCTURES

A structure is a method of creating a single data object from one or more variables. The variables within a structure are called members. This method allows for a collection of members to be referenced from a single name. Some high-level languages refer to this type of object as a record or based-variable. Unlike an array, the variables contained within a structure do not need to be of the same type.

A structure declaration has the following form:

```
struct structure_tag_name {           struct structure_tag_name {
   type member_1;                        type member_1;
   type member_2;           or           type member_2;
        ...                                   ...
   type member_x;                        type member_x;
};                                    } structure_var_name;
```

Once a structure template has been defined, the *structure_tag_name* serves as a common descriptor and can be used to declare structures of that type throughout the program. Declared below are two structures, *var1* and *var2*, and an array of structures *var3*:

```
struct structure_tag_name var1, var2, var3[5];
```

Structure templates can contain all sorts of variable types, including other structures, pointers to functions, and pointers to structures. It should be noted that when a template is defined, no memory space is allocated. Memory is allocated when the actual structure variable is declared.

Members within a structure are accessed using the member operator (.). The member operator connects the member name to the structure that it is associated with:

```
structure_var_name.member_1    structure_var_name.member_x
```

Like arrays, structures can be initialized by following the structure name with a list of initializers in braces:

```
struct DATE {
        int month;          // declare a template for a
        int day;            // date structure
        int year;
}
// declare a structure variable and initialize it..
struct DATE date_of_birth = { 2, 21, 1961 };
```

This yields the same result as these assignments:

```
date_of_birth.month = 2;
date_of_birth.day = 21;
date_of_birth.year = 1961;
```

Since structures themselves are a valid type, there is no limit to the nesting of members within a structure. For example, if a set of structures is declared as

```
struct LOCATION {
        int x;
        int y;      // this is the location coordinates x and y
};

struct PART {
        char part_name[20];         // a string for the part name
        long int sku;               // a SKU number for the part
        struct LOCATION bin;        // its location in the warehouse
} widget;
```

To access the location of a "widget," you would provide a reference like this:

```
x_location = widget.bin.x;
                      // the x coordinate of the bin of the widget
y_location = widget.bin.y;
                      // the y coordinate of the bin of the widget
```

To assign the location of the "widget," the same rules apply:

```
widget.bin.x = 10;     // the x coordinate of the bin of the widget
widget.bin.y = 23;     // the y coordinate of the bin of the widget
```

A structure can be passed to a function as a parameter as well as returned from a function. For example, the function

```
struct PART new_location( int x, int y)
{
   struct PART temp;

   temp.part_name = "";      // initialized the name to NULL
   temp.sku = 0;             // zero the sku number
   temp.bin.x = x;           // set the location to the passed x and y
   temp.bin.y = y;
```

```
    return temp;          // and then returns the structure to the caller
}
```

would return a PART structure with the *bin.x* and *bin.y* location members assigned to the parameters passed to the function. The *sku* and *part_name* members would also be cleared before the structure was returned. The function above could then be used in an assignment like

```
widget = new_location(10,10);
```

The result would be that the *part_name* and *sku* would be cleared and the *widget.bin.x* and *widget.bin.y* values would be set to 10.

1.10.2 ARRAYS OF STRUCTURES

As with any other variable type, arrays of structures can also be declared. The declaration of an array of structures appears as follows:

```
struct PART {
        char part_name[20];         // a string for the part name
        long int sku;               // a SKU number for the part
        struct LOCATION bin;        // its location in the warehouse
} widget[100];
```

The access of a member is still the same. The only difference is in the indexing of the structure variable itself. So, to access a "particular widget's location," a reference like this may be used:

```
x_location = widget[12].bin.x;
              // the x coordinate of the bin of the widget 12
y_location = widget[12].bin.y;
              // the y coordinate of the bin of the widget 12
```

In this example, there is a character string, *part_name*, which can be accessed as strings normally are:

```
widget[12].part_name;     // the name of widget 12

      or

widget[12].part_name[0];
            // the first character in the name of widget 12
```

Arrays of structures can be initialized by following the structure name with a list of initializers in braces; there simply needs to be an initializer for each structure element within each array element:

```
struct DATE {
      int month;
      int day;
      int year;
}
```

```
struct DATE birthdates[3] = {  2, 21, 1961,
                               8,  8, 1974,
                               7, 11, 1997 };
```

1.10.3 POINTERS TO STRUCTURES

Sometimes it is desirable to manipulate the members of a structure in a generalized fashion. One method of doing this is to use a pointer to reference the structure, for example, passing a pointer to a structure to a function instead of passing the entire structure.

A pointer to a structure is declared as follows:

```
struct structure_tag_name *structure_var_name;
```

The pointer operator (*) states that *structure_var_name* is a pointer to a structure of type *structure_tag_name*. Just as with any other type, a pointer must be assigned a value that points to something tangible, like a variable that has already been declared. A variable declaration guarantees memory has been allocated for a purpose.

The following code would be used to declare the structure variable, *widget*, and a pointer to a structure variable, *this_widget*. The final line in the example assigns the address of *widget* to the pointer *this_widget*.

```
struct LOCATION {
      int x;
      int y;        // this is the location coordinates x and y
};

struct PART {
      char part_name[20];      // a string for the part name
      long int sku;            // a SKU number for the part
      struct LOCATION bin;     // its location in the warehouse
};

struct PART widget, *this_widget;
      // declare a structure and a pointer to a structure

  . . .

this_widget = &widget;
         // assign the pointer the address of a structure
```

When a pointer is used to reference a structure, the structure pointer operator, -> (minus-greater-than), is used to access the members of the structure through indirection:

```
this_widget->sku = 1234;
```

This could also be stated using the indirection operator to first locate the structure and then using the member operator to access the *sku* member:

```
(*this_widget).sku = 1234;
```

Since *this_widget* is a pointer to *widget*, both methods of assignment, shown above, are valid. The parentheses around *this_widget* are required because the member operator has a higher precedence than the indirection (*) operator. If the parentheses were omitted, the expression would be misinterpreted as

```
*(this_widget.partname[0])
```

which is the actual address of *widget* (&widget).

Structures can contain pointers to other structures but can also contain pointers to structures of the same type. A structure cannot contain itself as a member, because that would be a recursive declaration, and the compiler would lack the information required to resolve the declaration. Pointers are always the same size regardless of what they point to. Therefore, by pointing to a structure of the same type, a structure can be made "self-referential." A very basic example would be as follows:

```
struct LIST_ITEM {
        char *string;        // a text string
        int position;        // its position in a list
        struct LIST_ITEM *next_item; // a pointer to another
                                     // structure of the same type

} item, item2;

item.next_item = &item2;
        // assign the pointer with the address of the item2
```

Now

```
item.next_item->position
```

is the *position* member of the structure pointed to by *next_item*. This would be equivalent to

```
item2.position
```

Self-referential structures are typically used for data manipulations like linked-lists and quick sorts.

1.10.4 UNIONS

A union is declared and accessed much like a structure. A union declaration has the following form:

```
union union _tag_name {            union union _tag_name {
        type member_1;                          type member_1;
        type member_2;      or                  type member_2;
             ...                                ...
        type member_x;                          type member_x;
};                                 } union _var_name;
```

The primary difference between a union and a structure is in the way the memory is allocated. The members of a union actually share a common memory allocated to the largest member of that union:

```
union SOME_TYPES {
      char character;
      int integer;

      long int long_one;
} my_space;
```

In this example, the total amount of memory allocated to *my_space* is equivalent to the size of the long int *long_one* (4 bytes). If a value is assigned to the long int,

```
my_space.long_one = 0x12345678L;
```

then the value of *my_space.character* and *my_space.integer* are also modified. In this case, their values would now be

```
my_space.character = 0x12;
my_space.integer = 0x1234;
```

Unions are sometimes used as a method of preserving valuable memory space. If there are variables that are used on a temporary basis, and there is never a chance that they will be used at the same time, a union is a method for defining a "scratch pad" area of memory.

More often, a union is used as a method of extracting smaller parts of data from a larger data object. This is shown in the previous example. The position of the actual data depends on the data types used and how a particular compiler handles numbers larger than type char (8 bits). The example above assumes big-endian (most significant byte first) storage. Compilers vary in how they store data. The data could be byte order swapped, word order swapped, or both! This example could be used as a test for a compiler in order to find out how the data is organized in memory.

Union declarations can save steps in coding to convert the format of data from one organization to another. Shown below are two examples where two 8-bit input ports are combined in one 16-bit value. The first method uses shifting and combining; the second method uses a union:

```
#include <stdio.h>
#include <MEGA8535.h>    // register definition header file for
                         // an Atmel ATMEGA8535
void main(void)
{
    unsigned int port_w;

    while(1)
    {
          port_w = PINA;    // get port A into the 16 bit value
          port_w <<= 8;     // shift it up..
          port_w |= PINB;   // now combine in port B
```

```
            printf("16-Bits = %04X\n",port_w);
        }                        // put the combined value out to
    }                            // the standard output
```

Now the same results using a **union** declaration:

```
    #include <stdio.h>
    #include <MEGA8535.h>  // register definition header file for
                           // an Atmel ATMEGA8535

    void main(void)
    {
                           // declare the two types of data in
                           // a union to occupy the same space..
        union {
            unsigned char port_b[2];
            unsigned int port_w;
        } value;

        while(1)
        {
            value.portb[0] = PINA;   // get port A
            value.portb[1] = PINB;   // get port B
                                  // the union eliminates the
                                  // need to combine the data
                                  // manually
            printf("16-Bits = %04X\n",value.port_w);
        }                         // put the combined value out to
    }                             // the standard output
```

1.10.5 TYPEDEF OPERATOR

The C language supports an operation that allows for creating new type names. The **typedef** operator allows for a name to be declared synonymous with an existing type. For example,

```
        typedef  unsigned char byte;
        typedef  unsigned int word;
```

Now the aliases "byte" and "word" can be used to declare other variables that are actually of type **unsigned char** and **unsigned int.**

```
        byte var1;   // this is the same as an unsigned char
        word var2;   // this is the same as an unsigned int
```

This method of alias works for structures and unions as well:

```
        typedef struct
        {
            char name[20];
            char age;
            int home_room_number;
        } student;
```

```
student Bob;        // these allocate memory in the form of
student Sally;      // a structure of type student
```

The **#define** statement is sometimes used to perform this alias operation through a text substitution in the compiler's preprocessor. **typedef** is evaluated by the compiler directly and can work with declarations, castings, and usages that would exceed the capabilities of the preprocessor.

1.10.6 BITS AND BITFIELDS

Bits and bitfields are often used when memory space is at a premium. Some compilers support a type **bit,** which is automatically allocated by the compiler and is referenced as a variable of its own. For example,

```
bit running; // the compiler allocates a single bit of
             // storage for this flag..

running = 1; // the only two possibilities for value are 1 and 0
running = 0;

if(running)
      ;      // the value can be tested as well as assigned
```

Unlike bits, bitfields are more common for larger, more generalized systems and are not always supported by embedded compilers. Bitfields are associated with structures because of the form of their declaration:

```
struct structure_tag_name {           struct structure_tag_name {
     unsigned int bit_1:1;                  unsigned int bit_1:1;
     unsigned int bit_2:1;    or            unsigned int bit_2:1;
         ...                                    ...
     unsigned int bit_15:1;                 unsigned int bit_15:1;
};                                    } struct_var_name;
```

A bitfield is specified by a member name (of type **unsigned int** only), followed by a colon (:) and the number of bits required for the value. The width of the member can be from 1 to the size of type **unsigned int** (16 bits). This allows several bitfields within a structure to be represented in a single unsigned integer memory location.

```
struct {
      unsigned int      running : 1;
      unsigned int      stopped : 1;
      unsigned int      counter : 4;
} machine;
```

These bitfields can be accessed by member name, just as you would for a structure:

```
machine.stopped = 1;      // these are single bits, so only
                          // 1 and 0 are allowed..
machine.running = 0;
```

```
machine.counter++;          // this is 4-bit value, so
                            // 0-15 are allowed..
```

Sometimes in embedded systems bitfields are used to describe I/O port pins:

```
typedef struct
{
        bit_0: 1;
        bit_1: 1;
        bit_2: 1;
        bit_3: 1;
        bit_4: 1;
        bit_5: 1;
        bit_6: 1;
        bit_7: 1;
} bits;
```

The following **#define** labels the contents of memory at location 0x38. (Remember, **#define** is treated as a textual substitution directive by the preprocessor.) This allows the programmer to access memory location 0x38 by the name *PORTB*, as if it were a variable, throughout the program:

```
// PORTB is the contents of address 0x38, a bitfield "bits"
#define PORTB (*(bits *) 0x38)
```

The *bits* bitfield allows the individual bits of I/O port *PORTB* to be accessed independently (sometimes called "bit banged"):

```
// bitfields can be used in assignments
PORTB.bit_3 = 1;
PORTB.bit_5 = 0;

// bitfields can also be used in conditional expressions
if(PORTB.bit_2 || PORTB.bit_1)
        PORTB.bit_4 = 1;
```

1.10.7 SIZEOF OPERATOR

The C language supports a unary operator called **sizeof.** This operator is a compile-time feature that creates a constant value related to the size of a data object or its type. The forms of the **sizeof** operation are as follows:

```
sizeof( type_name ) // type_name could be keyword int,
                    // char, long, etc.

sizeof( object )    // object could be a variable, array,
                    // structure, or union variable name
```

These operations produce an integer that reveals the size (in bytes) of the type or data object located in the parentheses. For example, assuming these declarations,

```
int value,x;
long int array[2][3];
```

```
struct record
{
      char  name[24];
      int     id_number;
} students[100];
```

here are some of the resulting possibilities:

```
x = sizeof(int);       // this would set x=2, since an int is 2 bytes
x = sizeof(value);     // this would set x=2, since value is an int
x = sizeof(long);      // this would set x=4, since a long is 4 bytes

x = sizeof(array);
            // x = sizeof(long)*array_width*array_length= 4*2*3 = 24

x = sizeof(record);
            // x = 24+2 for the character string plus the integer

x = sizeof(students);
               // x = 100 Elements * (24 characters + sizeof(int))
               // x = 100*(24+2) = 2600 !!
```

1.11 MEMORY TYPES

The architecture of a microprocessor may require that variables and constants be stored in different types of memory. Data that will not change should be stored in one type of memory, while data that must be read from and written to repetitively in a program should be stored in another type of memory. A third type of memory can be used to store variable data that must be retained even when power is removed from the system. When special memory types such as pointers and register variables are accessed, additional factors must be considered.

1.11.1 CONSTANTS AND VARIABLES

The AVR microcontroller was designed using Harvard architecture, with separate address spaces for data (SRAM), program (FLASH), and EEPROM memory. The CodeVisionAVR® and other compilers implement three types of memory descriptors to allow easy access to these very different types of memory.

The default or automatic allocation of variables, where no memory descriptor keyword is used, is in SRAM. Constants can be placed in FLASH memory (program space) with the **flash** or **const** keywords. For variables to be placed in EEPROM, the **eeprom** keyword is used.

When declarations are made, the positions of the **flash** and **eeprom** keywords become part of the meaning. If **const, flash,** or **eeprom** appear first, this states to the compiler that the actual allocation of storage or the location of data is in that memory area. If the type is declared followed by the **flash** or **eeprom** keyword, this indicates that it is a variable that *references* FLASH or EEPROM, but the variable itself is physically located in SRAM. This scenario is used when declaring pointers into FLASH or EEPROM.

The following declarations will place physical data directly in program memory (FLASH). These data values are all constant and cannot be changed in any way by program execution:

```
flash int  integer_constant = 1234 + 5;
flash char char_constant = 'a';
flash long long_int_constant1 = 99L;
flash long long_int_constant2 = 0x10000000;
flash int  integer_array1[] = {1,2,3};

// The first two elements will be 1 and 2, the rest will be 0.
flash int  integer_array2[10] = {1,2};

flash int  multidim_array[2,3] = {{1,2,3},{4,5,6}};
flash char string_constant1[] = "This is a string constant";
const char string_constant2[] = "This is also a string constant";

flash struct {
          int  a;
          char b[3], c[3];
} sf = {{0x000a},{0xb1,0xb2,0xb3},{0xb1,0xb2,0xb3}};
```

EEPROM space is a non-volatile yet variable area of memory. Variables can be placed in EEPROM memory simply by declaration:

```
eeprom int cycle_count;      // allocates an integer space in EEPROM
eeprom char ee_string[20];  // allocates a 20-byte area in EEPROM

eeprom struct  {
      char a;
      int  b;
      char c[15];
} se;                  // allocates 18-byte structure "se" in EEPROM
```

These permanent (FLASH) and semi-permanent (EEPROM) memory areas have many system-specific uses in the embedded world. FLASH space is an excellent area for non-changing data. The program code itself resides in this region. Declaring items such as text strings and arithmetic look-up-tables in this region directly frees up valuable SRAM space.

If a string is declared with an initializer like

```
char mystring[30] = "This string is placed in SRAM";
```

30 bytes of SRAM will be allocated, and the text "This string is placed in SRAM" is physically placed in FLASH memory with the program. On startup, this FLASH-resident data is copied to SRAM and the program works from SRAM whenever accessing *mystring*. This is a waste of 30 bytes of SRAM unless the string is intended for alteration by the program during run time. To prevent this loss of SRAM space, the string could be stored in FLASH memory directly by the declaration:

```
flash char mystring[30] = "This string is placed in SRAM";
```

The EEPROM area is called non-volatile, meaning that when power is removed from the microprocessor the data will remain intact, but it is semi-permanent in that the program can alter the data located in this region. EEPROM also has a life—it has a maximum number of write cycles that can be performed before it will electrically fail. This is due to the way that EEPROM memory itself is constructed, a function of electro-chemistry. In many cases, this memory area will have a rating of 10,000 write operations, maximum. Newer technologies are being developed all the time, increasing the number of write operation to the hundreds of thousands, even millions. There are no limitations on the number of times the data can be read.

It is important to understand this physical constraint when designing software that uses EEPROM. Data that needs to be kept and does not change frequently can be stored in this area. This region is great for low-speed data logging, calibration tables, runtime hour meters, and software setup and configuration values.

1.11.2 POINTERS

Pointers to these special memory regions are each handled differently during program execution. Even though the pointers can point to FLASH and EEPROM memory areas, the pointers themselves are always stored in SRAM. In these cases, the allocations are normal (**char, int,** and so on) but the type of memory being referenced must be described so that the compiler can generate the proper code for accessing the desired region. The **flash** and **eeprom** keywords in these cases are used to expound on the name, like this:

```
                /* Pointer to a string that is located in SRAM */
char *ptr_to_ram = "This string is placed in SRAM";

                /* Pointer to a string that is located in FLASH */
char flash *ptr_to_flash = "This string is placed in FLASH";

                /* Pointer to a string that is located in EEPROM */
char eeprom *ptr_to_eeprom = "This string is placed in EEPROM";
```

1.11.3 REGISTER VARIABLES

The SRAM area of the AVR microcontroller includes a region called the Register File. This region contains I/O ports, timers, and other peripherals, as well as some "working" or "scratch pad" area. To instruct the compiler to allocate a variable to a register or registers, the storage class modifier **register** must be used.

```
register int abc;
```

The compiler may choose to automatically allocate a variable to a register or registers, even if this modifier is not used. In order to prevent a variable from being allocated to a register or registers, the **volatile** modifier must be used. This warns the compiler that the variable may be subject to outside change during evaluation.

```
volatile int abc;
```

The **volatile** modifier is frequently used in applications where variables are stored in SRAM while the microcontroller sleeps. (Sleep is a stopped, low-power mode typically used in battery applications.) This allows the value in SRAM to be valid each time the microcontroller is awakened and returned to normal operation.

Global variables that have not been allocated to registers are stored in the General or Global Variable area of SRAM. Local variables that have not been allocated to registers are stored in dynamically allocated space in the Data Stack or Heap Space area of SRAM.

sfrb and sfrw

The I/O ports and peripherals are located in the register file. Therefore, special instructions are used to indicate to the compiler the difference between an SRAM location, an I/O port, or another peripheral in the register file.

The **sfrb** and **sfrw** keywords indicate to the compiler that the **IN** and **OUT** assembly instructions are to be used to access the AVR microcontroller's I/O registers:

```
/* Define the Special Function Registers (SFR)*/

sfrb PINA=0x19;             // 8-bit access to the SFR input port A

sfrb TCNT1L=0x2c;           // 8-bit access to lower part of timer
sfrb TCNT1H=0x2d;           // 8-bit access to upper part of timer
sfrw TCNT1=0x2c;    // 16-bit access to the timer (unsigned int)

void main(void)
{
        unsigned char a;

        a=PINA;             // Read PORTA input pins
        TCNT1=0x1111;       // Write to TCNT1L & TCNT1H registers

        while(1)
            ;
}
```

The bit-level access to the I/O registers is allowed by using bit selectors appended after the name of the I/O register. Because bit-level access to I/O registers is done using the **CBI, SBI, SBIC,** and **SBIS** assembly language instructions, the register address must be in the 0x00 to 0x1F range for declarations using **sfrb,** and in the 0x00 to 0x1E range for declarations using **sfrw.** For example,

```
sfrb PORTA=0x1b;
sfrb DDRA=0x18;
sfrb PINA=0x19;

void main(void)
```

```
{
        DDRA.0   = 1;       // set bit 0 of Port A as output
        DDRA.1   = 0;       // set bit 1 of Port A as input
        PORTA.0  = 1;       // set bit 0 of Port A to a 1

        while(1)
        {
              if (PINA.1)          // test bit 1 input of Port A
                    PORTA.0 = 0;
        }
}
```

Usually the **sfrb** and **sfrw** keywords are found in a header files included at the top of the program with the **#include** preprocessor directive. These header files are typically related to a particular processor. The header files provide predetermined names for the I/O and other useful registers in the microcontroller being used in a particular application.

Listed below is a typical include file that might be provided with a compiler (MEGA8515.h):

```
// I/O registers definitions for the ATMEGA8515
      sfrb OSCCAL=4;
      sfrb PINE=5;
      sfrb DDRE=6;
      sfrb PORTE=7;
      sfrb ACSR=8;
      sfrb UBRRL=9;
      sfrb UCSRB=0xa;
      sfrb UCSRA=0xb;
      sfrb UDR=0xc;
      sfrb SPCR=0xd;
      sfrb SPSR=0xe;
      sfrb SPDR=0xf;
      sfrb PIND=0x10;
      sfrb DDRD=0x11;
      sfrb PORTD=0x12;
      sfrb PINC=0x13;
      sfrb DDRC=0x14;
      sfrb PORTC=0x15;
      sfrb PINB=0x16;
      sfrb DDRB=0x17;
      sfrb PORTB=0x18;
      sfrb PINA=0x19;
      sfrb DDRA=0x1a;
      sfrb PORTA=0x1b;
      sfrb EECR=0x1c;
      sfrb EEDR=0x1d;
```

```
sfrb EEARL=0x1e;
sfrb EEARH=0x1f;
sfrw EEAR=0x1e;    // 16-bit access
sfrb UBRRH=0x20;
sfrb UCSRC=0x20;
sfrb WDTCR=0x21;
sfrb ICR1L=0x24;
sfrb ICR1H=0x25;
sfrw ICR1=0x24;    // 16-bit access
sfrb OCR1BL=0x28;
sfrb OCR1BH=0x29;
sfrw OCR1B=0x28;  // 16-bit access
sfrb OCR1AL=0x2a;
sfrb OCR1AH=0x2b;
sfrw OCR1A=0x2a;  // 16-bit access
sfrb TCNT1L=0x2c;
sfrb TCNT1H=0x2d;
sfrw TCNT1=0x2c;  // 16-bit access
sfrb TCCR1B=0x2e;
sfrb TCCR1A=0x2f;
sfrb SFIOR=0x30;
sfrb OCR0=0x31;
sfrb TCNT0=0x32;
sfrb TCCR0=0x33;
sfrb MCUCSR=0x34;
sfrb MCUCR=0x35;
sfrb EMCUCR=0x36;
sfrb SPMCR=0x37;
sfrb TIFR=0x38;
sfrb TIMSK=0x39;
sfrb GIFR=0x3a;
sfrb GICR=0x3b;
sfrb SPL=0x3d;
sfrb SPH=0x3e;
sfrb SREG=0x3f;
#pragma used-

// Interrupt vectors definitions

#define EXT_INT0 2
#define EXT_INT1 3
#define TIM1_CAPT 4
#define TIM1_COMPA 5
#define TIM1_COMPB 6
#define TIM1_OVF 7
```

```
#define TIM0_OVF 8
#define SPI_STC 9
#define USART_RXC 10
#define USART_UDRE 11
#define USART_TXC 12
#define ANA_COMP 13
#define EXT_INT2 14
#define TIM0_COMP 15
#define EE_RDY 16
#define SPM_RDY 17
```

1.12 REAL-TIME METHODS

Real-time programming is sometimes misconstrued as some sort of complex and magical process that can only be performed on large machines with operating systems like Linux or Unix. Not so! Embedded systems can, in many cases, perform on a more real-time basis than a large system.

A simple program may run its course over and over. It may be able to respond to changes to the hardware environment it operates in, but it will do so in its own time. The term "real-time" is used to indicate that a program function is capable of performing all of its functions in a regimented way within a certain allotment of time. The term may also indicate that a program has the ability to respond immediately to outside (hardware input) stimulus.

The AVR, with its rich peripheral set, has the ability to not only respond to hardware timers, but to input changes as well. The ability of a program to respond to these real-world changes is called interrupt or exception processing.

1.12.1 USING INTERRUPTS

An interrupt is just that, an exception, change of flow, or interruption in the program operation caused by an external or internal hardware source. An interrupt is, in effect, a hardware-generated function call. The result is that the interrupt will cause the flow of execution to pause while the interrupt function, called the interrupt service routine (ISR), is executed. Upon completion of the ISR, the program flow will resume, continuing from where it was interrupted.

In an AVR, an interrupt will cause the status register and program counter to be placed on the stack, and, based on the source of the interrupt, the program counter will be assigned a value from a table of addresses. These addresses are referred to as vectors.

Once a program has been redirected by interrupt vectoring, it can be returned to normal operation through the machine instruction **RETI** (RETurn from Interrupt). The **RETI** instruction restores the status register to its pre-interrupt value and sets the program counter to the next machine instruction following the one that was interrupted.

There are many sources of interrupts available on the AVR microcontroller. The larger the AVR, the more sources that are available. Listed below are some of the possibilities. These definitions are usually found in a header file, at the top of program, and are specific to a given microprocessor. These would be found in a header for an ATMega128, Mega128.h:

```
// Interrupt vectors definitions

#define EXT_INT0 2
#define EXT_INT1 3
#define EXT_INT2 4
#define EXT_INT3 5
#define EXT_INT4 6
#define EXT_INT5 7
#define EXT_INT6 8
#define EXT_INT7 9
#define TIM2_COMP 10
#define TIM2_OVF 11
#define TIM1_CAPT 12
#define TIM1_COMPA 13
#define TIM1_COMPB 14
#define TIM1_OVF 15
#define TIM0_COMP 16
#define TIM0_OVF 17
#define SPI_STC 18
#define USART0_RXC 19
#define USART0_DRE 20
#define USART0_TXC 21
#define ADC_INT 22
#define EE_RDY 23
#define ANA_COMP 24
#define TIM1_COMPC 25
#define TIM3_CAPT 26
#define TIM3_COMPA 27
#define TIM3_COMPB 28
#define TIM3_COMPC 29
#define TIM3_OVF 30
#define USART1_RXC 31
#define USART1_DRE 32
#define USART1_TXC 33
#define TWI 34
#define SPM_RDY 35
```

This list contains a series of vector indices associated with a name describing the interrupt source. To create an ISR, the function that is called by the interrupt system, the ISR, is declared using the reserved word **interrupt** as a function type modifier.

```
interrupt [EXT_INT0] void external_int0(void)
```

```
{
      /* Called automatically on external interrupt 0 */
}
```

or

```
interrupt [TIM0_OVF] void timer0_overflow(void)
{
      /* Called automatically on TIMER0 overflow */
}
```

The **interrupt** keyword is followed by an index, which is the vector location of the interrupt source. The vector numbers start with [1], but since the first vector is the reset vector, the actual interrupt vectors available to the programmer begin with [2].

ISRs can be executed at any time, once the interrupt sources are initialized and the global interrupts are enabled. The ISR cannot return any values, since technically there is no "caller," and it is always declared as type **void.** It is for this same reason that nothing can be passed to the ISR.

Interrupts in an embedded environment can genuinely create a real-time execution. It is not uncommon in a peripheral-rich system to see an empty *while(1)* loop in the *main()* function. In these cases, *main()* simply initializes the hardware and the interrupts perform all the necessary tasks when they need to happen.

```
// I/O register definitions for ATMEGA8515
#include <MEGA8515.h>

// quartz crystal frequency [Hz]
#define XTAL 4000000

// LED blink frequency [Hz]
#define Blink_Rate 2

// TIMER1 overflow interrupt service routine
// occurs every 0.5 seconds

interrupt [TIM1_OVF] void timer1_overflow(void)
{
      // preset again TIMER1
      TCNT1=0x10000-( XTAL /1024/ Blink_Rate);

      PORTB.0 ^= 1;      // toggle the LSB of PORTB
}

void main(void)
{
      // set the I/O ports and initialize TIMER1
```

```
        DDRB=0x01;    // PORTB.0 pin is an output

        TCCR1A=0;     // TIMER1 is disconnected from pin OC1, no PWM
        TCCR1B=5;     // TIMER1 clock is xtal/1024

        TCNT1=0x10000-(XTAL/1024/ Blink_Rate);   // preset TIMER1

        TIFR=0;       // clear any TIMER1 interrupt flags

        TIMSK=0x80; // enable TIMER1 overflow interrupt

        GIMSK=0;      // all other interrupt sources are disabled

        #asm("sei");        // global enable interrupts

        while (1)
              ;
    // the rest is done by the "timer1_overflow" function
}
```

In this example, *main()* is used to initialize the system and enable the TIMER1 overflow interrupt. The interrupt function *timer1_overflow* blinks the LED every half second. Note that *main()* sits in a *while(1)* loop while the blinking goes on.

1.12.2 REAL-TIME EXECUTIVES

A Real-Time Executive (RTX) or Real-Time Operating System (RTOS) is a program, commonly referred to as a "kernel," that coordinates the management and execution of multiple sub-programs or "tasks." An RTX strictly coordinates program runtime operations, while an RTOS is usually associated with extra functionality such as file management and other generalized I/O operations. There are many real-time executives available for the AVR. Some are in the public domain and are free of charge. Others are commercially licensed and supported but are still inexpensive. In this section, we are only going to touch on the RTX, specifically the Progressive Resources' PR_RTX that is readily available and simple to configure and use.

An RTX utilizes a time-based interrupt (i.e., TIMER0) to guarantee that various subprogram executions happen on time and as often as is required to obtain the desired operational results. The beauty of an RTX is that the program operation, from a timing perspective, is defined in a header block and the tasks are usually configured with a call into an RTX initialization function. Characteristics such as the number of tasks allowed to be executing within the system, how often the operation is to switch from one task to another, and whether tasks are to run in a round-robin or top-to-bottom (sometimes called preemptive or priority) fashion, are defined by the header. Since each subprogram stands somewhat alone, each is given its own stack space to work in—keeping the processes truly independent. For example,

```
// PR_RTX Control Block *************************************
// System crystal frequency [Hz]
#define PR_Xtal 14745600L

// Task switch time [microseconds]
#define PR_Switch_Time 5000

// Operating mode 1 = round robin
// 2 = priority
#define PR_op_mode 2

// Set low point in RAM for the task stacks.
#define PR_Stack_Ram_Low 0x60

// Define the stack sizes for the default task. These
// are optional defines. If either is not present, the default
// task gets the same default stack sizes as the other tasks.
#define PR_default_task_data_stack 40
#define PR_default_task_system_stack 80

// Define the maximum number of tasks for the project.
#define PR_Max_Tasks 10

// Define the prescaler divisor type. See docs for details.
#define PR_Divisor_Type_1

#include "PR_RTX_Control.c"
```

Once the operational characteristics are defined, the RTX is initialized with definitions of what tasks you wish to have run and the priority you would like to have the tasks assigned to. Priority is important in that if for some reason a task could not complete its function within the allocated amount of time, the RTX can take the operation back to the highest priority task and start again. This allows for the important and generally time-critical tasks to get processor time no matter what is going on with lower priority or less important tasks. In the following example, two LEDs are being flashed by two independent tasks. You will see that each task is simply written as a C function of type **void** and that each function contains a *while(1)* loop. This is because the functions do not return to a caller—they are each running as stand-alone programs.

```
void blnkgrn (void)
{
   while (1)
   {
      PORTC ^= 0x01;        //change green LED state
      PR_Wait_Ticks(40);    //wait 200 ms
   }
}
```

```
void blnkred (void)
{
   unsigned char x = 255;   //start slow blink
   PORTC |= 0x2;            //start with light on

   while (1)
   {
      PORTC ^= 0x02;       //change red LED state
      PR_Wait_Ticks(x);    //wait 500 ms
      x = 10;              //change to fast blink after first blink
   }

}

void PR_Init_Tasks(void)
{
   PR_Create_Task(0,blnkred,0,0,1); //start task 0
   PR_Create_Task(1,blnkgrn,0,0,1); //start task 1
}

void main(void)
{
   // Declare your local variables here
   // Input/Output Ports initialization

   // Port C
   PORTC=0x00;
   DDRC=0x03;

   PR_RTX();               //Start RTX

   // The main program's while(1) loop becomes
   // the 'default' or lowest priority task.
   while(1)
   {
        // Other operations can be done here
        // if desired.
   }
}
```

The *while(1)* loop in the function *main()* becomes the "default" or lowest-level task. Since each task is responsible for releasing time back to the system using the *PR_Wait_Ticks()* function, that unused system time is turned over to *main()*. In a typical application development, the slow operations such as *printf()* or *scanf()* would be performed in *main()* and global variables would be used to communicate needed information to the other tasks in the program.

1.12.3 STATE MACHINES

State machines are a common method of structuring a program such that it never sits idle waiting for input. State machines are generally coded in the form of a **switch/case** construct, and flags are used to indicate when the process is to move from its current state to its next state. State machines also offer a better opportunity to change the function and flow of a program without a rewrite, simply because states can be added, changed, and moved without impacting the other states that surround it.

State machines allow for the primary logical operation of a program to happen somewhat in background. Since typically very little time is spent actually processing each state, more free CPU time is left available for time-critical tasks like gathering analog information, processing serial communications, and performing complex mathematics. The additional CPU time is often devoted to communicating with humans: user interfaces, displays, keyboard services, data entry, alarms, and parameter editing.

Figures 1–1 and 1–2 show an example state machine used to control an "imaginary" traffic light.

The state machine in this example uses PORTB to drive the red, green, and yellow lights in the North-South and East-West directions. Note that in *main()* the *delay_ms()* function is used to control the time. This keeps the example simple. In real life, this time could be used for a myriad of other tasks and functions, and the timing of the lights could be controlled by an interrupt from a hardware timer.

The state machine *Do_States()* is called every second, but it executes only a few instructions before returning to *main()*. The global variable *current_state* controls the flow of the execution through *Do_States()*. The *PED_XING_EW*, *PED_XING_NS*, and *FOUR_WAY_STOP* inputs create exceptions to the normal flow of the machine by altering either the timing, the path of the machine, or both.

The example system functions as a normal stoplight. The North-South light is green when the East-West light is red. The traffic is allowed to flow for thirty-second periods in each direction. There is a five-second yellow warning light during the change from a green-light to a red-light condition. If a pedestrian wishes to cross the flowing traffic, pressing the *PED_XING_EW* or *PED_XING_NS* button will cause the time remaining for traffic flow to be shortened to no longer than ten seconds before changing directions. If the *FOUR_WAY_STOP* switch is turned on, the lights will convert over to all four flashing red. This conversion will happen only during the warning (yellow) light transition to make it safe for traffic.

User interfaces, displays, keyboard services, data entry, alarms, and parameter editing can also be performed using state machines. The more "thinly sliced" a program becomes through the use of flags and states, the more real-time it is. More things are being dealt with continuously without becoming stuck waiting for a condition to change.

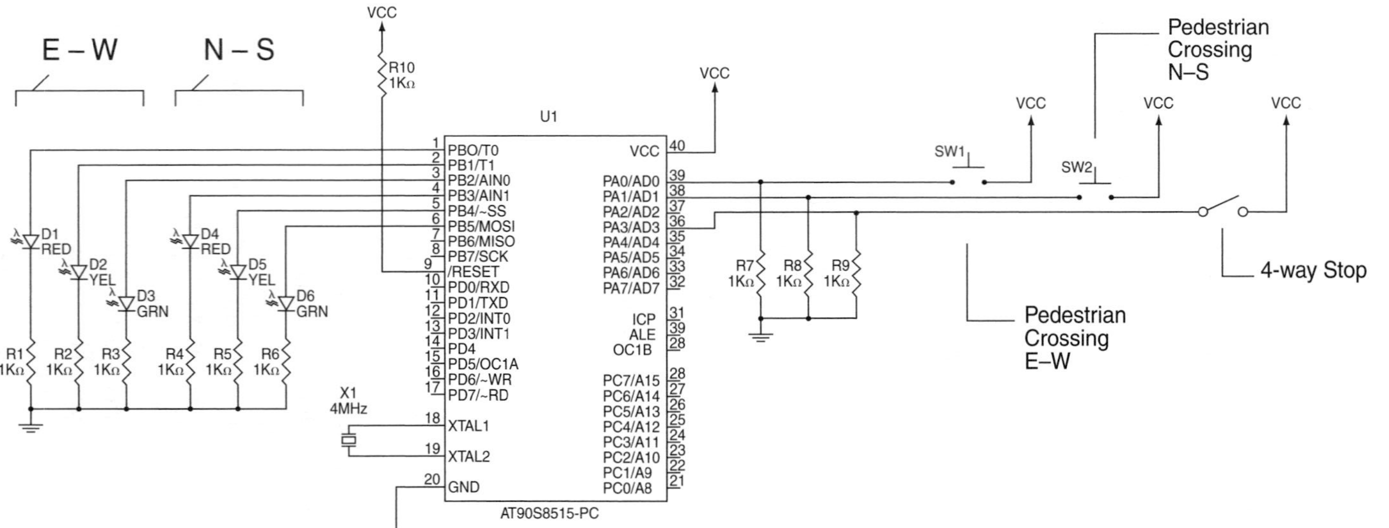

Figure I–I *Imaginary Traffic Light Schematic*

```
#include <Mega8515.h>
#include <delay.h>

#define   EW_RED_LITE PORTB.0    /* definitions to actual outputs */
#define   EW_YEL_LITE PORTB.1    /* used to control the lights */
#define   EW_GRN_LITE PORTB.2
#define   NS_RED_LITE PORTB.3
#define   NS_YEL_LITE PORTB.4
#define   NS_GRN_LITE PORTB.5

#define   PED_XING_EW PINA.0    /* pedestrian crossing push button */
#define   PED_XING_NS PINA.1    /* pedestrian crossing push button */
#define   FOUR_WAY_STOP PINA.3     /* switch input for 4-Way Stop */

char time_left;          // time in seconds spent in each state
int current_state;       // current state of the lights
char flash_toggle;       // toggle used for FLASHER state

// This enumeration creates a simple way to add states to the machine
// by name. Enumerations generate an integer value for each name
// automatically, making the code easier to maintain.

enum { EW_MOVING , EW_WARNING , NS_MOVING , NS_WARNING , FLASHER };

// The actual state machine is here..
void Do_States(void)
{
    switch(current_state)
    {
      case  EW_MOVING:          // east-west has the green!!
            EW_GRN_LITE = 1;
            NS_GRN_LITE = 0;
            NS_RED_LITE = 1;    // north-south has the red!!
            EW_RED_LITE  = 0;
            EW_YEL_LITE = 0;
            NS_YEL_LITE = 0;

            if(PED_XING_EW || FOUR_WAY_STOP)
            {      // pedestrian wishes to cross, or
                   // a 4-way stop is required
                   if(time_left > 10)
                         time_left = 10;    // shorten the time
            }
            if(time_left != 0)              // count down the time
            {
                   --time_left;
                   return;                  // return to main
```

Figure 1–2 *Imaginary Traffic Light Software (Continues)*

```
        }                    // time expired, so..
        time_left = 5;       // give 5 seconds to WARNING
        current_state = EW_WARNING;
                                         // time expired, move
        break;                           // to the next state
    case  EW_WARNING:
        EW_GRN_LITE = 0;
        NS_GRN_LITE = 0;
        NS_RED_LITE = 1;     // north-south has the red..
        EW_RED_LITE  = 0;
        EW_YEL_LITE = 1;     // and east-west has the yellow
        NS_YEL_LITE = 0;

        if(time_left != 0)               // count down the time
        {
              --time_left;
              return;                    // return to main
        }                    // time expired, so..
        if(FOUR_WAY_STOP)    // if 4-way requested then start
              current_state = FLASHER; // the flasher
        else
        {                          // otherwise..
              time_left = 30;      // give 30 seconds to MOVING
              current_state = NS_MOVING;
        }                          // time expired, move
        break;                     // to the next state

    case  NS_MOVING:
        EW_GRN_LITE = 0;
        NS_GRN_LITE = 1;
        NS_RED_LITE = 0;     // north-south has the green!!
        EW_RED_LITE  = 1;    // east-west has the red!!
        EW_YEL_LITE = 0;
        NS_YEL_LITE = 0;

        if(PED_XING_NS || FOUR_WAY_STOP)
        {     // if a pedestrian wishes to cross, or
              // a 4-way stop is required..
              if(time_left > 10)
                    time_left = 10;    // shorten the time
        }
        if(time_left != 0)               // count down the time
        {
              --time_left;
              return;                    // return to main
        }                    // time expired, so..
        time_left = 5;       // give 5 seconds to WARNING
```

Figure 1–2 *Imaginary Traffic Light Software (Continues)*

```
            current_state = NS_WARNING;     // time expired, move
            break;                          // to the next state

      case  NS_WARNING:
            EW_GRN_LITE = 0;
            NS_GRN_LITE = 0;
            NS_RED_LITE = 0;   // north-south has the yellow..
            EW_RED_LITE  = 1;
            EW_YEL_LITE = 0;   // and east-west has the red..
            NS_YEL_LITE = 1;

            if(time_left != 0)              // count down the time
            {
                  --time_left;
                  return;                   // return to main
            }                  // time expired, so..
            if(FOUR_WAY_STOP)  // if 4-way requested then start
                  current_state = FLASHER; // the flasher
            else
            {                         // otherwise..
                  time_left = 30;     // give 30 seconds to MOVING
                  current_state = EW_MOVING;
            }                         // time expired, move
            break;                    // to the next state

      case  FLASHER:
            EW_GRN_LITE = 0;   // all yellow and
            NS_GRN_LITE = 0;   // green lites off
            EW_YEL_LITE = 0;
            NS_YEL_LITE = 0;

            flash_toggle ^= 1;        // toggle LSB..
            if(flash_toggle & 1)
            {
                  NS_RED_LITE = 1;    // blink red lights
                  EW_RED_LITE  = 0;
            }
            else
            {
                  NS_RED_LITE = 0;    // alternately
                  EW_RED_LITE  = 1;
            }
            if(!FOUR_WAY_STOP)        // if no longer a 4-way stop
                  current_state = EW_WARNING;
            break;                    // then return to normal
```

Figure 1–2 *Imaginary Traffic Light Software (Continues)*

```
        default:
                current_state = NS_WARNING;
                break;      // set any unknown state to a good one!!

    }
}

void main(void)
{
    DDRB = 0xFF;          // portb all out
    DDRA = 0x00;          // porta all in

    current_state = NS_WARNING;    // initialize to a good starting
                                   // state (as safe as possible)
    while(1)
    {
        delay_ms(1000);     // 1 second delay.. this time could
                            // be used for other needed processes

        Do_States();        // call the state machine, it knows
                            // where it is and what to do next
    }
}
```

Figure 1–2 *Imaginary Traffic Light Software (Continued)*

1.13 PROGRAMMING STYLE, STANDARDS, AND GUIDELINES

Using the C language to write source code is just part of the entire software development process. There are many considerations that fall outside of writing the code and the desired operation of a program. These considerations include:

- Readability and maintainability of the software
- A documented development process
- Project management
- Quality control and meeting outside requirements such as ISO9001 and ISO9003
- Configuration management and revision control
- Design-rule and code-style requirements of your company or organization
- Verification and validation processes to meet medical and industrial requirements
- Hazard analysis

As you begin to develop code for products and services that are released into the marketplace, these aspects will become as much or more a part of the development process as writing the actual operating software. Many companies have software "style guides" that define how the software should be physically structured. Items such as header block format, bracket and parentheses placement, naming conventions for variables and definitions, and rules for variable types and usage will be outlined. This may sound a bit ominous, but once you begin writing using a defined style and development criteria, you will find it easier to collaborate and share effort with others in your organization, and you will have fewer errors in your code from the outset, as well.

Organizations such as MISRA (Motor Industry Software Reliability Association) have created documents that show how following some basic rules and guidelines during software development can greatly improve the safety and reliability of the developed software—and reduce some of the annoying "got-yah" type errors during the development process. You can find more information about these guidelines at

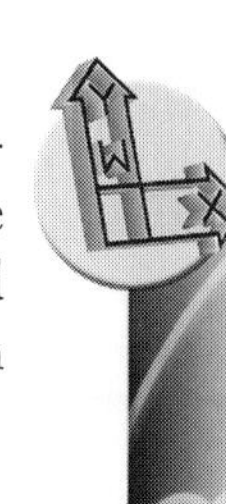

http://www.misra.org.uk/

1.14 CHAPTER SUMMARY

This chapter has provided a foundation for you to begin writing C language programs.

The beginning concepts demonstrated the basic structure of a C program. Variables, constants, enumerations, their scope and construction, both simple as well as in arrays, structures, and unions, have been shown to be useful in defining how memory is to be allocated and how the data within that memory is to be interpreted by a C program.

Expressions and their operators, including I/O operations, were discussed to provide a basis for performing arithmetic operations and determining logical conditions. These operations and expressions were also used with control constructs such as **while** and **do/while** loops, **for** loops, and **switch/case** and **if/else** statements to form functions as well as guide the flow of execution in a program.

The advanced concepts of real-time programming using interrupts executives and state machines were explored to demonstrate how to streamline the execution of programs, improve their readability, and provide a timely control of the processes within a C language project.

1.15 EXERCISES

1. Define the terms *variable* and *constant* (Section 1.4).

*2. Create an appropriate declaration for the following (Section 1.4):

   A. A constant called "x" that will be set to 789.

   B. A variable called "fred" that will hold numbers from 3 to 456.

   C. A variable called "sensor_out" that will contain numbers from –10 to +45.

   D. A variable array that will have ten elements, each holding numbers from –23 to 345.

   E. A character string constant that will contain the string "Press here to end".

EXERCISE

F. A pointer called "array_ptr" that will point to an array of numbers ranging from 3 to 567.

G. Use an enumeration to set "uno"," dos","tres" to 21, 22, 23, respectively.

3. Evaluate the following (Section 1.6):

A. unsigned char t; t = 0x23 * 2; // t = ?

B. unsigned int t; t = 0x78 / 34; // t = ?

C. unsigned char x; x = 678; // x = ?

D. char d; d = 456; // d = ?

E. enum {start = 11, off, on, gone}; // gone = ?

F. x = 0xc2; y = 0x2; z = x ^ y; // z = ?

G. e = 0xffed; e = e >> 4; // e = ?

H. e = 27; e += 3; // e = ?

I. f = 0x90 | 7; // f = ?

J. x = 12; y = x + 2; // y = ?   x = ?

*4. Evaluate as true or false as if used in a conditional statement (Section 1.6):

For all problems: x = 0x45; y = 0xc6

A. (x == 0x45)

B. (x | y)

C. (x > y)

D. (y – 0x06 == 0xc)

5. Evaluate the value of the variables after the fragment of code runs (Section 1.7):

```
unsigned char loop_count;
unsigned int value = 0;
for (loop_count = 123; loop_count < 133;loop_count++)
        value += 10;
                //value = ??
```

*6. Evaluate the value of the variables after the fragment of code runs (Section 1.7):

```
unsigned char cntr = 10;
unsigned int value = 10;

do
{
        value++;
```

```
} while (cntr < 10);
        // value = ??     cntr = ??
```

7. Evaluate the value of the variables after the fragment of code runs (Section 1.7):

```
unsigned char cntr = 10;
unsigned int value = 10;

while (cntr < 10)
{
      value++;
}
          // value = ??     cntr = ??
```

8. Given: unsigned char num_array[ ] = {1,2,3,4,5,6,7,8,9,10,11,12}; (Section 1.9):

```
// num_array[3] = ??
```

9. Given: unsigned int *ptr_array; (Section 1.9):

```
unsigned int num_array[ ] = {10,20,30,40,50,60,70};
unsigned int x,y;
ptr_array = num_array;
x = *ptr_array++;
y = *ptr_array++;
    // x = ??   y = ??   ptr_array = ??
```

*10. Write a fragment of C code to declare an appropriate array and then fill the array with the powers of 2 from 2^1 to 2^6 (Section 1.7).

11. What does the acronym MISRA stand for?

"*" preceding an exercise number indicates that the question is answered or partially answered in the appendix.

1.16 LABORATORY ACTIVITIES

1.

   A. Create, compile and test a program to print "C Rules and Assembler Drools!!" on the standard output device.

   B. Modify the program to use a **for** loop to print the same phrase four times.

   C. Repeat using a **while** loop to print the phrase three times.

   D. And yet again using a **do/while** to print the phrase five times.

2. Create a program to calculate the capacity of a septic tank. The three dimensions height, width, and length should be read from three parallel ports A, B, and C (in feet) and the result printed to the standard output device. The maximums the program should handle are 50 feet of width, 50 feet of length, and 25 feet of depth (it's a big tank!). Use appropriate variable sizes and use casting where required to correctly calculate the capacity and print the result as follows (x, y, z, and qq should be the actual numbers):

   The capacity of a tank x feet long by y feet high by z feet wide is qq cubic feet of "stuff."

   The program should run once, print the results, and stop.

3. Modify the program in Activity 2 so that if any of the three inputs exceeds the allowed range, the program says

   ```
   The input exceeds the program range.
   ```

4. Create a program to print the powers of 2 from 2^0 to 2^{12}.

5. Use a **switch/case** construct to create a program that continuously reads port A and provides output to port B according to the following table:

Input	*Output*
0x00	0b00000000
0x01	0b00001111
0x02	0b11110000
0x03	0b01010101
0x04	0b10101010
0x05	0b11000011
0x06	0b00111100
0x07	0b10000001
other	0b11111111

6. Create a program to determine the order of storage of long variables according to the suggestion in the "Structures and Unions" section (Section 1.10) relative to unions (near the middle of the section).

7. Create a program that can search an array of structures for specific values. Declare an array of structures containing the members "*month*", "*day*", and "*year*". Initialize the array to contain one and only one instance of your own birthday. However, the month, day, and year of your birthday must show up in other places in the array, just not all at the same point. The program must search the array to find your birthday and then print a message to indicate at what index in the array the birthday resides.

8. Modify the program in Activity 7 to search the array of structures using pointers.

9. Create a program to use the standard input function *getchar()* to retrieve one character and compare its value to the value placed on port A. If they match, print a message to that effect on the standard output device using *printf()*. Otherwise, output a 0x01 to Port B if port A is higher in value than the character received, or output an 0x80 if port A is lower.
10. Use an external hardware interrupt to print a message to the standard output device each time an interrupt occurs. Also print the number of times the interrupt has occurred. You *may not* use a global variable to count the interrupt occurrences.

The Atmel RISC Processors

2.1 OBJECTIVES

At the conclusion of this chapter, you should be able to:

- Define, describe the uses of, and identify the memory spaces and registers of the RISC microcontrollers
- Allocate the microcontroller resources to a project
- Apply interrupts to C programs
- Write C programs to complete simple I/O tasks
- Apply the major peripherals of the AVR microcontrollers to solve problems in interfacing to electronic devices
- Write C programs to accomplish serial communication
- Read and determine the results of AVR assembly language programs

2.2 INTRODUCTION

This chapter provides most of the hardware background needed by an engineer to successfully apply the Atmel AVR RISC microcontrollers to a problem. It is based on the premise that in order to be successful, designers must be as intimately familiar with the hardware as they are with the programming language.

The chapter begins with the basic overview of the processor, its architecture, and its functions. Next, the memory spaces are discussed in detail so that you can identify and understand where your variables are stored and why. The uses of interrupts are next discussed, along with the most common built-in peripherals. These topics are then combined into the process of allocating processor resources for a project.

A variety of the peripherals common to many of the Atmel microcontrollers are discussed. Many of the Atmel parts have additional peripherals or extended functionality in their peripherals. Consult the specification for your particular part to discover the full range of peripheral functionality for your particular microcontroller. Each of the peripherals is shown as it applies to projects. Lastly, assembly language is described as it pertains to understanding the microcontroller's functions.

2.3 ARCHITECTURAL OVERVIEW

The microcontrollers being considered are the Atmel AVR RISC microcontrollers. This one sentence says volumes about the architecture of the devices.

First, they are *RISC* devices. RISC stands for "reduced instruction set computing" and means that the devices are designed to run very fast through the use of a reduced number of machine-level instructions. This reduced number of instructions contributes to the speed due to the fact that, with a limited number of machine instructions, most can run in a single cycle of the processor clock. In terms of MIPS (millions of instructions per second), this means that an AVR processor using an 8 MHz clock can execute nearly 8 million instructions per second, a speed of nearly 8 MIPS.

Secondly, the devices are *microcontrollers*. A microcontroller is a complete computer system optimized for hardware control that encapsulates the entire processor, memory, and all of the I/O peripherals on a single piece of silicon. Being on the same piece of silicon means that speed is once again enhanced, because internal I/O peripherals take less time to read or write than external devices do.

Being optimized for hardware control means that the machine-level instruction set provides convenient instructions to allow easy control of I/O devices—there will be instructions to allow setting, clearing, or interrogating a single bit of a parallel I/O port or a register. The ability to set or clear a bit is typically used to turn a hardware device off or on. In a processor not optimized for hardware control, the setting, clearing, or reading of the individual bits of an I/O port (or any other register) would require additional AND, OR, XOR, or other instructions to affect a single bit of the port.

As an example, consider the following code fragment that sets the third bit of Port A high:

```
PORTA = PORTA | 0x4;       //set bit 2 high
```

While this line of code is correct for use with either a standard microprocessor or a microcontroller, the actual results are different for the two devices.

The following snippet shows some assembly code generated for a standard microcomputer and for a microcontroller to execute the line shown above:

```
Standard microcomputer:
        IN    R30,0x1B
        MOV   R26,R30
        LDI   R30,LOW(4)
        OR    R30,R26
        OUT   0x1B,R30
```

```
Microcontroller:
        SBI   0x1B,2
```

As these examples demonstrate, the standard microprocessor shown takes five instructions to set the bit high while the microcontroller takes only one. This equates to a five times speed increase for port operations in the microcontroller and a five times decrease in code size for these operations. Microcontrollers allow for reduced code size and increased execution speed by providing instructions that are directly applicable to hardware control.

Overall, the architecture of the Atmel AVR devices is that of a microcomputer. It is organized into a CPU (central processing unit), a memory, and an I/O section. The CPU is masked from our vision, but the memory and I/O sections are very visible and need to be understood by the applications designer. Each of these is discussed below.

2.4 MEMORY

The memory section of the Atmel RISC AVR processors is based on the Harvard model, in which various portions of the memory are separated to allow faster access and increased capacity. The CPU has a separate interface for the FLASH code memory section, the data memory section, and the EEPROM memory section, if one is present.

The actual memory being discussed is that of the medium-sized microcontrollers such as the ATMega16. Check the specification of your particular microcontroller to see the exact memory configuration for your part.

2.4.1 FLASH CODE MEMORY

The FLASH code memory section is a block of FLASH memory that starts at location 0x0000 and is of a size that is dependent on the particular Atmel AVR microcontroller in use. The FLASH code memory is *non-volatile* memory (retains its data even when power is removed), and it is used to store the executable code and constants, because they must remain in the memory even when power is removed from the device. The code memory space is 16 bits wide at each location to hold the machine-level instructions that are typically a single 16-bit word. Figure 2–1 shows the memory map for the FLASH section of memory.

Although the FLASH memory can be programmed and reprogrammed with executable code, there is no provision to write to the FLASH by means of an executable program; it must be programmed by external means. Consequently, it is viewed as read-only memory from the programmer's perspective and is therefore used only to store constant type variables, along with the executable code. Constants are automatically promoted to **int size** variables when they are stored in FLASH code memory space because of the memory width.

2.4.2 DATA MEMORY

The data memory of the Atmel AVR processor typically contains three separate areas of read/write (R/W) memory. The lowest section contains the thirty-two general-purpose working registers, followed by the sixty-four I/O registers, which are then followed by the internal SRAM.

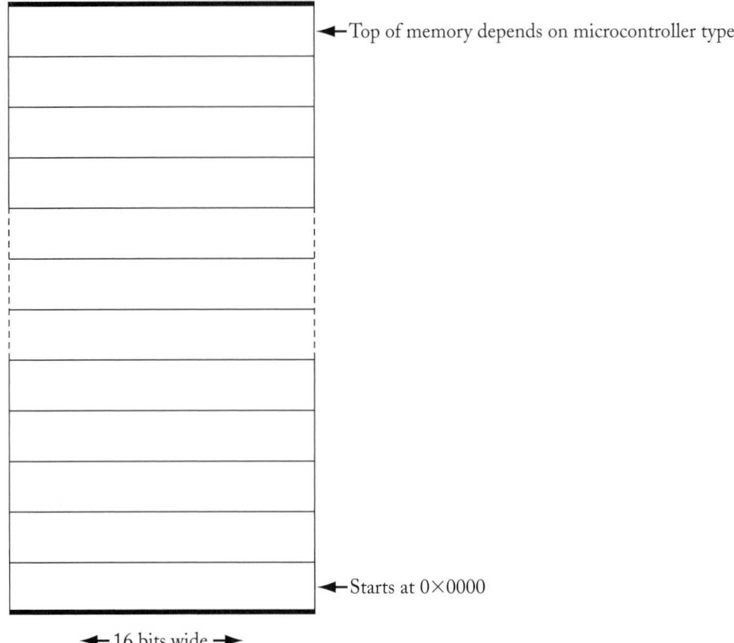

Figure 2–1 *FLASH Code Memory*

The general-purpose working registers are just that: they are used for the storage of local variables and other temporary data used by the program while it is executing, and they can even be used for the storage of global variables. The sixty-four I/O registers are used as the interface to the I/O devices and peripherals on board the microcontroller. And the internal SRAM is used as a general variables storage area and also for the processor stack.

The memory map for the data memory area is shown in Figure 2–2.

Registers

The general-purpose working registers occupy the lowest thirty-two cells in the data memory. These registers are used much like the storage locations in a calculator, in that they store temporary or intermediate results. Sometimes they are used to store local variables, sometimes global variables, and sometimes the pointers into memory that are used by the processor. In short, the processor uses these thirty-two working registers as it executes the program. The use of the thirty-two working registers is controlled by the C compiler and is typically out of the programmer's control unless assembly language is being used.

I/O Registers

In an ATMega16, the I/O working registers occupy the next highest sixty-four bytes in the data memory space (refer to Figure 2–2). Each of these registers provides access to the control registers or the data registers of the I/O peripherals contained within the microcontroller.

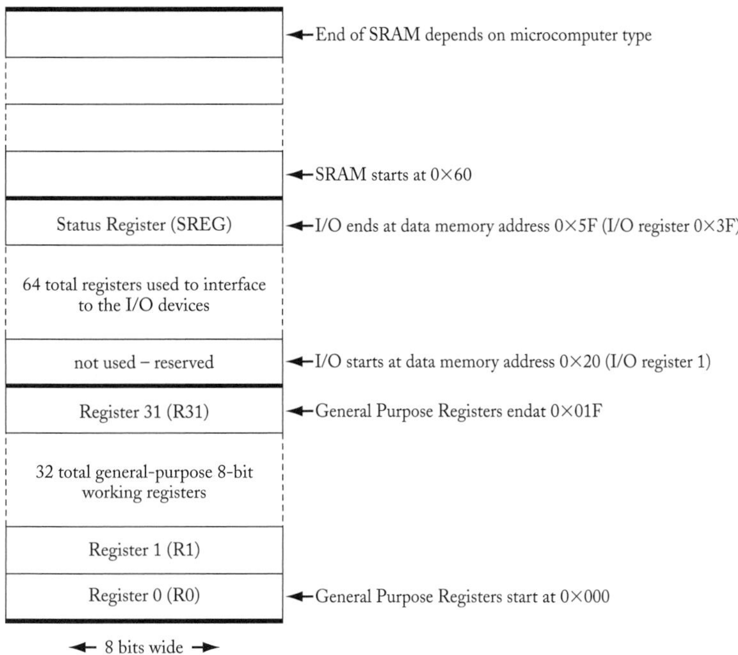

Figure 2–2 *Data Memory of the ATMega16*

The programmer uses the I/O registers extensively to provide interface to the I/O peripherals of the microcontroller. Other Atmel devices such as the ATMega128 have more I/O registers then will fit in the 64-byte space. These devices occupy space up to 0x100 to accommodate the larger number of I/O registers.

In Chapter 1, "Embedded C Language Tutorial," you have already seen examples using the I/O registers called PORTA and DDRA. These I/O registers were used in the examples relative to writing to and reading from a parallel port. Figure 2–3 shows a selection of the I/O registers, emphasizing those associated with the parallel ports.

I/O Register Name	***I/O Register Address***	***SRAM Address***	***Description***
DDRA	0×1A	0×3A	Data Direction Register for Port A
DDRB	0×17	0×37	Data Direction Register for Port B
PORTA	0×1B	0×3B	Output Latches for Port A
PINB	0×16	0×36	Input Pins for Port B
SREG	0×3F	0×5F	Processor Status Register

Figure 2–3 *Example I/O registers*

Each I/O register has a name, an I/O address, and an SRAM address. A C language programmer will most commonly use the I/O register name. The two different numeric addresses are important when writing in assembly language because some of the instructions relate to the SRAM address and some relate to the I/O address. For further information about the AVR instruction set and assembly language, see Section 2.12, "The AVR RISC Assembly Language Instruction Set."

The I/O register names are much more convenient for the C language programmer to use than the addresses. However, the C language compiler does not inherently know what these names are or know what addresses to associate with the names. Each program contains a **#include** header file that defines the names and the associated addresses for the C language compiler. The header files were detailed in the Chapter 1, "Embedded C Language Tutorial." The C language program in Figure 2–4 demonstrates these concepts.

```
#include <mega16.h>
main()
{
      DDRA = 0xFF;     /*all bits of Port A are output*/
      DDRB = 0x00;     /*all bits are input*/
      while (1)
      {
        PORTA = PINB;   /*read port B and write to Port A*/
      }
}
```

Figure 2–4 *Port Initialization using Register Names*

This program reads the pins of Port B using the PINB I/O register and writes the results to the output latches of Port A using the PORTA register. The C compiler gets the addresses to use with these register names from the **#include** header file in the first line of the program, in this example, mega16.h.

In summary, the C language programmer uses the I/O registers as the interface to the I/O hardware devices within the microcontroller. Subsequent sections of this chapter describe the use and function of each of the I/O peripherals in the microcontroller and their associated I/O registers.

SRAM

The SRAM section of memory is used to store variables that do not fit into the registers and to store the processor *stack*. Figure 2–5 shows the SRAM section of data memory.

As Figure 2–5 shows, there are no special areas or divisions of memory within the SRAM. The addresses shown in this figure are those for the SRAM section; their physical addresses will not be the same. Data is usually stored starting at the bottom of the SRAM, and the processor *stack* or *stacks* start at the top of memory and utilize memory from the top down.

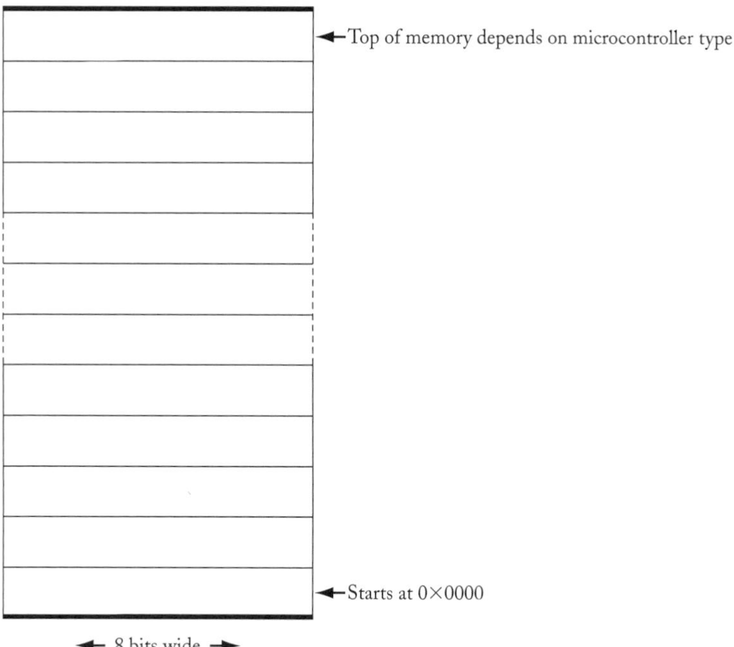

Figure 2–5 *SRAM Section of Data Memory*

The processor uses the stack area of memory as temporary storage for function return addresses, intermediate results from calculations, and any other short-term temporary data storage. The CodeVisionAVR C language compiler actually implements two stacks: The *system stack* starts at the top of the SRAM area and is used to store return addresses and processor control information. The *data stack* starts below the system stack and works its way down through memory, in a manner similar to the system stack, and is used to store temporary data, such as the local variables used in a function.

While the use of the SRAM memory for data storage is fairly straightforward, its use for the stack is not. The stack may be thought of as an area of memory set aside for temporary storage, much as a spot on your desk would be set aside for the storage of papers. New data (or a new piece of paper) is *pushed* onto the stack, and old data (or the top piece of paper) is *popped* off the stack in a *LIFO* (last in, first out) manner. The *stack pointer* (SP) holds the address of the memory cell that is currently available for use.

The stack pointer keeps track of the next available location to store data onto the stack. It works as follows:

1. When new data is ready to be placed onto the stack, the processor uses the SP to find where in memory to place the data. For example, if the SP contains the value 0x200, the data will be placed in SRAM memory location 0x200.

2. The SP is then decremented to location 0x1FF, and the processor begins to execute the next instruction. In this manner, the SP always contains the location of the next memory cell of the stack that is available to be used.

At some later point, the processor may want to retrieve the last piece of data pushed onto the stack. The retrieval process works as follows:

1. When the processor is ready to retrieve or pop data off the stack, it reads the SP and immediately increments the SP contents by one. Since the stack pointer is always pointing to the next available (empty) location on the stack, incrementing the stack pointer means that the address now in the stack pointer is the address of the last *used* location that contains data.
2. The processor uses the address in the SP to *pop* or retrieve the data from the stack. Since the data has been retrieved from this location, its address is left in the SP as being the next available location on the stack. The next byte pushed onto the deck will overwrite the data that was present at that location.

Stack operations are summarized in Figure 2–6.

The major concept relative to the stack is that as data is pushed onto the stack, the stack uses progressively more memory in the SRAM area, starting at the top of SRAM and working downward. As data is popped off the stack, the memory that has been previously used is released, and the available stack location moves up in memory.

Likewise, the data stack is located in SRAM below the system stack. The exact location is set by the programmer in the compiler configuration settings. The data stack works very much like the system stack, using a data stack pointer just like the system stack pointer, growing downward through memory as it is used and back upward as it is released.

Global variables may also be stored in the SRAM memory area starting above the I/O register area. The SRAM is used in an upward direction until space is made available for the global variables. It is the programmer's responsibility to ensure that the system stack does not come down low enough in SRAM to overwrite the data stack and that the data stack does not come down low enough in memory to overwrite the global variables area. In most cases, the compiler takes care of these limits, but ultimately it is the responsibility of the programmer.

2.4.3 EEPROM MEMORY

The EEPROM section of memory is an area of read/write memory that is non-volatile. It is typically used to store data that must not be lost when power is removed and then reapplied to the microcontroller. The EEPROM space starts at 0x000 and goes up to a maximum value that is dependent on the specific microcontroller in use.

Although the EEPROM section of memory may be both read and written, it is seldom used for general variable storage. The reason is that EEPROM memory is very slow to write; it

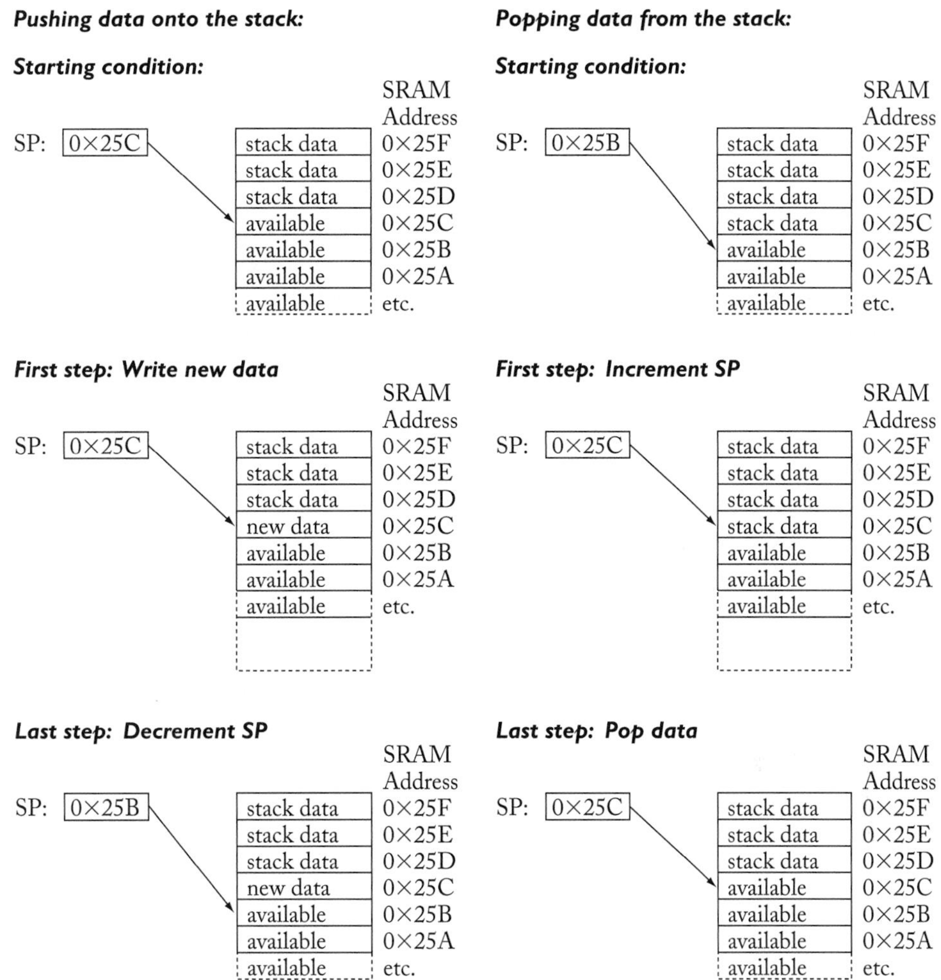

Note: All of the cells marked "available" actually contain some data. Either it comes from whatever was there at power up, or it may be data previously on the stack. But the data is no longer needed and so the cells are available to write new data into.

Figure 2–6 *Stack Operations*

can take up to one millisecond to complete a write operation on one byte of memory. The extensive use of this memory for variable storage would slow the processor down appreciably. Also, EEPROM memory can withstand only a limited number of write cycles, as was mentioned in Chapter 1. So, for these reasons, EEPROM memory is usually reserved for those variables that must maintain their value in the event of a power loss.

Chapter 2 Example Project: Part A

This is the first part of a project example that will be used to further demonstrate the concepts presented in Chapter 2. The example project will center on a data collection system for a stock car according to the following scenario:

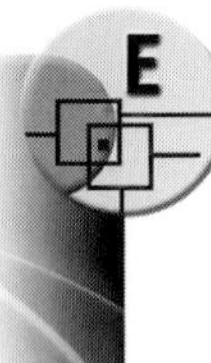

Assume that your boss hands you a data collection system based on an Atmel ATMega16 microcontroller. He explains that this system was designed to collect data for a drag racer over their fifteen-second run, but that he would like you to modify the system to collect data once per second over an entire two-minute lap for his stock car. The collected data is then uploaded to a PC for analysis. The system was originally designed to record engine rpm based on a pulse from one spark plug, drive shaft rpm from a magnetic sensor reading a magnet attached to the drive shaft, and engine temperature from an RTD thermocouple attached at the cylinder head. Your boss explains that he tried to have the programming modified once before, but, unfortunately, the inept programmer succeeded only in erasing the entire program. Happily for you, this means that all of the conditioning circuitry for the engine signals exists and that you need only to provide a program to collect the data and send it to the PC.

The first task for you is to verify that the processor has sufficient resources to do the job. First, we verify that the memory is big enough to hold the data and processor variables. The data is to be saved every second for two minutes, so the number of sets of data is as follows:

2 * 60 seconds/minute * 1 data set/second = 120 sets of data

So what you now need to know is the number of bytes in each set of data. According to your boss, engine rpm is in the range of 2000 to 10,000, shaft rpm is in the range of 1000 to 2000, and engine temperature is in the range of 100°F to 250°F. Your boss specifies that he wants rpm accurate to 1 rpm and temperature accurate to 1°F.

Therefore, you will be storing two integer-sized numbers (two bytes each for a total of four bytes) for the two rpm readings and one byte-size number for the temperature in each set of data. The total storage needed is as follows:

120 sets of data * 5 bytes/set = 600 bytes

In addition to the data storage, you will need about twenty-five bytes for a *system stack* and forty bytes for a *data stack*. (As you progress in your programming prowess, you will learn to judge the size of stack required; in general, the more math you do and the more functions you nest, the more stack you will need.) Checking the ATMega16 specification, you discover that the microcontroller has 1024 bytes of SRAM memory, and so the memory resources are sufficient for the task. Normally, at this point, you would also consider whether the peripheral resources are sufficient as well. In this case, however, you don't know about the peripherals yet, and besides, the system worked for drag racers, so the peripheral resources must be there.

2.5 RESET AND INTERRUPT FUNCTIONS

Interrupts, as discussed in Chapter 1, "Embedded C Language Tutorial," are essentially hardware-generated function calls. The purpose of this section is to describe how the interrupts function and how *reset*, which is a special interrupt, functions.

Interrupts, as their name suggests, *interrupt* the flow of the processor program and cause it to branch to an *interrupt service routine* (ISR) that does whatever is supposed to happen when the interrupt occurs. Interrupts are useful for those situations in which the processor must immediately respond to the interrupt or in those cases where it is very wasteful for the processor to poll for an event to occur. Examples of the need for immediate response include using interrupts to keep track of time (the interrupt may provide a tick for the clock's time base) or an emergency off button that *immediately* stops a machine when an emergency occurs.

Examples of the other useful applications of interrupts include devices like keypads or other input devices. In terms of the microcontroller's ability to process instructions, we humans are very, very slow. It would be wasteful for the processor to poll a keypad in hopes that one of us slow humans has pressed a key. In this instance, the pressing of a key might cause an interrupt and the microcontroller would then briefly pause to see if the key press required it to do anything. If the key press were the first of several required to cause an action to occur, the processor would return to its tasks until enough keys had been pressed to require action. Using interrupts frees the processor from polling the keypad constantly at a rate that is much, much faster than we can press the keys.

All of the interrupts, including *reset,* function in exactly the same manner. Each interrupt has a vector address assigned to it low in program memory. The compiler places the starting address of the associated interrupt service routine and a relative jump instruction at the vector location for each interrupt. When the interrupt occurs, the program completes executing its current instruction and then branches to the vector location associated with that interrupt. The program then executes the relative jump instruction to the ISR and begins executing it.

When the interrupt occurs, the *return address* (the address of the next instruction to be executed when the interrupt is finished) is stored on the *system stack*. The last instruction in the interrupt service routine is an RETI assembly language instruction, which is a *return from interrupt*. This instruction causes the *return address* to be popped off the stack and program execution to continue from the point at which it was interrupted.

In the AVR processors, all interrupts have the same priority. There is no allowance for one interrupt to interrupt another interrupt, that is, one interrupt cannot have priority over another interrupt. It is possible, however, for two interrupts to occur simultaneously. An arbitration scheme, sometimes referred to as *priority*, is provided to determine which interrupt executes in these cases. In cases where two interrupts occur simultaneously, the interrupt with the lowest-numbered vector will be executed first. Refer to the vector table in the next section or in the specification for a specific processor.

2.5.1 INTERRUPTS

Figure 2–7 shows the first six interrupt sources available in an ATMega16. The choices for interrupt sources and their vectors will depend on the processor being used.

Starting Address	Vector Number	Source	Description
0x0000	1	Reset	External reset, Power up, Watchdog timeout
0x0002	2	External Interrupt 0	Occurs on a hardware signal on pin INT0
0x0004	3	External Interrupt 1	Occurs on a hardware signal on pin INT1
0x0006	4	Timer 2 Compare	Occurs on a Timer 2 compare match
0x0008	5	Timer 2 Overflow	Occurs on an overflow of Timer 2

Figure 2–7 *Sample Interrupt Vectors*

Interrupts must be initialized before they become active, or usable. Initializing interrupts is a two-step process: The first step is to unmask the interrupts that are to be active, and the second step is to globally enable all of the unmasked interrupts.

Unmasking amounts to setting a 1 in the control register bit that corresponds to the interrupt that is to be used. For instance, the *general interrupt control register* (GICR) register is shown in Figure 2–8.

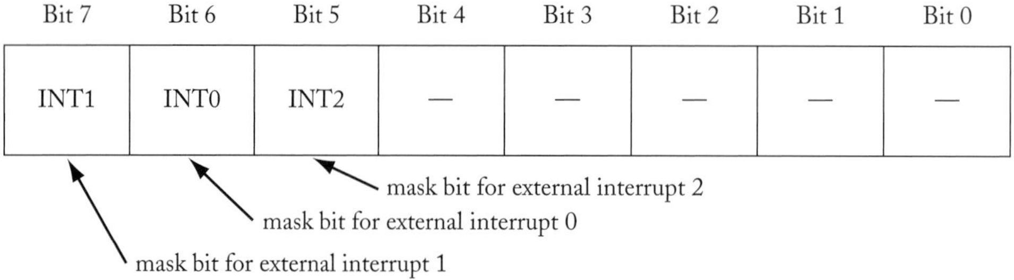

Figure 2–8 *GICR*

GICR is used to enable the external interrupts. As an example, setting the INT0 bit will enable external interrupt 0 (an external hardware signal applied to the pin that is also named INT0), or setting the INT1 bit will enable external interrupt 1. Setting both bits will enable both interrupts. The mask bits are used as exactly that, a mask.

When an appropriate hardware signal occurs on INT1, for instance, that signal is logically ANDed with the INT1 bit in GICR. If the result is a 1, the interrupt may be allowed to occur (see the following paragraphs), but if the INT1 bit in GICR is set to 0, then the result of the logical AND is always 0, and the interrupt will never be allowed to occur. Setting the interrupt mask bit is necessary for an interrupt to be allowed to occur.

The second step to enable an interrupt is to set the global interrupt enable bit in the *status register* (SREG) of the processor. This is accomplished as follows:

```
#asm("sei")
```

This line of code is inserted in a C language program at the point where you want to enable the global interrupt bit. This line uses a compiler directive, "`#asm`", to insert in the program the assembly language instruction *SEI*, the instruction to set the global interrupt enable bit.

When an interrupt occurs, the interrupt signal is first logically ANDed with the mask bit from the control register appropriate to that particular interrupt, and the result is then logically ANDed with the global interrupt enable bit. If the result of all of this is a logic 1, then the interrupt is allowed to occur and the processor vectors to the appropriate address in the interrupt vector table. The program/hardware example in Figure 2–9 illustrates the use of external interrupt 0.

```
#include <mega16.h>
 interrupt [EXT_INT0] void ext_int0_isr(void)
 {
   PORTC = PORTC ^ 0x1;   /*toggle the LSB of port C*/
 }

 void main(void)
 {
   DDRC=0x01;      /*port C LSB set for output*/

   GICR=0x40;     /*set the INT0 bit to enable external interrupt 0*/
   MCUCR=0x02;     /*set the ISC01 to activate on a falling edge*/

   #asm("sei")     /*enable the global interrupt enable bit*/

   while (1)
       ;     /*do nothing - the interrupt does it all*/
 }
```

Figure 2–9 *Interrupt Example Program*

This program toggles the LED attached to Port C, bit 0, each time the pushbutton attached to the INT0 pin is pressed. Figure 2–10 shows the hardware setup for this program.

Looking at the program critically, we see that the **#include** statement in the first line defines the registers and interrupt vectors for an ATMega16 microcontroller. The next four lines are the interrupt service routine (ISR) for External Interrupt 0. Notice that the first line of the ISR, the function declaration, begins with the reserved word **interrupt** to tell the compiler that the following function is an interrupt service routine. The compiler uses this information to know that the register contents must be saved and restored by the function (by pushing

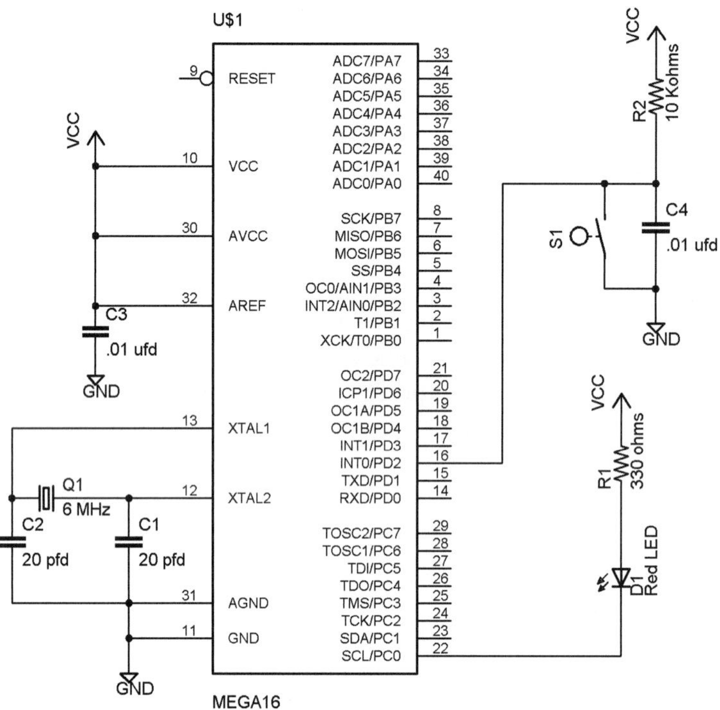

Figure 2–10 *Interrupt Program Hardware*

them onto the data stack at the start of the function and by popping them off the data stack at the end of the function) and that the function must end with an **RETI** instruction rather than an **RET** instruction.

The "`[EXT_INT0]`" definition comes from the **#include** file and amounts to the number 2, because External Interrupt 0 has 2 as its vector number. The compiler uses this information to place the relative jump to the ISR function at the appropriate place in the interrupt vector table. The balance of the function declaration is a normal one.

The third line of the ISR is the active line in the ISR. It uses an exclusive OR to toggle the least significant bit of Port A each time it is executed.

In the first line of the *main()* function, the least significant bit of Port C is set for output. The following line sets the interrupt mask bit in GICR to allow External Interrupt 0 to function when the global interrupt enable bit is set. The next line sets the Interrupt Sense bits in the *MCU control register* (MCUCR) so that External Interrupt 0 is triggered by a falling edge applied to INT0. The Interrupt Sense bits for external interrupts 0 and 1 are shown along with their definitions in Figure 2–11.

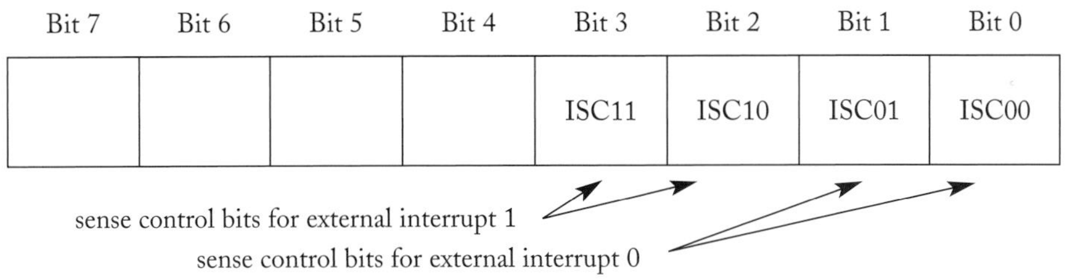

ISCx1	*ISCx0*	*Interrupt function*
0	0	Interrupt x is triggered by a low level
0	1	Interrupt x is triggered by any change in logic level
1	0	Interrupt x is triggered by a falling edge
1	1	Interrupt x is triggered by a rising edge

Figure 2–11 *MCUCR Interrupt Sense Control Bits*

The "`#asm ("sei")`" line is an assembly language instruction, *S*et *E*nable *I*nterrupt, which sets the global interrupt enable bit so that a falling edge applied to INT0 will cause the interrupt to occur and toggle the LED. The "`while (1)`" in the *main()* function does nothing except keep the processor busy until the interrupt occurs.

2.5.2 RESET

Reset is the lowest numbered interrupt. It is also a special interrupt in that it always takes precedence over all executing interrupts and any code that may be running. Three sources can cause a reset to occur: a logic low applied to the external reset pin for greater than 50 ns, as a part of the power-on sequence of the microcontroller, and by a timeout of the watchdog timer. Reset is used to preset the microcontroller to a known state so that it can start executing the program located at location 0x0000 in code memory.

The microcontroller state following a reset will vary slightly depending on the processor being used, but it is essentially the following:

- All peripherals (including the watchdog timer) are disabled.
- All parallel ports are set to input.
- All interrupts are disabled.

By disabling all the peripherals and interrupts, the microcontroller can begin executing the program without unexpected jumps due to interrupts or peripherals that might cause unexpected or erratic behavior. Setting the parallel ports to *input mode* ensures that a port and an external device will not be trying to drive a port pin to opposite levels, which could damage the port pin.

Watchdog Timer and Reset

The watchdog timer is a safety device. It is designed to cause a processor reset in the case where the processor is lost or confused or doing anything other than running the program it is supposed to be running.

The watchdog timer is a timer that causes a processor reset to occur if it is allowed to time out. When a program is operating normally, the program consistently reloads or resets the watchdog timer to prevent a timeout. As long as the program is operating normally and resets the watchdog timer, the processor reset will not occur, but if the program gets lost or gets hung somewhere, the timeout will occur and the processor will be reset. The theory is that a reset of the processor will get the program back to operating normally.

Figure 2–12 shows the bits of the *watchdog timer control register*, WDTCR.

Bit 7	Bit 6	Bit 5	Bit 4	Bit 3	Bit 2	Bit 1	Bit 0
			WDTOE	WDE	WDP2	WDP1	WDP0

Bit(s)	***Name(s)***	***Description***
WDTOE	Watchdog Timer Off Enable	Allows disabling the WDT
WDE	Watchdog Timer Enable	Enables the Watchdog Timer
WPx	Watchdog Prescaler Bit x	Sets the timeout period for the WDT

Figure 2–12 *Watchdog Timer Control Register, WDTCR*

The oscillator that clocks the watchdog timer is separate from the system clock. Its frequency is dependent on the voltage applied to the processor. With 5 V applied to Vcc, the frequency is approximately 1 MHz. This creates a situation in which, although the Watchdog Prescaler Bits in WDTCR are set to determine the timeout, the actual time it takes to time out will also depend on the Vcc applied to the processor. Figure 2–13 shows the approximate timeout periods determined by the watchdog prescaler bits at two levels of Vcc.

The times shown in Figure 2–13 are approximate, because the oscillator frequency is approximate and depends on the exact voltage applied to Vcc. The program and circuit in Figure 2–14 demonstrates the use of the watchdog timer.

The program and hardware in Figure 2–14 demonstrate the use of the watchdog timer. When the program starts, it hangs in the first *while()* loop waiting for the pushbutton to be

WDP2	WDP1	WDP0	Time out @ 5V Vcc	Time out @ 3V Vcc
0	0	0	16 ms	17 ms
0	0	1	33 ms	34 ms
0	1	0	65 ms	68 ms
0	1	1	130 ms	140 ms
1	0	0	260 ms	260 ms
1	0	1	520 ms	550 ms
1	1	0	1000 ms	1000 ms
1	1	1	2100 ms	2200 ms

Note: Times shown are approximate.

Figure 2–13 *Watchdog Timeout Period Selection*

```
#include <mega16.h>
static unsigned int waiter; /*variable to time blinks*/

void main (void)
{
   DDRC = 0x1;  /*set PORTC.0 for output*/

   /*wait for button to be pressed*/
   while (PINA.0 == 1)
   {
      ;   /*wait for PB to be pressed*/
   }

   WDTCR = 0x0b;  /*enable WDT with 130 ms timeout*/

   /*stay in this loop while button is held*/
   while(PINA.0 == 0)
   {
      #asm("wdr");  /*reset WT continuously*/
      ++waiter;  /*increment waiter variable*/
      if (waiter == 50000)
      {
         PORTC.0 = PORTC.0 ^ 1; /*complement port bit*/
         waiter = 0;  /*reset waiter for next delay time*/
      }
   }

while(1);  /*lock up to allow WDT to time out*/
}
```

Figure 2-14 *WDT Program Example (Continues)*

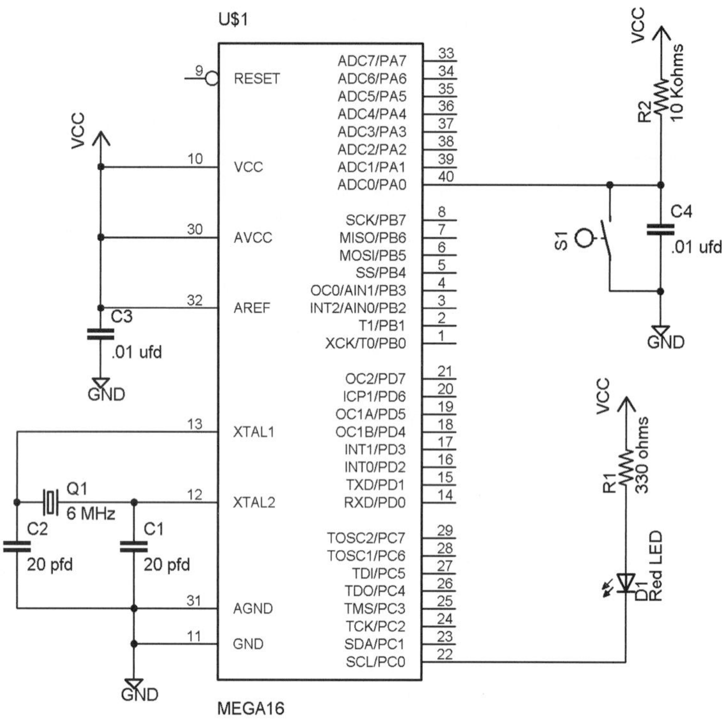

Figure 2-14 *(Continued)*

pressed (pressing the pushbutton provides a logic 0 to PINA.0). At this time, the program falls out of the *while()* loop and enables the WDT by setting WDTCR to 0x0b, setting Watchdog Enable bit high, and selecting the timeout period. At this point, the WDT is active and needs to be reset at least every 120 ms to avoid having it reset the processor.

As long as the pushbutton is held down, the program now remains in the second *while()* loop. In this loop, the WDT is continuously reset by the assembly language instruction *wdr*. The WDT must be reset using the embedded assembly language instruction because there is no equivalent instruction in C. The LED blinks continuously while program is in this loop.

When the user releases the pushbutton, the program falls out of the second *while()* loop and enters the *while(1)* loop, where it is held forever. Since the WDT is no longer being reset, approximately 130 ms after entering the *while(1)*, the WDT times out and resets the processor. This can be verified by noting that the processor resets and restarts the bootloader or by noting that the program is now back in the first *while()* loop waiting for the switch to be pressed.

In order to prevent the program from accidentally disabling the WDT, it takes a two-step process to disable the WDT once it is enabled. Specifically, the WDTCR must be first loaded with 0x18 (setting the both WDE and WDTOE bits high simultaneously). Then the WDE bit must be cleared as the next instruction. An example is shown below:

```
WDTCR = 0x18;   /*set WDE and WDTOE simultaneously*/
WDTCR = 0c00;   /*set WDE to 0 immediately*/
```

Normally the watchdog timer, once enabled, is never disabled because the purpose of enabling it to start with is to protect the processor against errant or erratic processing.

2.6 PARALLEL I/O PORTS

The parallel I/O ports are the most general-purpose I/O devices. Each of the parallel I/O ports has three associated I/O registers: the *data direction register*, DDRx (where "x" is A, B, C, and so on depending on the specific processor and parallel port being used), the *port driver register*, typically called PORTx, and the *port pins register*, PINx.

The data direction register's purpose is to determine which bits of the port are used for input and which bits are used for output. The input and output bits can be mixed as desired by the programmer. A processor reset clears all of the data direction register bits to logic 0, setting all of the port's bits for input. Setting any bit of the data direction register to a logic 1 sets the corresponding port bit for output mode. For instance, setting the least significant two bits of DDRA to a logic 1 and the other bits to a logic 0 sets the least significant two bits of port A for output and the balance of the bits for input.

Writing to the output bits of port A is accomplished as follows:

```
PORTA = 0x2;    /*sets the second bit of port A*/
                /*and clear the other seven bits*/
```

Reading from the input bits of port A is accomplished as follows:

```
x = PINA;   /*reads all 8 pins of port A*/
```

In this latter example, "x" would contain the value from all of the bits of port A, both input and output, because the PIN register reflects the value of all of the bits of the port.

Input port pins are floating, that is, there is not necessarily a pull-up resistor associated with the port pin. The processor can supply the pull-up resistor, if desired, by writing a logic 1 to the corresponding bit of the port driver register as shown below:

```
DDRA = 0xC0;   /*upper 2 bits as output, lower 6 as input*/
PORTA = 0x3;   /*enable internal pull-ups on lowest 2 bits*/
```

In general, although this varies by the specific processor, the port pins are capable of sinking 20 mA, but they can source much less current. This means that the ports can directly drive LEDs, provided that the port pin is sinking the current as shown in Figure 2–10.

Figure 2–15 shows a program that uses the least significant four bits of port C as input and the most significant four bits as output. Pull-up resistors are enabled on the input pins of the least significant two bits of the port.

Additional information and examples relative to the I/O ports are shown in Section 1.5, "I/O Operations," in Chapter 1, "Embedded C Language Tutorial."

```
#include <mega16.h>

void main (void)
{
   DDRC = 0xf0;  /*set upper 4 bits of PORTC as output*/
   PORTC = 0x03; /*enable pull-ups on the 2 least significant*/
                 /*bits of PORTC*/

   while(1)
   {
      PORTC = (PINC << 4);  /*read the lower 4 bits of POTRC*/
                            /*shift them 4 bits left and output*/
                            /*them to the upper 4 bits of PORTC*/
   }
}
```

Figure 2–15 *Parellel Port Example*

EXAMPLE

Chapter 2 Example Project: Part B

This is the second part of the Chapter 2 example program to create a system to collect operational data for a stock car racer. In the first part, it was determined that the system has sufficient memory resources to do the task. In this part, the overall code structure and user interface will be created.

It is important to determine which peripheral resources are to be used for each measurement, input, and output. If you were creating the data collection system from scratch, you could determine the resources to use, and, as you assigned the resources, you would be sure that the processor had sufficient resources. In this case, however, you need to determine which peripherals are being used for each measurement so you can write your program accordingly.

Investigation, and perhaps a little circuit tracing, has shown that the system connections are as follows:

- The Start button is connected to the INT1 pin on the Mega163. Pressing it pulls INT1 low.
- The Upload button is connected to Port A, bit 0. Pressing it pulls the bit low.
- The engine rpm pulses are connected to the ICP (input capture pin).
- The drive shaft rpm pulses are connected to the INT0 pin.
- The engine temperature signal is connected to the ADC3 (analog-to-digital converter input #3) pin.

For now, all that you need to be concerned with are the Start and Upload buttons. The other signals will be handled in Part C of the example project.

You will now want to plan the overall structure of the program. Thinking back to Section 1.12, "Real-Time Methods," in Chapter 1, it would make sense to apply those methods to this project. In general, this program is supposed to record data at specific intervals and eventually upload the data to a PC for analysis. Using real-time programming methods allows us to keep the measurements running in real time (using interrupts) and then to simply record the current values when the one-second interval has elapsed. In other words, the measurements are kept as current as possible all the time, whether they are being recorded or not. When the time comes to record the data, the recording function simply grabs the latest data and records it.

So you will eventually need interrupt service routines for the two rpm measurements and the one temperature measurement. Additionally, you must handle the Start and Upload buttons. A simple way to handle the job is for the program to record 120 sets of data and then stop; pressing the Start button simply clears the data set counter so it then records another 120 sets of data at the time you want. The advantage here is that pressing the Start button will always start a new set of data in the case where someone started the recording process errantly. The Upload button would send the data whenever it is pressed if the 120 sets were complete.

The following program skeleton seems to fit the needs and will be filled in as the Chapter 2 example proceeds:

```
#include <mega16.h>
#include <stdio.h>

unsigned char data_set_cntr;   /*number of data sets recorded*/
unsigned int e_rpm[120], s_rpm[120];   /*arrays for 120 sets of data*/
unsigned char temp[120];
unsigned int current_e_rpm, current_s_rpm;   /*most-current values*/
unsigned char current_temp;

/* External Interrupt 0 service routine*/
interrupt [EXT_INT0] void ext_int0_isr(void)
{
/*place code here to handle drive shaft RPM measurement*/
}

/* External Interrupt 1 service routine*/
interrupt [EXT_INT1] void ext_int1_isr(void)
{
   data_set_cntr = 0;    /*clear counter to start counting*/
}

/* Timer 0 overflow interrupt service routine*/
```

EXAMPLE

```
interrupt [TIM0_OVF] void timer0_ovf_isr(void)
{
/* place code here to produce 1 second intervals*/
   if (data_set_cntr < 120)  /* code to record a set of data */
                            /*if less than 120 sets are done*/
}

/* Timer 1 input capture interrupt service routine*/
interrupt [TIM1_CAPT] void timer1_capt_isr(void)
{
/* Place code here to handle engine RPM measurement*/
}

#define ADC_VREF_TYPE 0x00
/* ADC interrupt service routine*/
interrupt [ADC_INT] void adc_isr(void)
{
/* Place code here to measure engine temperature*/
}

void main(void)
{
/* Input/Output Ports initialization*/
PORTA=0x01;       /*enable pull up on bit 0 for upload switch*/
DDRA=0x00;        /*all input*/
PORTB=0x00;
DDRB=0x00;        /*all input*/
PORTC=0x00;
DDRC=0x00;        /*all input*/
PORTD=0x4C;       /*enable pull ups on Port D, Bits 2,3, & 6*/
DDRD=0x00;        /*all input*/

/* Timer/Counter 0 initialization*/
/* Initialization to be added later*/

/* Timer/Counter 1 initialization*/
/* Initialization to be added later*/

/* External Interrupt(s) initialization*/
GICR=0xC0;        /* both external interrupts enabled*/
MCUCR=0x0A;       /* set both for falling edge triggered*/
MCUCSR=0x00;      /* clear the interrupt flags in case they are*/
GIFR=0xC0;        /* errantly set*/

/* Timer(s)/Counter(s) Interrupt(s) initialization*/
TIMSK=0x21;      /* timer interrupts enabled for later use*/

/* UART initialization*/
/*initialization to be added later*/
```

```
/* Analog Comparator disabled as it is not used*/
ACSR=0x80;
SFIOR=0x00;

/* ADC initialization*/
/* ADC initialization to be added later*/

/* Global enable interrupts*/
#asm("sei")

while (1)
      {
      if (!PINA.0) send_data();   /*send data to PC if switch pressed*/
      };
}
```

This code structure sets up the operation of the entire system. INT1 is connected to the Start button; when the button is pressed, it resets the data set counter to 0. Timer 0 will be set (details to follow in the next part of the example) to produce an interrupt that will provide a one-second interval. Within the Timer 0 interrupt service routine, the data is recorded at a one-second interval whenever the data counter is less than 120. So the scheme is as follows: pressing the Start button sets the data counter to 0, and data is then recorded until 120 sets have been recorded. In the meantime, all of the data is kept up to date in a real-time fashion by the various interrupts. The last piece, uploading to the PC, is handled in the *while(1)* loop in *main()*. When the Upload button is pressed, the data is sent out serially.

After further sections of this chapter, more pieces of the example program will be added.

EXAMPLE

2.7 TIMER/COUNTERS

Timer/counters are probably the most commonly used complex peripherals in a microcontroller. They are highly versatile, being able to measure time periods, to determine pulse width, to measure speed, to measure frequency, or to provide output signals. Example applications might include measuring the rpm of a car's engine, timing an exact period of time, such as that necessary to time the speed of a bullet, producing tones to create music or to drive the spark ignition system of a car, or providing a pulse-width or variable-frequency drive to control a motor's speed. In this section timer/counters will be discussed in the generic sense, and then typical timer/counters in the AVR microcontrollers will be discussed.

Although used in two distinctly different modes, timing and counting, timer/counters are simply binary up-counters. When used in timing mode, the binary counters are counting time periods applied to their input, and in counter mode, they are counting the events or pulses or something of this nature. For instance, if the binary counters had 1-millisecond pulses as their input, a time period could be measured by starting the counter at the beginning of an event and stopping the counter at the end of the event. The ending count in the counter would be the number of milliseconds that had elapsed during the event.

When a timer/counter is used as a counter, the events to be counted are applied to the input of the binary counter, and the number of events occurring are counted. For instance, the counter could be used to count the number of cans of peas coming down an assembly line by applying one pulse to the input of the counter for each can of peas. At any time, the counter could be read to determine how many cans of peas had gone down an assembly line.

The AVR microcontrollers provide both 8-bit and 16-bit timer/counters. In either case, an important issue for the program is to know when the counter reaches its maximum count and *rolls over*. In the case of an 8-bit counter, this occurs when the count reaches 255, in which case the next pulse will cause the counter to roll over to 0. In the case of a 16-bit counter, the same thing occurs at 65,535. The rollover events are extremely important for the program to be able to accurately read the results from a timer/counter. In fact, rollovers are so important that an interrupt is provided that will occur when a timer/counter rolls over.

Some AVR microcontrollers have two 8-bit timers (typically, Timer 0 and Timer 2) and one 16-bit timer (typically, Timer 1), although this configuration will vary depending on the exact type of AVR microcontroller being used. The following sections will discuss the timer prescaler and input selector, a feature common to all of the timers, and then discuss each of the common timers in turn. The most common uses for each timer/counter will be discussed, although many timers have more functions than are discussed in this text. For any specific processor, check the specifications to determine all of the various functions possible with the timer/counters.

2.7.1 TIMER/COUNTER PRESCALERS AND INPUT SELECTORS

Timer/counter units may use a variety of internal frequencies derived from the system clock as their input, or they may get their input from an external pin. The timer counter *control register* (TCCRx) associated with the timer contains the *counter select* bits (CSx2, CSx1, CSx0) that control which input is used with a specific counter. Figure 2–16 shows the prescaler and input selector configuration for a timer counter control register as used in most AVR microcontrollers.

The following code fragment shows, as an example, how to initialize Timer 0 to use the system clock divided by 8 as its clock source (the counter select bits are the three least significant bits of TCCR0):

```
TCCR0 = 0x2;           /*Timer 0 uses clock/8*/
```

2.7.2 TIMER 0

Timer 0 is typically an 8-bit timer, but this varies by specific processor type. It is capable of the usual timer/counter functions but is most often used to create a time base or *tick* for the program. *Timer counter control register 0,* TCCR0, controls the function of Timer 0 by

Timer 0 Input Sources

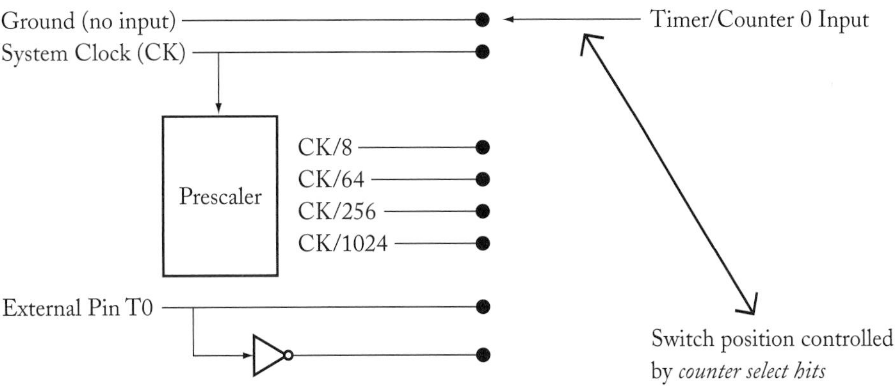

Figure 2–16 *Timer 0 Prescaler/Selector*

selecting the clock source applied to Timer 0. Figure 2–17 shows the clock prescaler bit definitions for TCCR0. Other bits of TCCR0 control additional functions of Timer 0 in a manner similar to the control bits for Timer 1 presented later.

Bit 7	Bit 6	Bit 5	Bit 4	Bit 3	Bit 2	Bit 1	Bit 0
					CS02	CS01	CS00

← Other bits →

CS02	**CS01**	**CS00**	**Interrupt function**
0	0	0	Stop, Timer 0 is stopped
0	0	1	System Clock, CK
0	1	0	System Clock / 8, CK/8
0	1	1	System Clock / 64, CK/64
1	0	0	System Clock / 256, CK/256
1	0	1	System Clock / 1024, CK/1024
1	1	0	External Pin T0, counts a falling edge
1	1	1	External Pin T0, counts a rising edge

Figure 2–17 *TCCR0 Prescaler Definitions*

A program *tick*, like the tick of a clock, provides a highly accurate timing event. The overall scheme is that a number is selected and loaded into the timer. The timer counts from this number up to 255 and rolls over. Whenever it rolls over, it creates an interrupt. The interrupt service routine reloads the same number into the timer, executes any time-critical activities that may be required, and then returns to the program. The cycle then repeats, with the counter counting up from the number that was loaded to 255, and rolls over, creating another interrupt. The interrupt, then, is occurring on a regular basis when each time period has elapsed. The number loaded into the counter determines the length of the period. The lower the number, the longer it will take the timer to reach to 255 and roll over, and the longer the period of the tick will be.

As an example, assume that a program is to toggle the state of an LED every 0.5 seconds. The LED is attached to the least significant bit of port C as shown in Figure 2–18.

For a timer to be used as a tick, the first necessary task is to determine the number that is loaded into the timer each time the interrupt occurs. In this example, we want the LED to toggle every 0.5 seconds. One obvious solution would be for the interrupt to occur every

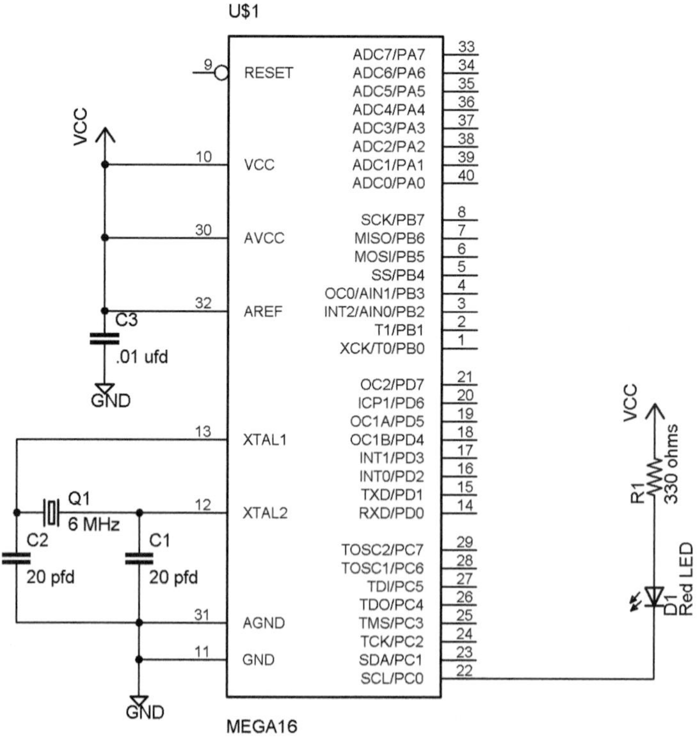

Figure 2–18 *LED Toggle Hardware*

0.5 seconds. In the case of Timer 0, the slowest setting of the clock prescaler is system clock /1024.

6 MHz/1024 = 5.859 kHz, which has a period of 1/5.859 kHz = 171 µs

This shows that every 171 microseconds another clock pulse will be applied to Timer 0. Timer 0 is an 8-bit timer/counter, and so it can count up to 256 such periods before it rolls over. So the total time that is possible with this hardware is as follows:

256 * 171 µs = 43.69 ms

Forty-four milliseconds is not sufficient to time 500 millisecond events. In order to accomplish longer periods, a global counter variable is placed in the interrupt service routine of the tick. So, for instance, if the tick occurs every 50 milliseconds, in this example we would want the counter to count up to 10 before toggling the LED (10 * 50 ms = 500 ms).

The choice of a reload number is up to the programmer. Usually, it is desirable that the reload number produce an even and easy-to-use time delay period, such as 1 millisecond or 10 milliseconds. In the case of the 6 MHz clock shown in the example, using the divide-by-eight prescaler will give a clock period of 1.33 microseconds for each clock pulse applied to the counter. Given a 1.33-microsecond clock applied to an 8-bit counter, the maximum time possible using the counter alone would be as follows:

1.33 µs * 256 counts = 340 µs

340 microseconds is not a very even period of time to work with, but using 225 counts would give a timeout period of 300 microseconds. Therefore, the counter reload number would be as follows:

256 – 225 = 31

Rollover occurs at count number 256, and it is desirable for the counter to count 225 counts before it rolls over; therefore, the reload number in this case is 31. This means that the interrupt service routine will be executed once every 300 microseconds (1.33 µs/clock cycle * 225 clock cycles/interrupt cycle = 300 µs/interrupt cycle). A global variable will be used to count to 1666 to produce the entire 500 milliseconds time period (300 µs * 1666 ≅ 500 ms). The entire program is shown in Figure 2–19.

Figure 2–19 illustrates all of the concepts discussed above. The timer/counter register itself, *timer counter 0* (TCCNT0), is reloaded with the value 31 each time the ISR executes so that it can once again count up 225 steps to reach 255 and roll over.

A global variable called "`timecount`" is used to keep track of the number of times that the interrupt service routine is executed by incrementing it each time the ISR is executed. "`timecount`" is both incremented and checked inside the expression of the if statement. When "`timecount`" reaches 1666, the most significant bit of port A is toggled, and "`timecount`" is reset to zero to count up for the next 500-millisecond period. This tick could also handle any other event that occurs on an even number of 300 microsecond increments in the same manner.

```
#include <mega16.h>
static unsigned int time_count;   /*.5 second counter*/

// Timer 0 overflow interrupt service routine
interrupt [TIM0_OVF] void timer0_ovf_isr(void)
{
   TCNT0 = 31;  /*set for 300 us timeout*/
   ++time_count;     /*increment .5 second counter*/
   if (time_count == 1666)
   {
      PORTC.0 = (PORTC.0 ^ 1); /*complement port bit to blink LED*/
      time_count = 0;  /*reset time counter for next .5 second*/
   }
}

void main (void)
{
   DDRC = 0x01;  /*set lsb for output*/

   /*set timer 0 prescaler to clk/8*/
   TCCR0=0x02;
   TCNT0=0x00;
   OCR0=0x00;

   /* unmask  Timer 0 overflow interrupt*/
   TIMSK=0x01;

   /* enable interrupts */
   #asm("sei")

   while(1)
   {
      ;  /*do nothing in here */
   }
}
```

Figure 2–19 *LED Blinker Program*

2.7.3 TIMER 1

The 16-bit timer (Timer 1 from an ATMega16 is used as an example) is a much more versatile and complex peripheral than the typical 8-bit timers. In addition to the usual timer/counter, Timer 1 contains one 16-bit input capture register and two 16-bit output compare registers. The input capture register is used for measuring pulse widths or capturing times. The output compare registers are used for producing frequencies or pulses from the timer/counter to an output pin on the microcontroller. Each of these modes will be discussed in the sections that follow. Remember, however, that although each mode is discussed separately, the modes may be, and often are, mixed together in a program.

Timer 1 is also conceptually very different from Timer 0. Timer 0 is usually stopped, started, reset, and so on in its normal use. Timer 1, on the other hand, is usually left running. This creates some considerable differences in its use. These differences will be discussed in detail in the sections that follow, covering the special uses of Timer 1.

Timer 1 Prescaler and Selector

In spite of its many special features, Timer 1 is still a binary up-counter whose count speed or timing intervals depend on the clock signal applied to its input, just as Timer 0 was. As with all peripherals in the microcontroller, Timer 1 is controlled through a control register.

Timer/counter control register 1, TCCR1 (the ATMega16 timer control register for Timer 1), is actually composed of two registers, TCCR1A and TCCR1B. TCCR1A controls the compare modes and the pulse width modulation modes of Timer 1. These will be discussed later in the section. TCCR1B controls the prescaler and input multiplexer for Timer 1, as well as the input capture modes. Figure 2–20 shows the bit definition for the bits of TCCR1B.

Bit 7	Bit 6	Bit 5	Bit 4	Bit 3	Bit 2	Bit 1	Bit 0
ICNC1	ICES1		WGM 13	WGM 12	CS12	CS11	CS10

Bit	**Function**
ICNC1	Input Capture Noise Canceller (1 = enabled)
ICES1	Input Capture Edge Select (1 = rising edge, 0 = falling edge)
WGM 1x	Output waveform control. See TCCR1A
CS12	Counter Input Select Bits Exactly the same definition as for Timer 0
CS11	
CS10	

Figure 2–20 *TCCR1B Bit Definitions*

TCCR1B *counter select bits* control the input to Timer 1 in exactly the same manner as the counter select bits of Timer 0. In fact, the three select bits provide clock signals that are absolutely identical to those of Timer 0.

Timer 1 Input Capture Mode

Measuring a time period with Timer 0 involves starting the timer at the beginning of the event, stopping the timer at the end of the event, and finally reading the time of the event from the timer counter register. The same job with Timer 1 is handled differently because Timer 1 is always running. To measure an event, the time on Timer 1 is captured or held at the beginning of the event, the time is also captured at the end of the event, and the two are subtracted to find the time that it took for the event to occur. You would do much the same if you are trying to find out how long it took you walk from your history class to the bookstore. You would note the time when you left history class, again note the time when you arrived at the bookstore, and subtract the two times to determine how long it took you get to the bookstore. In Timer 1, these tasks are managed by the *input capture register* (ICR1).

ICR1 is a 16-bit register (made up of ICR1H and ICR1L) that will capture the actual reading of Timer 1 when the microcontroller receives a certain signal. The signal that causes a capture to occur can be either a rising or a falling edge applied to the *input capture pin,* ICP, of the microcontroller. As shown in Figure 2–20, the choice of a rising or falling edge trigger for the capture is controlled by the *input capture edge select bit,* ICES1. Setting ICES1 will allow ICR1 to capture the Timer 1 time on a rising edge, and clearing it will allow ICR1 to capture the time on a falling edge.

As is probably obvious by now, since there is only one capture register available to Timer 1, the captured contents must be read out as soon as they are captured to prevent the next capture from overwriting and destroying the previous reading. In order to accomplish this, an interrupt is provided that occurs whenever new data is captured into ICR1. Each time the capture interrupt occurs, the program must determine whether the interrupt signals the beginning or the ending of an event that is being timed so that it can treat the data in ICR1 appropriately.

Timer 1 also provides an input noise canceller feature to prevent miscellaneous unwanted spikes in the signal applied to the ICP from causing a capture to occur at the wrong time. When the noise canceller feature is active, the ICP must remain at the active level (high for a rising edge, or low for a falling edge) for four successive samples before the microcontroller will treat the trigger as legitimate and capture the data. This prevents a noise spike from triggering the capture register. Setting the *input capture noise canceller bit,* ICNC1, in TCCR1B enables the noise canceller feature (refer to Figure 2–20).

The hardware and software shown in Figures 2–21 and 2–22, respectively, demonstrate the use of the input capture register. The goal of this hardware and software is to measure the period of a square wave applied to the ICP of the microcontroller and to output the result, in milliseconds, on port C.

The software in Figure 2–22 has several features worth noting. A **#define** statement is used to connect a meaningful name with the output port.

The ISR for the Timer 1 overflow does nothing more than increment an overflow counter when the overflow occurs during a period measurement. The number of overflows is used in the calculation for the period.

The *Input Capture Event* ISR occurs every time the input waveform is a rising edge. At this instant, the ISR reads the input capture register to get the time that was captured when the

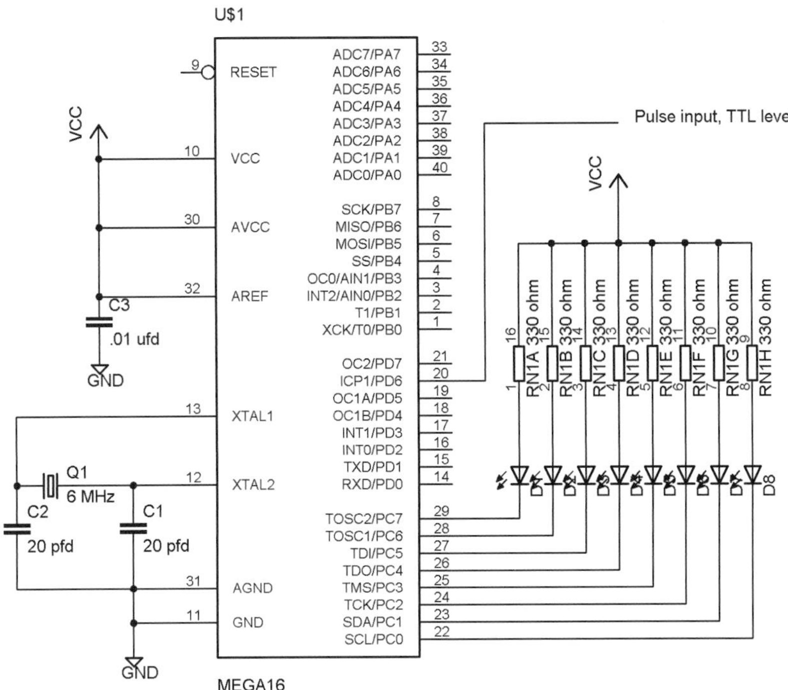

Figure 2–21 *Period Measurement Hardware*

```
#include <mega16.h>

/*define Port C as output for pulse width*/
#define period_out PORTC

unsigned char ov_counter;    /*counter for timer 1 overflow*/
unsigned int starting_edge, ending_edge;   /*storage for times*/
unsigned int clocks;    /*storage for actual clock counts in the pulse*/

/*Timer 1 overflow ISR*/
interrupt [TIM1_OVF] void timer1_ovf_isr(void)
{
   ++ov_counter;    /*increment counter when overflow occurs*/
}

/*Timer 1 input capture ISR*/
interrupt [TIM1_CAPT] void timer1_capt_isr(void)
{
```

Figure 2–22 *Period Measurement Software (Continues)*

```
    /*combine the two 8-bit capture registers into the 16-bit count*/
    ending_edge = 256*ICR1H + ICR1L; /*get end time for period*/
    clocks = (unsigned long)ending_edge
             + ((unsigned long)ov_counter * 65536)
             - (unsigned long)starting_edge;
    period_out = ~(clocks / 750);  /*output milliseconds to Port C*/
    /*clear overflow counter for this measurement*/
    ov_counter = 0;
    /*save end time to use as starting edge*/
    starting_edge = ending_edge;
}

void main(void)
{
    DDRC=0xFF;   /*set Port C for output*/
    TCCR1A = 0;   /*disable all waveform functions*/
    TCCR1B = 0xC2;/*Timer 1 input to clock/8, enable input capture*/
    TIMSK = 0x24; /*unmask timer 1 overflow and capture interrupts*/

    #asm("sei") /*enable global interrupt bit*/

    while (1)
    {
        ; /*do nothing here*/
    }
}
```

Figure 2–22 *Period Measurement Software (Continued)*

rising edge occurred. This number is the actual count of clock ticks that was present when the input capture event occurred.

Because ICR1 is composed of two separate 8-bit registers (ICR1H and ICR1L), they must be combined into an integer-sized number to use in the period calculation. This is accomplished by:

```
ending_edge = 256*ICR1H + ICR1L; /*get end time for period*/
```

Multiplying the 8-bit result is exactly the same as shifting the number form ICR1H eight bits to the left, making the more significant byte of the integer count. The result of this multiplication is added to the value from ICR1L to form the complete integer result. This number is the timer count at the *end* of the period being timed. The count from the start of the period is the same as the ending count from the previous period.

The total number of clock ticks that occurred during the measurement period may then be calculated as follows:

```
clocks = (unsigned long) ending_edge
         + ((unsigned long) ov_counter * 65536)
         - (unsigned long) starting_edge;
```

The second line of the equation accounts for any overflows that occurred during the measurement period.

The timer is being clocked by Fclk/8 (6 MHz / 8 = 750 kHz), which means that for every millisecond that has elapsed, 750 counts have occurred. Finally, dividing the total number of counts by 750 produces the actual number of milliseconds in the measurement period.

Timer 1 Output Compare Mode

The output compare mode is used by the microcontroller to produce output signals. The outputs may be square or asymmetrical waves, and they may be varying in frequency or symmetry. Output compare mode, for instance, would be appropriate if you attempt to program a microcontroller to play your school's fight song. In this case, the output compare mode would be used to generate the musical notes that make up the song.

Output compare mode is sort of the antithesis of input capture mode. In input capture mode, an external signal causes the current time in a timer to be captured or held in the input capture register. In output compare mode, the program loads an *output compare register*. The value in the output compare register is compared to the value in the timer/counter register, and an interrupt occurs when the two values match. This interrupt acts as an alarm clock to cause a processor to execute a function relative to the signal it is producing, exactly when it is needed.

In addition to generating an interrupt, output compare mode can automatically set, clear, or toggle a specific output port pin. For Timer 1, the output compare modes are controlled by *timer counter control register 1A*, TCCR1A. Figure 2–23 shows the bit definitions for TCCR1A that relate to output compare operation.

Bit 7	Bit 6	Bit 5	Bit 4	Bit 3	Bit 2	Bit 1	Bit 0
COM1A1	COM1A0	COM1B1	COM1B0	FOC1A	FOC1B	WGM11	WGM10

COM1A0 & COM1A1 Control the compare mode function for compare resister A
COM1B0 & COM1B1 Control the compare mode function for compare resister B

Control Bit Definitions:

COM1x1	*COM1x0*	*Function. ('x' is 'A' or 'B' as appropriate)*
0	0	No Output
0	1	Compare match toggles the OC1x line
1	0	Compare match clears the OC1x line to 0
1	1	Compare match sets the OC1x line to 1

Figure 2–23 *TCCR1A Bit Definitions for Output Compare Mode*

As shown in Figure 2–23, compare mode control bits determine what action will be taken when a match occurs between the compare register and the timer register. The associated output pin can be unaffected, toggled, set, or cleared. The match also causes an interrupt to occur. The goal of the interrupt service routine is to reset or reload the compare register for the *next* match that must occur.

By way of example, assume the microcontroller needs to produce a 7.5 kHz square wave. This scheme works something like this:

1. When the first match occurs, the output bit is toggled, and the interrupt occurs.
2. In the interrupt service routine, the program will calculate when the next match should occur and load that number into the compare register. In this case, the 7.5 kHz waveform has a period of 133.3 microseconds with 0.667 microseconds in each half of the waveform. So the time from one toggle to the next is 0.667 microseconds. Using the frequency of the clock applied to the timer, the program calculates the number of clock ticks that will occur in 0.667 microseconds.
3. This number of clock ticks is added to the existing contents of the compare register and reloaded into the compare register to cause the next toggle and to repeat the calculation and reload cycle.

An example of hardware and software demonstrating these concepts is shown in Figures 2–24 and 2–25, respectively. Note that no external hardware is required because the signal is produced entirely by the microcontroller.

In the program in Figure 2–25, notice first the initializations that occur in *main()*. Register DDRD is set to 0x20 so that the output compare bit, OC1A, that relates to output compare register A, OCR1A, is set for output mode, so that the output signal can appear on the bit. TCCR1A and TCCR1B set the prescaler to clock /8 (in this case clock = 6 MHz, so clock /8 = 750 kHz) and set the output compare mode for output compare register A to toggle the output bit, OC1A, when a match occurs. And finally, the compare interrupt is unmasked by setting the OCIE1A bit in the *timer interrupt mask register*, TIMSK.

Information such as that shown in Figure 2–26 is used to calculate the number that is added to the compare register contents to determine the point of the next toggle. In this case, one-half of the waveform is 66.7 microseconds long. Since the clock applied to the counter is 750 kHz (clock/8), the number of clock cycles between match points (waveform toggle points) is given by the following:

length of time between toggle points/clock cycle time = interval number

or, for this example,

66.7 μs/1.33 μs per clock cycle = 50 clock cycles per match point

And so, for this example, each time the compare match interrupt occurs, 50 is added to the current contents of the compare register to determine when the next match and toggle of the output waveform will occur.

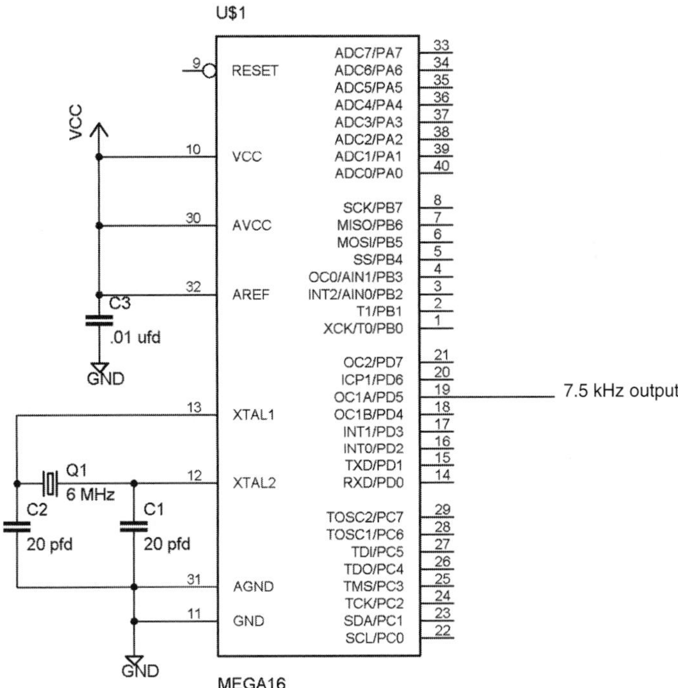

Figure 2–24 *7.5 kHz Hardware*

One additional important point to consider relative to the output compare registers, and specifically relative to the calculation of the next match point number, is the situation in which adding the interval number to the current contents of the compare register results in a number bigger than 16 bits. For example, if the output compare register contains a number 65,000, and the interval is 1000, then

65,000 + 1000 = 66000 (a number greater than 65,535)

As long as unsigned integers are used for this calculation, those bits greater than 16 will be truncated and so the actual result will be

65,000 + 1000 = 464 (drop the 17th bit from 66000 to get 464)

This will work out perfectly, since the *output compare register* is a 16-bit register and the timer/counter is a 16-bit device as well. The timer/counter will count from 65,000 up to 65,536 (a total of 536 counts) and then an additional 464 counts to reach the match point. The 536 counts plus the 464 counts is exactly 1000 counts, as desired. In other words, in both the timer/counter and the compare register, rollover occurs at the same point, and as long as unsigned integers are used for the related math, rollover is not a problem.

```
#include <mega16.h>

/* Timer 1 output compare A interrupt service routine*/
interrupt [TIM1_COMPA] void timer1_compa_isr(void)
{
   /*OC1A is really OCR1AH and OCR1AL but CVAVR provides*
   /*a psuedo register OCR1A that is the integer result*/
   /*of combining the two 8-bit registers*/
   OCR1A = OCR1A + 50;     /*set to next match (toggle) point*/
}

void main(void)
{
   DDRD=0x20;       /*set OC1A bit for output*/

   TCCR1A=0x40;     /*set prescaler to Clock/8*/

   /*enable output compare mode to toggle OC1A pin on match*/
   TCCR1B=0x02;

   /*unmask output compare match interrupt for register A*/
   TIMSK=0x10;

   #asm("sei")      /*set global interrupt bit*/

   while (1)
   {
      ;      /*do nothing*/
   }
}
```

Figure 2–25 *7.5 kHz Software*

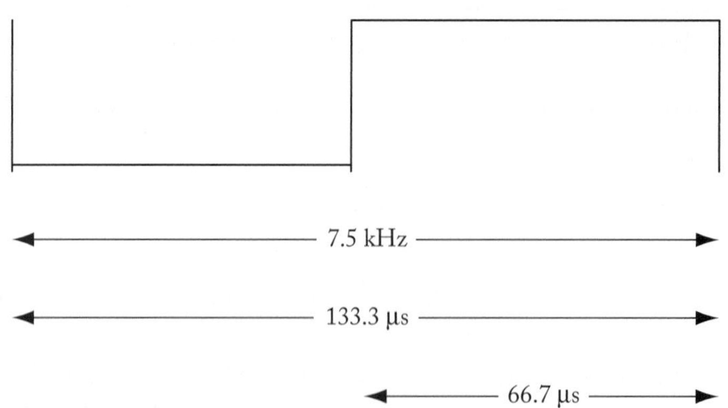

Figure 2–26 *7.5 kHz Waveform*

Timer 1 Pulse Width Modulator Mode

Pulse width modulation (PWM) mode is one of a number of methods of providing digital-to-analog conversion. PWM is the scheme in which the duty cycle of a square wave output from the microcontroller is varied to provide a varying DC output by filtering the actual output waveform to get the average DC. Figure 2–27 illustrates this principle.

As is shown in Figure 2–27, varying the duty cycle, or proportion of the cycle that is high, will vary the average DC voltage of the waveform. The waveform is then filtered and used to control analog devices, creating a digital-to-analog converter (DAC). Examples of PWM control schemes are shown in Figure 2–28.

In Figure 2–28A, the RC circuit provides the filtering. The time constant of the RC circuit must be significantly longer than the period of the PWM waveform. Figure 2–28B shows an LED whose brightness is controlled by the PWM waveform. Note that in this example, a logic 0 will turn the LED on, and so the brightness will be inversely proportional to the PWM. In this case, our eyes provide the filtering because we cannot distinguish frequencies above about 42 Hz, which is sometimes called the flicker rate. In this case, the frequency of the PWM waveform must exceed 42 Hz, or we will see the LED blink.

The final example, Figure 2–28C, shows DC motor control using PWM. The filtering in this circuit is largely a combination of the mechanical inertia of the DC motor and the inductance of the windings. It simply cannot physically change speed fast enough to keep up

Figure 2–27 *PWM Waveforms*

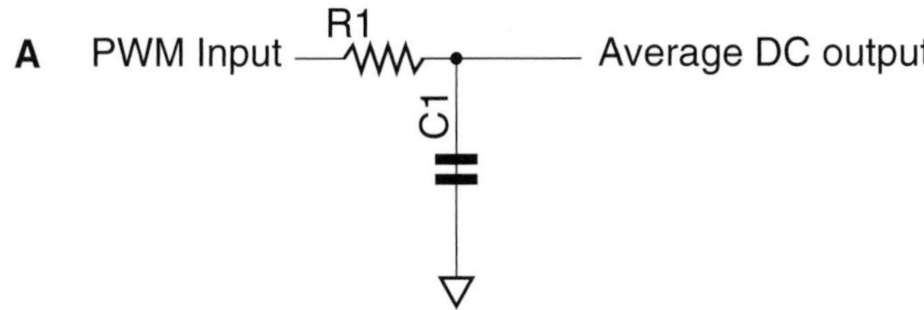

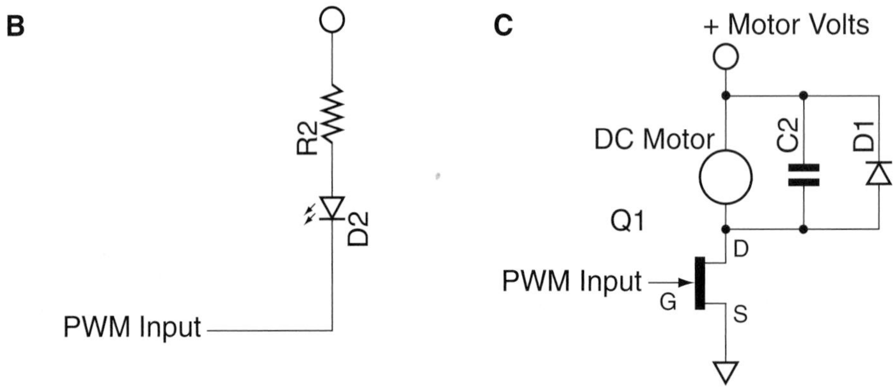

Figure 2–28 *PWM Examples*

with the waveform. The capacitor also adds some additional filtering, and the diode is important to suppress voltage spikes caused by switching the current on and off in the inductive motor.

One method of creating PWM with Timer 1 would be to use the output compare register and, each time a match occurs, vary the increment number being reloaded to create the PWM waveform. However, Timer 1 provides a built-in method to provide PWM without the need for the program to be constantly servicing the compare register to create a pulse width modulator.

Timer 1 changes its mode of operation in order to provide PWM. When operating in PWM mode, Timer 1 counts both up and down, making it difficult to use any of the other Timer 1 modes with PWM mode. During PWM operations, Timer 1 counts from zero up to a *top* value and back down again to zero. The top value is determined by the *resolution* desired. PWM is provided in 8-bit, 9-bit, or 10-bit resolution as determined by the Waveform Generation Mode (WGM) *select bits* in TCCR1A (refer to Figure 2–23). Figure 2–29 shows the effect of these bits on PWM.

Figure 2–29 shows that the resolution chosen will determine the top value for the counter to count up to and down from, and it will also affect the *frequency* of the resulting PWM

PWM Select Bits		PWM Resolution	Timer Top Value
WGM11	WGM10		
0	0	PWM Disabled	
0	1	8 - bit	255 (0xff)
1	0	9 - bit	511 (0x1ff)
1	1	10 - bit	1023 (0x3ff)

Figure 2–29 *Waveform Generator Mode Select Bits*

waveform. For example, choosing 9-bit resolution will result in a top count of 511 and the PWM frequency as calculated below (given a 6 MHz system clock with a ÷ 8 prescaler):

$$f_{PWM} = f_{system\ clock} / (prescaler * 2 * top\ count) =$$

$$6\ Mhz / (8 * 2 * 511) = 733.8\ Hz$$

Resolution is the fineness, or accuracy, of PWM control. In 8-bit mode, PWM is controlled to 1 part in 256; in 9-bit mode, PWM is controlled to 1 part in 512; and in 10-bit mode, PWM may be controlled to 1 part in 1024. In PWM, resolution must be weighed against frequency to determine the optimum choice as higher resolutions produce lower frequency PWM signals.

The actual duty cycle being output in PWM mode depends on the value loaded into the output compare register for the timer/counter. In normal PWM mode, as the counter counts down, it sets the output bit on a match to the output compare register, and as it counts up, it clears the output bit on a match to the output compare register. In this manner, loading the output compare register with the value equal to, say, 20 percent of the top value will produce a 20 percent duty cycle waveform. It is also possible to provide the inverted PWM for applications such as controlling the brightness of an LED connected in a current-sink mode directly to the PWM output pin—loading the output compare register to 75 percent of the top value in the inverted mode will produce a 75 percent low duty cycle square wave.

Figures 2–30 and 2–31 show the hardware and software, respectively, of an example program using PWM. This program provides four different duty cycles (10, 20, 30, and 40 percent at a frequency close to 2 kHz) based on the logic levels applied to port A.

Setting the exact frequency output requires juggling the timer prescaler and the resolution to get as close to the desired frequency as possible. The table in Figure 2–31 shows the frequency output (calculated using the formula presented previously) for all possible combinations of the prescaler and resolution, given the 8 MHz system clock. From the table, we can pick the combination that provides the frequency closest to the one we desire. If the problem had required a particular resolution of the PWM, the choices from the table would be limited by the resolution, and the frequency might not have been as close.

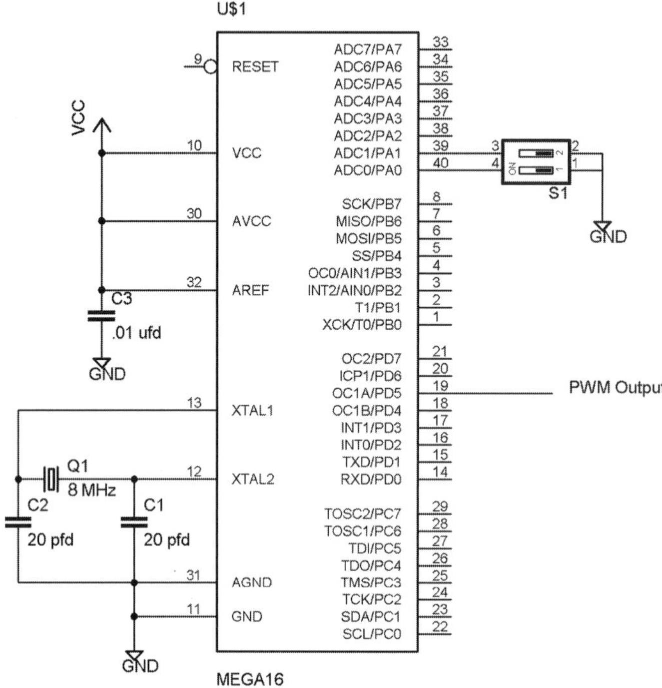

Figure 2–30 *PWM Hardware*

For this example, Figure 2–31 shows that 8-bit resolution and system clock/8 would give 1.96 kHz, which is very close to the desired frequency. Figure 2–31 also demonstrates that the frequency choices for any given combination of crystal, prescaler, and PWM resolution are fairly limited. If a project requires a very specific PWM frequency, the system clock crystal will likely have to be chosen to allow this frequency to be generated at the desired resolution.

The program shown in Figure 2–32 uses a **switch/case** statement to select the duty cycle of the PWM output. A **#define** statement is used to mask port A so that only the two lower bits are used as input, and port A is set to 0x3 so that the internal pull-ups are enabled on the same lower two bits. This avoids the need for external pull-up resistors.

TCCR1A and TCCR1B are initialized to select the PWM mode and the prescaler to provide the desired frequency output.

Processor time in many microcontroller applications is precious. In this program, an **if** statement is used along with the variable "`old_sw`" so that new data is written to the pulse width modulator only if the switches are actually changed, thereby preventing continuous writes to the output compare register.

		PWM Mode	
PRESCALER	**8-Bit (Top = 255)**	**9-Bit (Top = 511)**	**10-Bit (Top = 1023)**
System Clock	f_{PWM} = 15.7 kHz	f_{PWM} = 7.8 kHz	f_{PWM} = 3.91 kHz
System Clock / 8	f_{PWM} = 1.96 kHz	f_{PWM} = 978 Hz	f_{PWM} = 489 Hz
System Clock / 64	f_{PWM} = 245 Hz	f_{PWM} = 122 Hz	f_{PWM} = 61 Hz
System Clock / 256	f_{PWM} = 61 Hz	f_{PWM} = 31 Hz	f_{PWM} = 15 Hz
System Clock / 1024	f_{PWM} = 15 Hz	f_{PWM} = 8 Hz	f_{PWM} = 4 Hz

$f_{PWM} = f_{system\ clock} / (\text{prescaler} * 2 * \text{top count})$

Figure 2–31 *Frequency Selection Chart for an 8 MHz System Clock*

```
#include <mega16.h>

/*use a 'define' to set the two lowest bits of port C as the*/
/*control input.*/
#define PWM_select (PINA & 3)

void main(void)
{
   unsigned int old_sw;    /*storage for past value of input*/

   PORTA = 0x03; /*enable the internal pull-ups for in the input bits*/
   DDRD = 0x20; /*OC1A as output*/
   TCCR1A = 0x91;/*compare A for non-inv. PWM and 8 bit resolution*/
   TCCR1B = 0x02;   /*clock / 8 prescaler*/

   while (1)
   {
      if (PWM_select != old_sw)
      {
         old_sw = PWM_select; /*save toggle switch value*/
         switch (PWM_select)
         {
            case 0: OCR1A = 25;   /*10 % duty cycle desired*/
                    break;
            case 1: OCR1A = 51;   /*20% duty cycle desired*/
                    break;
```

Figure 2–32 *PWM Software (Continues)*

```
                case 2: OCR1A = 76;  /*30% duty cycle desired*/
                        break;
                case 3: OCR1A = 102; /*40% duty cycle desired*/
                        break;
                default:
            }
        }
    }
}
```

Figure 2–32 *PWM Software (Continued)*

2.7.4 TIMER 2

Timer 2 is usually (depending on a specific microcontroller in use) an 8-bit timer/counter with output compare and PWM features similar to Timer 1.

The most interesting difference with Timer 2 is that it can use a crystal separate from the system clock as its clock source. Selection of the external crystal as the clock source for Timer 2 is accomplished by setting the AS2 bit in the *asynchronous status register* (ASSR). The bit definitions for the bits of ASSR are shown in Figure 2–33.

Setting the AS2 bit allows Timer 2 to use the external crystal as its clock source. This means that the clock source for Timer 2 is running asynchronously to the microcontroller system clock. The other three bits in the ASSR register are used by the programmer to ensure that data is not written into the Timer 2 registers at the same moment that the hardware is updating the Timer 2 registers. This is necessary because the oscillator of Timer 2 is asynchronous to the system oscillator, and it is possible to corrupt the data in the Timer 2 registers by writing data to the registers as they attempt to update.

A single control register, TCCR2, controls the operation of Timer 2. The TCCR2 bit definitions are shown in Figure 2–34.

Using, for instance, a 32.768 kHz crystal allows Timer 2 to function as the time base for a real-time clock. Using Timer 2 in this way will allow the microcontroller to keep accurate time when necessary. Simply allow T2 to run with the 32.768 crystal and it will overflow exactly 128 times each second.

Bit 7	Bit 6	Bit 5	Bit 4	Bit 3	Bit 2	Bit 1	Bit 0
				AS2	TCN2UB	OCR2UB	TCR2UB

Figure 2–33 *ASSR Bit Definitions*

Bit 7	Bit 6	Bit 5	Bit 4	Bit 3	Bit 2	Bit 1	Bit 0
FOC2	WGM20	COM21	COM20	WGM21	CS22	CS21	CS20

Bit	Description
FOC2	Force a Compare match in non-PWM mode.
WGM20	Setting this bit enables the Timer 2 PWM function.
COM21	These two bits set the output compare mode function. The bit definitions are identical to the COM1x1 and COM1x0 bits of timer 1.
COM20	
WGM21	Set to enable counter clear on compare match.
CS22	Counter prescaler select bits for Timer 2. See AVR spec. for details.
CS21	
CS20	

Figure 2–34 *TCCR2 Bit Definitions*

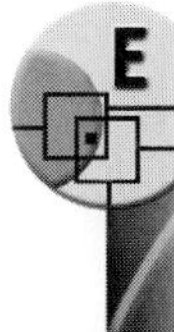

EXAMPLE

Chapter 2 Example Project: Part C

This is the third part of the Chapter 2 example system to collect operational data for a stock car. This part will be concerned with using the timers to measure the one-second recording interval and the two rpm measurements.

ONE-SECOND RECORDING INTERVAL USING TIMER 0

Using a timer to produce a "tick" or delay interval has been discussed. In this case, you are trying to produce a long (in microcontroller terms) time interval, and you have several other interrupt-dependent processes running as well. So it would be well to use as slow a clock as practical for the tick clock so as to minimize the amount of time used in the tick routine.

Further investigation of the Mega16 system you have been assigned to use shows that the system clock is 8 MHz. The Timer 0 prescaler allows division ratios of 1, 8, 64, 256, and 1024. Choosing a prescaler value of 64 produces a 125 kHz clock applied to timer 0.

Now you can figure out the number of clock pulses that must elapse for 1 second to have passed. In this case it is easy, and 125,000 clock pulses must occur for 1 second to have elapsed. Timer 0 is an 8-bit counter and can only count up 255, so you select to have it count 250 counts per tick. 125000 / 250 = 500, meaning that the timer interrupt must occur 500 times for 1 second to have elapsed.

The following code is placed in *main()* to initialize Timer0:

```
/* Timer/Counter 0 initialization*/
TCCR0=0x03;        /*clock set to clk/64*/
TCNT0=0x00;        /*start timer0 at 0 */
```

The following code is placed above *main()* as the interrupt service routine for the Timer 0 overflow:

```
int time_cntr = 0;     /*global variable for number of Timer 0*/
                       /*overflows*/

/* Timer 0 overflow interrupt service routine*/
interrupt [TIM0_OVF] void timer0_ovf_isr(void)
{
  TCNT0 = 6;      /*reload timer 0 for next period*/
  if (time_cntr++ == 500) /*check for one second, increment counter*/
  {
    if (data_set_cntr < 120)  /*record data if less than 120 sets*/
    {
      e_rpm[data_set_cntr] = current_e_rpm;    /*record engine rpm*/
      s_rpm[data_set_cntr] = current_s_rpm;    /*record shaft rpm*/
      temp[data_set_cntr++] = current_temp;   /*record engine temp*/
    }
    time_cntr = 0;      /*reset counter for next 1 second interval */
  }
}
```

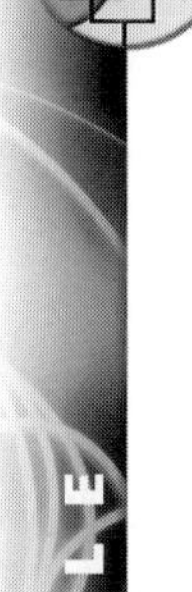

Notice that the "`data_set_cntr`" variable is post-incremented as a part of the storing of the current temperature.

ENGINE RPM MEASUREMENT USING TIMER 1

The input capture feature is used to capture the period of the engine rpm pulses. The period is used to calculate the rpm. Note that one pulse occurs for every two engine revolutions in a 4-cycle engine such as this one.

The system clock is 8 MHz in this system, so using a prescaler of 8 will provide clock pulses of 1 MHz with a convenient period of 1 microsecond. These clock pulses drive Timer 1, and the *input capture register* (ICR1) will be used to capture the Timer 1 count each time a falling edge occurs on the ICP of the ATMega16. When this occurs, an interrupt will be generated. In the interrupt service routine, the current captured count is compared to the previously captured time to determine the period of the pulses, which is used to calculate the engine rpm.

The following code is placed in *main()* to initialize Timer 1 to capture the time on the falling edge:

```
/* Timer(s)/Counter(s) Interrupt(s) initialization*/
TIMSK=0x21;   /* timer0 overflow, ICP interrupts enabled*/

/* Timer/Counter 1 initialization*/
TCCR1A=0x00;
TCCR1B=0x02;        /*prescaler = 8, capture on falling edge*/
TCNT1H=0x00;        /*start at 0*/
TCNT1L=0x00;
```

And the following code makes up the ISR for the capture event:

```
unsigned int previous_capture_time;   /*saved time from previous*/
                                      /*capture*/

/* Timer 1 input capture interrupt service routine*/
interrupt [TIM1_CAPT] void timer1_capt_isr(void)
{
  unsigned int current_capture_time, period; /*current time and*/
                                             /*period*/

  current_capture_time = (256* ICR1H) + ICR1L;   /*get captured time*/
  if (current_capture_time > previous_capture_time) /*check rollover*/
    period = current_capture_time - previous_capture_time;
  else
    period = 0xFFFF - current_capture_time + previous_capture_time;
  current_e_rpm = (unsigned long)120E6 / (unsigned long)period;
  previous_capture_time = current_capture_time;   /*save for next*/
                                                  /*calculation*/
}
```

In this code, a global variable, "`previous_capture_time`", is initialized to retain the value from the previous capture. The ISR function reads the current capture time and uses it, along with the previous captured time, to calculate the rpm. Notice the **if** statement that checks for rollover; this is another method to allow for the case in which the 16-bit timer/counter rolls over from 0xFFFF to 0x0000 during the elapsed period. The last statement in the ISR saves the "`current_capture_time`" for use as the "`previous_capture_time`" the next time a pulse occurs.

The rpm is being calculated as follows:

> RPM = 1E6 micro-seconds/second * 1 pulse / period in micro-seconds * 2 revolutions per pulse * 60 seconds per minute

Note that casting is used to ensure accuracy with the large numbers. Also, combining constants shortens the calculation formula.

DRIVE SHAFT RPM MEASUREMENT USING TIMER 1

This measurement gets a little bit more involved, because there is not a second capture register in the Mega16 and the shaft rpm signal is connected to INT0. You can create your own capture register within the INT0 ISR by reading the Timer 1 count when the interrupt occurs. Then the rest of the function works very much like the one above for the engine rpm:

```
unsigned int previous_shaft_capture_time; /*saved time from previous*/
                                          /*capture*/

/* External Interrupt 0 service routine*/
interrupt [EXT_INT0] void ext_int0_isr(void)
{
```

EXAMPLE

```
    unsigned int current_capture_time, period;     /*current time and*/
                                                   /*period*/

    current_capture_time = TCNT1;   /*get Timer 1 time*/
    /*check for roll-over*/
    if (current_capture_time > previous_shaft_capture_time)
         period = current_capture_time -
         previous_shaft_capture_time;
    else
        period = 0xFFFF - current_capture_time +
          previous_shaft_capture_time;
    current_s_rpm = (unsigned long)60E6 / (unsigned long)period;
    previous_shaft_capture_time = current_capture_time;   /*save for*/
                                                   /*next calculation*/
}
```

The calculation is the same, except that there is one pulse per revolution instead of two. Also note that some of the local variables have the same name. Since they are local, this is acceptable, but it is not acceptable with some programing standards.

2.8 SERIAL COMMUNICATION USING THE USART

Serial communication is the process of sending multiple bits of data over a single wire. It is an offshoot of the original telegraph, in which the bits were the dots and dashes of Morse code. The bits of a serial byte are separated by time so that the receiving device can determine the logic levels of each bit.

The USART is used to communicate from the microcontroller to various other devices. Examples of such devices include the terminal tool in CodeVisionAVR (used for troubleshooting and program debugging), other microcontrollers that need to communicate to control a system, or a PC that is communicating with the microcontroller to complete a task.

The usual form of serial communication, and the form being discussed here, is *asynchronous* serial communication. It is asynchronous in the sense that a common clock signal is not required at both the transmitter and the receiver in order to synchronize the data detection. Asynchronous serial communication uses a start bit and a stop bit added to the data byte to allow the receiver to determine the timing of each bit.

Figure 2–35 shows the elements of a standard asynchronous serial communication byte. This figure shows both the waveform and a definition for each bit of the serial word. The serial transmit line idles at a logic 1 and drops to zero to indicate the beginning of the start bit. The start bit takes one full bit time and is followed by the eight bits of the data byte that appear on the serial line in inverse order, that is, the least significant bit appears first and the most significant bit appears last. The stop bit follows the most significant data bit and is a logic 1, the same level as the idle state.

The falling edge of the start bit begins the timing sequence in the serial receiver. Beginning from the falling edge of the start bit, the receiver waits 1.5 bit times before sampling the

Logic 1

Logic 0

State	Idle	Idle	Active	Active	Active	Active	Active	Active	Active	Active	Active	Active	Idle
Logic Level	1	1	0	1	0	0	0	0	0	1	0	1	1
Bit Type			Start	Data	Data	Data	Data	Data	Data	Data	Data	Stop	
Data Bit Number				0	1	2	3	4	5	6	7		

Figure 2–35 *Serial Byte Format*

receiving line to get the first data bit. After that, the receiver waits 1 bit time per bit, thereby sampling each successive data bit in the center of its time period for maximum reliability.

Fortunately, all of the timing relative to the serial byte formatting, the sampling of the serial bits, and the addition of the start stop bits are handled automatically by the *universal synchronous asynchronous receiver-transmitter* (USART, pronounced "you sart"). None of these issues directly affects the programmer.

The only issue confronting the programmer is to ensure that the serial communication parameters of both the transmitter and the receiver match. Specifically, this includes setting the correct number of data bits (usually 8), determining whether or not to include a parity bit (usually not), and setting the baud rate. The baud rate is the speed of serial communication and determines the timing of the bits. Baud rate is defined as the inverse of the time per bit (baud rate = 1/bit time).

The serial waveform and other parameters relating to the serial communication discussed above are all involved with the *information* being transmitted. The serial information is independent of the *medium* being used to carry the serial information. The medium may consist of a wire, an infrared beam, a radio link, or other means.

The most common medium is defined as RS-232. It was developed in order to provide a greater distance for reliable serial communication using wire to carry the signal. RS-232 is an inverted scheme, in that a logic 1 is represented by a negative voltage more negative than −3 V and a logic 0 is represented by a positive voltage more positive than +3 V. Using a different nonzero voltage for each logic level allows some hardware error checking, because a broken line will present 0 V at the receiver and so may be detected. Most microcontroller-to-PC communication is RS-232.

CodeVisionAVR, like most C language compilers, provides built-in library functions to handle the common serial communication tasks. These tasks usually involve communicating with a terminal device, for example, a PC executing a terminal program to transmit serially the characters typed on the keyboard and to display in a window the characters received serially from the microcontroller. CodeVisionAVR provides a built-in terminal emulator in the development environment as a convenience to the programmer. The standard library functions are included in the header file stdio.h, which may be included in the C language program.

Using the header file with its built-in functions makes serial communication very easy. The example program shown in Figure 2–36 prints the message "Yabba Dabba Do" on a terminal device.

As the simple serial example shows, other than initializing the serial port (the first two lines of *main()*), serial communication is extremely easy using the built-in library functions. For more details on the *printf()* function, see Chapter 3, "Standard I/O and Preprocessor Functions."

```
#include <mega16.h>
#include <stdio.h>   /* Standard Input/Output functions*/

void main(void)
{
   UCR=0x18;          /*serial port initialized*/
   UBRR=0x33;         /*set baud rate*/

   printf ("\n\rYabba Dabba Do"); /*print phrase*/

   while (1)
      ;        /*do nothing else*/
}
```

Figure 2–36 *Simple Serial Example*

The USART interface consists of three registers. These are the *USART control and status register*, the *USART baud rate register*, and the *USART data register*. These registers may be either 8 or 16 bits in length depending on the complexity of the USART contained in the processor being used. (Note: in the simpler processors, the USART may be called a UART due to the fact that the peripheral cannot handle synchronous communications.)

In the ATMega16 for example, for normal asynchronous operation, the USART control and status register is a 16-bit register consisting of two 8-bit registers, *UCSRA* and *UCSRB,* and the USART baud rate register is a 16-bit register consisting of *UBRRH* and *UBRRL,* which are the high and low bytes, respectively, of *UBRR*. For other modes of operation an additional USART control and status register, *UCSRC,* is also available.

Regardless of the processor type, the UDR is actually two registers sharing a single I/O address. One is a read-only register and one is a write-only register. The read-only register contains any serial byte received, and the write-only register contains any serial byte to be transmitted. So, when a program reads the UDR, it is reading the receiver UDR to get data that has been received serially. When the program writes to the transmitter UDR, it is writing data to be serially transmitted.

Figure 2–37 shows the bit definitions for the *USART control and status register*, part B, from an ATMega16.

UCSRB is used to initialize and set the function of the USART in the ATMega16. The most significant three bits are the mask bits for the three interrupts associated with the USART. *RXCIE* is the mask bit for the interrupt which occurs when a serial character is received by the USART, *TXCIE* is the mask bit for the interrupt that occurs when the USART has successfully completed transmitting a character, and *UDRIE* is the mask bit for the interrupt that occurs when the transmit UDR is ready to transmit another character. Refer to Figure 2–37 for other bits and how they affect the USART operation.

Bit 7	Bit 6	Bit 5	Bit 4	Bit 3	Bit 2	Bit 1	Bit 0
RXCIE	TXCIE	UDRIE	RXEN	TXEN	UCSZ2	RXB8	TXB8

Bit	**Description**
RXCIE	Mask bit for the Receive interrupt enable. Set to unmask the interrupt.
TXCIE	Mask bit for the Transmit interrupt enable. Set to unmask the interrupt.
UDRIE	Mask bit for the UART Data Register Empty interrupt enable. Set to unmask the interrupt.
RXEN	Set to enable the serial receiver.
TXEN	Set to enable the serial transmitter.
UCSZ2	With UCSZ1:0, sets number of data bits.
RXB8	9th bit received in 9-bit mode.
TXB8	9th bit transmitted in 9-bit mode.

Figure 2–37 *ATMega16 UCSRB*

In an ATMEGA16, UCSRA reflects the current status of the USART. The UCSRA contains the interrupt flag bits (most significant 3 bits) and other bits that indicate the results of USART activity. Figure 2–38 shows the bit definitions for the USR.

The USART status is important due to the fact that serial communication is always slower than parallel communication. During the 1.04 milliseconds that it takes to transmit a single serial byte at 9600 baud, a microcontroller using a system clock of 8 MHz can execute as many as 8000 instructions. So, in order for the microcontroller not to be waiting around for the serial port (remember Real-Time Processing from Chapter 1?), it is important for the program to be able to tell the state of the serial port. In the case of a received character, it takes the same 1.04 milliseconds from the time the start bit is received until the character has been completely received. After the eighth bit is received, the RXC bit is set in the USR to indicate that a serial byte has been received. The program uses the RXC bit to know when to read valid data from the receive UDR. The data must be read from the UDR before the next character is received, as it will overwrite and destroy the data in the UDR. This explains the provision for an interrupt to occur when a serial character has been received so that it may be read promptly without consuming large amounts of processor time polling to see if a byte has been received.

In a similar manner, the UDRE bit is used to indicate that the transmit UDR is empty and that another byte may be loaded into the transmit UDR to be transmitted serially. Again, this is necessary because the microcontroller can load out bytes much, much faster than the USART can transmit them. In order to keep the program from having to poll the USR

Bit 7	Bit 6	Bit 5	Bit 4	Bit 3	Bit 2	Bit 1	Bit 0
RXC	TXC	UDRE	FE	DOR	PE	U2X	MPCM

Bit	Description
RXC	Set to indicate receipt of a serial character.
TXC	Set to indicate that a serial character has been sent.
UDRE	Set to indicate that the transmit UDR is empty.
FE	Set to indicate a framing error.
DOR	Set to indicate an overrun error.
PE	Parity error.
U2X	Set to double transmission speed.
MPCM	Set to enable multi-processor communication mode.

Figure 2–38 *ATMega16 UCSRA*

continuously to see when it is available to send another byte, an interrupt is provided that indicates when the transmit UDR is empty.

The transmitter side of the USART is actually double buffered. That is, the UDR that the program writes to holds the data until the actual transmit register is empty. Then the data from the UDR is transferred to the transmit register and begins to serially shift out on the transmit pin. At this point, the UDR is available to accept the next data word from the program, the UDRE flag is set high and, if enabled, the UDRE interrupt occurs. Occasionally, it is necessary for the microcontroller to actually know when a byte has been sent. The TXC flag is provided to indicate that the transmit register is empty and that no new data is waiting to be transmitted. The program uses this flag, and its associated interrupt, when it is necessary to know exactly when the data has been sent.

The final register associated with the USART is the *USART baud rate register*, UBRR. This register determines the baud rate at which the serial port transmits and receives data. The number entered into the UBRR is determined according to the following formula:

```
UBRR = (System Clock / (16 * baudrate)) - 1
```

Using this formula, the UBRR number for the previous example (Figure 2–36) is calculated as follows:

```
UBRR = (System Clock / (16 * baudrate)) - 1
     = (8MHz / (16 * 9600 baud)) - 1 = 51 = 0x33
```

In this example, UBRRL is set to 0x33 and UBRRH is set to 0x00. The high byte of the value is placed in UBRRH, and the low byte of the value is placed into UBRRL.

Microcontrollers often need minimum amounts of serial communication but do not need all of the formatting options and other features included in the standard library functions. The standard library functions also consume large amounts of program memory. In these cases, the programmer must implement the serial communication. Figures 2–39 and 2–40 show hardware and software, respectively, for a simple system to implement serial communications between a microcontroller and a PC using RS-232.

Figure 2–39 shows a standard AVR processor with an attached MAX233 media driver to convert the TTL serial signals used by the microcontroller into the RS-232 signals used for serial communication to the PC. The hardware shown in this figure is used with the software shown in Figure 2–40 to make the serial communication example.

The serial communication example software is a program (Figure 2–40) that uses a **switch** construct to print one of three messages on the PC. The exact message will depend on whether an "a", a "b", or another key is pressed on the PC. The ASCII code for the character pressed is sent serially to the microcontroller. This example demonstrates both manual and interrupt handling of USART functions.

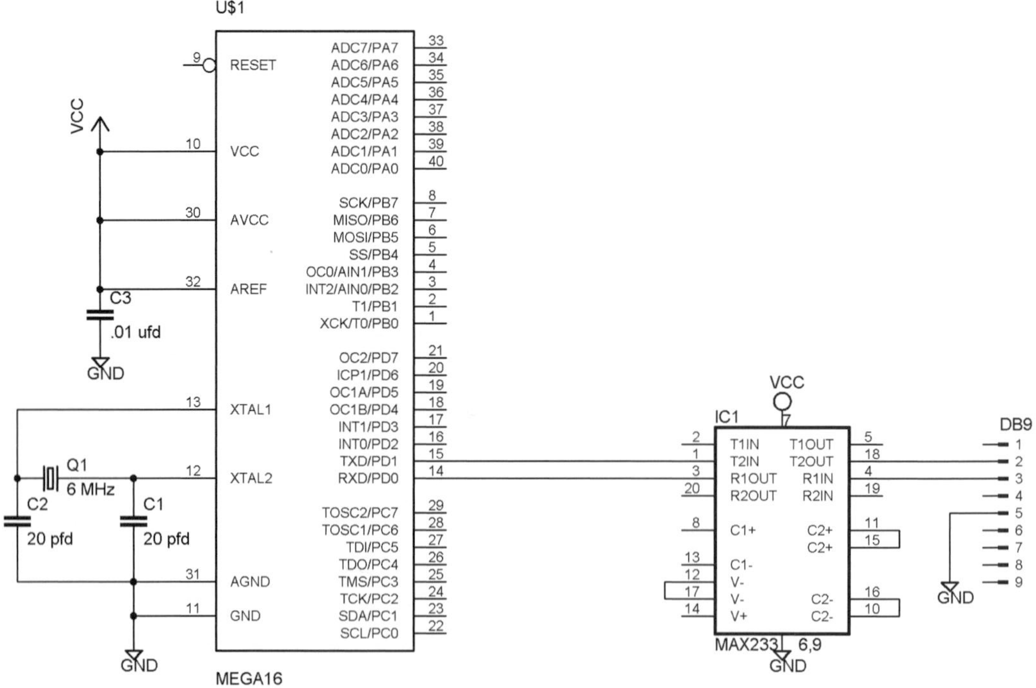

Figure 2–39 *Serial Communication Example Hardware*

```
#include <mega16.h>

unsigned char qcntr = 0,sndcntr = 0;    /*indexes into the que*/
unsigned char queue[50];        /*character queue*/

/*message arrays*/
char msg1[] = {"That was an a."};
char msg2[] = {"That was a b, not an a."};
char msg3[] = {"That was neither b nor a."};

/*this interrupt occurs whenever the */
/*USART has completed sending a character*/
interrupt [USART_TXC]void usart_transmit_isr (void)
{
   /*send next character and increment index*/
   if (qcntr != sndcntr)  UDR = queue[sndcntr++];
}

/*this function loads the queue and */
/*starts the sending process*/
void sendmsg (char *s)
{
   qcntr = 0;    /*preset indices*/
   sndcntr = 1;  /*set to one because first character already sent*/
   queue[qcntr++] = 0x0d;    /*put CRLF into the queue first*/
   queue[qcntr++] = 0x0a;
   while (*s) queue[qcntr++] = *s++;    /*put characters into queue*/
   UDR = queue[0];  /*send first character to start process*/
}

void main(void)
{
   char ch;  /* character variable for received character*/
   UCSRA=0x00;
   UCSRB=0x58;  /*enable receiver, transmitter and transmit interrupt*/
   UBRRH=0x00;  /*baud rate = 9600*/
   UBRRL=0x26;

   #asm("sei") /*global interrupt enable */

   while (1)
   {
      if (UCSRA & 0x80) /*check for character received*/
      {
         ch = UDR;    /*get character sent from PC*/
         switch (ch)
         {
```

Figure 2–40 *Serial Communication Example Software (Continues)*

```
                case 'a':
                    sendmsg(msg1);  /*send first message*/
                    break;
                case 'b':
                    sendmsg(msg2);  /*send second message*/
                    break;
                default:
                    sendmsg(msg3);  /*send default message*/
            }
        }
    }
}
```

Figure 2–40 *Serial Communication Example Software (Continued)*

The first five lines in *main()* declare the character variable and initialize the USART. UCSRB is loaded with 0x58 to enable the receiver, the transmitter, and the transmit interrupt. The UBRR is set to 0x33 to set the baud rate to 9600. Once the USART is set up and the interrupt is enabled, the program enters a *while(1)* loop that continuously checks for a received character using an **if** statement whose the expression will evaluate to TRUE after a character is received. Receiving the character will set the RXC bit in the USR that is being tested by the **if** statement. This is an example of manually polling the status of the serial port to determine when a character is received. TXC could be used in a similar manner to poll for the conclusion of a transmission.

When the **if** statement determines that a character has been received, the character is read from the UDR and used in a **switch** statement to determine which message to send. The message being transmitted (i.e., "That was an 'a'.") is composed of several characters, each of which must be sent serially to the PC along with a *carriage return* (*CR*) and a *line feed* (*LF*) character so that each message will start on a new line. The longest message is composed of 25 characters, and adding the CR and LF characters makes the total message 27 characters long. At 9600 baud (1.04 ms per serial byte), this message will take over 27 milliseconds to transmit. It is not appropriate for the microcontroller to wait around for 27 milliseconds, because it could be executing as many as 216,000 instructions (8 instructions per microsecond * 27 milliseconds * 1000 microseconds per milliseconds = 216,000 instructions) while this message is being sent. In order to free the microcontroller of the need to wait while the message is being sent, a FIFO queue is used in conjunction with the transmit interrupt to actually transmit the message.

A queue is a temporary holding device to hold bytes of data and return them on a first-in, first-out (FIFO) basis. In this case, the queue is mechanized as a variable array of data called, appropriately, *queue*. An index is used to indicate where new data is to be placed in the queue (in this case *qcntr*). As each new byte is added to the queue, the index is incremented so the next byte is added in sequence and so on until the queue holds all of the necessary data.

A separate index (in this case, *sndcntr*) is used to retrieve the data from the queue. As each byte is retrieved, this index is incremented. Finally, when the two indices are equal, the program knows that the queue has been emptied.

Actually, the CodeVisionAVR C language compiler can provide a transmitter queue, a receiver queue, or both using the "CodeWizard" code generator feature. This example is provided to demonstrate how the queue works for educational reasons.

Getting back to the example program, the *sendmsg()* function called from the switch statement puts the message into the queue and starts the transmit function. The switch statement passes a pointer (an address) to the appropriate message when it calls the function. The function first puts the CR and LF characters into the queue and then puts the message to be transmitted into the queue using the pointer. Finally, the function writes the first byte in the queue into the UDR to start the transmission process. After this character has been transmitted, the TXC interrupt occurs and the ISR loads the next character from the queue into the UDR, and so the cycle continues until the queue is empty, as indicated by the two indices being equal. A more elaborate form of the queue function with additional explanation may be found in Chapter 3, "Standard I/D and Preprocessor Functions."

2.9 ANALOG INTERFACES

In spite of the prevalence of digital devices, the world is still actually analog by nature. A microcontroller is able to handle analog data by first converting the data to digital form. An AVR microcontroller includes both an analog-to-digital conversion peripheral and an analog comparator peripheral. Each of these analog interfaces will be covered in this section, along with a brief background on analog-to-digital conversion.

Microcontrollers use analog-to-digital converters to convert analog quantities such as temperature and voltage (for example, a low-battery monitor), to convert audio to digital formats, and to perform a host of additional functions.

2.9.1 ANALOG-TO-DIGITAL BACKGROUND

Analog-to-digital conversion (as well as digital-to-analog conversion) is largely a matter of *proportion*. That is, the digital number provided by the analog-to-digital converter (ADC) relates to the proportion that the input voltage is of the full voltage range of the converter. For instance, applying 2 V to the input of an ADC with a full-scale range of 5 V will result in a digital output that is 40 percent of the full range of the digital output (2 V / 5 V = 0.4).

ADCs are available that have a variety of input voltage ranges and output digital ranges. The output digital ranges are usually expressed in terms of bits, such as 8 bits or 10 bits. The number of bits at the output determines the range of numbers that can be read from the output of the converter. An 8-bit converter will provide outputs from 0 up to $2^8 - 1$ or 255, and a 10-bit converter will provide outputs from 0 up to $2^{10} - 1$ or 1023.

In the previous example, in which 2 V was applied to a converter with a full-scale range of 5 V, an 8-bit converter would read 40 percent of 255, or 102. The proportion/conversion factor is summarized in the following formula:

$$\frac{V_{in}}{V_{fullscale}} = \frac{x}{2^n - 1}$$

In the formula above, "x" is the digital output and "n" is the number of bits in the digital output. Using this formula and solving for x, you can calculate the digital number read by the computer for any given input voltage. Using the formula within a program where you have x, you can use the formula to solve for the voltage being applied. This is useful when you might be trying to display the actual voltage on an LCD readout.

An important issue buried in this formula is the issue of resolution. The resolution of measurement, or the finest increment that can be measured, is calculated as follows:

$$V_{resolution} = \frac{V_{fullscale}}{2^n - 1}$$

For an 8-bit converter that has a full-scale input range of 5 V, the resolution would be calculated as follows:

$$V_{resolution} = 5V/(2^8 - 1) = 5V/255 = 20 \text{ mV} \quad \text{(approx.)}$$

Therefore, the finest voltage increment that can be measured in this situation is 20 mV. It would be inappropriate to attempt to make measurements that are, for example, accurate to within 5 mV with this converter.

2.9.2 ANALOG-TO-DIGITAL CONVERTER PERIPHERAL

The ADC peripheral in the AVR microcontrollers is capable of 10-bit resolution and can operate at speeds as high as 15 kSPS (kilo-samples per second). It can read the voltage on one of eight different input pins of the microcontroller, meaning that it is capable of reading from eight different analog sources.

Two registers control the analog-to-digital converter: The *ADC control and status register* (ADCSRA), controls the functioning of the ADC, and the *ADC multiplexer select register* (ADMUX), controls which of the eight possible inputs are being measured. Figure 2–41 shows the bit definitions for the ADC control and status register.

The ADC requires a clock frequency in the range of 50 kHz to 200 kHz to operate at maximum resolution. Higher clock frequencies are allowed but at decreased resolution. The ADC clock is derived from the system clock by means of a prescaler in a manner similar to the timers. The least significant three bits of ADCSRA control the prescaler division ratio. These bits must be set so that the system clock, when divided by the selected division ratio, provides an ADC clock between 50 kHz and 200 kHz. The selection bits and division ratios are shown in Figure 2–42.

Although it could be done by trial and error, the most direct method for choosing the ADC

Bit 7	Bit 6	Bit 5	Bit 4	Bit 3	Bit 2	Bit 1	Bit 0
ADEN	ADSC	ADATE	ADIF	ADIE	ADPS2	ADPS1	ADPS0

Bit	Description
ADEN	ADC Enable bit. Set to enable the ADC.
ADSC	ADC Start Conversion bit. Set to start a conversion.
ADATE	ADC Free Running Select bit. Set to enable free run mode.
ADIF	ADC Interrupt Flag bit. Is set by hardware at the end of a conversion cycle.
ADIE	ADC Interrupt Mask bit. Set to allow the interrupt to occur at the end of a conversion.
ADPS2	ADC Prescaler select bits.
ADPS1	
ADPS0	

Figure 2–41 *ADCSRA Bit Definitions*

preselector factor is to divide the system clock by 200 kHz and then choose the next higher division factor. This will ensure an ADC clock that is as fast as possible but under 200 kHz.

The ADC, like the serial USART, is somewhat slower than the processor. If the processor were to wait for each analog conversion to be complete, it would be wasting valuable time. As a result, the ADC is usually used in an interrupt-driven mode.

Although the discussion that follows uses the more common interrupt-driven mode, it is also possible for the ADC to operate in free-running mode, in which it continuously does conversions as fast as possible. When reading the ADC output in free-running mode, it is

ADPS2	ADPS1	ADPS0	Division Factor
0	0	0	2
0	0	1	2
0	1	0	4
0	1	1	8
1	0	0	16
1	0	1	32
1	1	0	64
1	1	1	128

Figure 2–42 *ADC Preselector Division Ratios*

necessary to disable the interrupts or stop the free-running conversions, read the result, and then re-enable the interrupts and free-running mode. These steps are necessary to ensure that the data read is accurate, in that the program will not be reading the data during the time that the processor is updating the ADC result registers.

The ADC is usually initialized as follows:

1. Set the three lowest bits of ADCSR for the correct division factor.
2. Set ADIE high to enable interrupt mode.
3. Set ADEN high to enable ADC.
4. Set ADSC to immediately start a conversion.

For a division factor of 8, the following lines of code would initialize the ADC to read the analog voltage on the ADC2 pin:

```
ADMUX = 2;            /*read analog voltage on ADC2*/
ADCSR = 0xcb;         /*ADC on, interrupt mode, /8, & started*/
```

The initialization above sets up the ADC, enables it, and starts the first conversion all at once. This is useful because the first conversion cycle after the ADC is enabled is an extra-long cycle to allow for the setup time of the ADC. The long cycle, then, occurs during the balance of the program initialization, and the ADC interrupt will occur immediately after the global interrupt enabled bit is set. Notice that the ADMUX register is loaded with the number of the ADC channel to be read.

Figures 2–43 and 2–44 show the hardware and software, respectively, for a limit detector system based on the analog input voltage to ADC channel 3. Briefly, the system lights the red LED if the input voltage exceeds 3 V, lights the yellow LED if the input voltage is below 2 V, or lights the green LED if the input voltage is within the range of 2 V to 3 V.

The limit detector program in Figure 2–44 shows a typical application for the ADC. The ADC is initialized and started in *main()* by setting ADCSRA to 0xCE. ADC channel 3 is selected by setting ADMUX to 3. This starts the process so the ADC interrupt will occur at the end of the first conversion. Checking the ADC output to see which LED to light and lighting the appropriate LED are all handled in the ADC interrupt ISR.

Notice that the 10-bit output from the ADC is read by reading the data from ADCW. ADCW is a pseudo-register provided by CodeVisionAVR that allows retrieving the data from the two ADC result registers, ADCL and ADCH, at once. Other compilers would more likely require the programmer to read both ADC registers (in the correct order, even) and combine them in the 10-bit result as a part of the program.

Also notice that the programmer has used the analog-to-digital conversion formula in the ISR. The compiler will do the math and create a constant that will actually be used in the program. You will find that this technique can be used to your advantage in several other places, such as loading the UBRR for the USART.

The ADC peripheral in the AVR microcontrollers varies somewhat according to the spe-

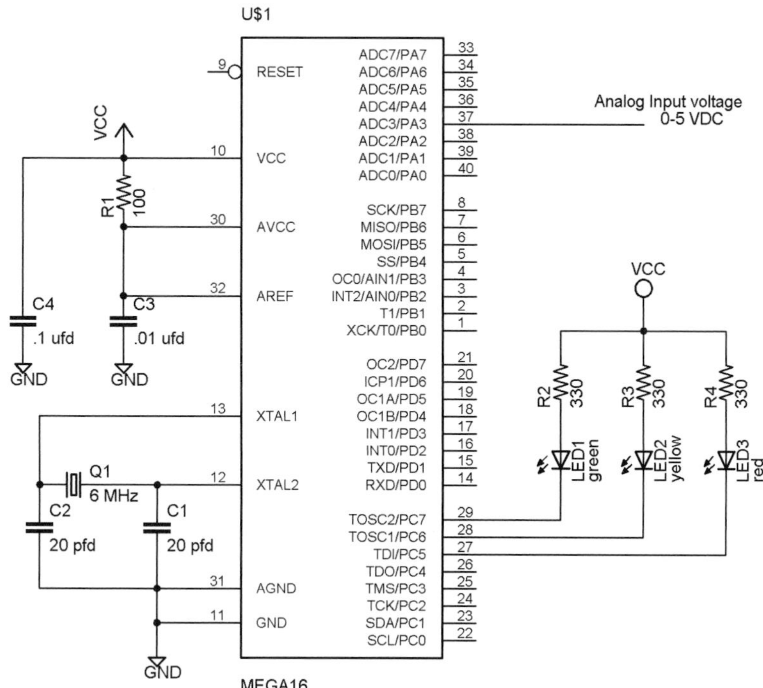

Figure 2–43 *ADC Example Hardware*

```
#include <mega16.h>

/*Define output port and light types*/
#define LEDs PORTC
#define red 0b11011111
#define green 0b01111111
#define yellow 0b10111111

/*ADC ISR, reads ADC, sets lights*/
interrupt [ADC_INT] void adc_isr(void)
{
   unsigned int adc_data;    /*variable for ADC results*/
   adc_data = ADCW;   /*read all 10 bits into variable*/
   if (adc_data >(3*1023)/5)
   {
      LEDs = red;         /*too high (>3V)*/
   }
   else if (adc_data < (2*1023)/5)
   {
```

Figure 2–44 *ADC Example Software (Continues)*

```
        LEDs = yellow;      /*too low (<2V)*/
    }
    else
    {
        LEDs = green;  /*must be just right - like Goldilock's porridge*/
    }
    ADCSRA = ADCSRA | 0x40;  /*start the next conversion   */
}

void main(void)
{
    DDRC = 0xe0;        /*most significant 3 bits for output*/
    ADMUX = 0x3;     /*select to read only channel 3*/
    ADCSRA=0xCE;      /*ADC on, /64, interrupt unmasked, and started*/

    #asm("sei")      /*global interrupt enable bit*/

    while (1)
    {
        ;      /*do nothing but wait on ADC interrupt*/
    }
}
```

Figure 2–44 *ADC Example Software (Continued)*

cific microcontroller in use. All of the ADCs require some noise suppression on the ADC Vcc connection (see Figure 2–43 or 2–46). Some also have a built-in noise canceller function, and some have the ability to control Vref internally. You will need to check the specification for your particular microcontroller when using the ADC.

2.9.3 ANALOG COMPARATOR PERIPHERAL

The analog comparator peripheral is a device that compares two analog inputs: AIN0, the positive analog comparator input, and AIN1, the negative analog comparator input. If AIN0 > AIN1, then the *analog comparator output* bit (ACO) is set. When the ACO bit changes state, either positive-going, negative-going, or both, it will cause an interrupt to occur, provided that the analog comparator interrupt is unmasked, or else the change of state may be set to cause an input capture to occur on timer/counter 1.

The *analog comparator control and status register* (ACSR) shown in Figure 2–45, is used to control the analog comparator functions.

As a simple example of analog comparator functioning, consider the system shown in Figure 2–46. A battery is used to power the system, so it is important to know when the battery voltage becomes dangerously low.

Bit 7	Bit 6	Bit 5	Bit 4	Bit 3	Bit 2	Bit 1	Bit 0
ACD	ACBG	ACO	ACI	ACIE	ACIC	ACIS1	ACIS0

Bit	**Description**
ACD	Analog Comparator Disable bit. Set to disable the analog comparator.
ACBG	Analog Comparator bandgap select.
ACO	Analog Comparator Output bit.
ACI	Analog Comparator Interrupt flag.
ACIE	Analog Comparator Interrupt mask bit.
ACIC	Analog Comparator Input Capture bit. Set to enable input capture on comparator change-of-state.
ACIS1 ACIS0	Analog Converter Comparator Mode Select bits. (See definitions below.)

Analog Compare Interrupt Mode Select Bit Definitions:

ACIS1	***ACIS0***	***Interrupt Mode***
0	0	Comparator interrupt on ACO toggle.
0	1	Reserved — do not use.
1	0	Comparator interrupt on ACO falling edge. (AIN1 becomes greater than AIN0.)
1	1	Comparator interrupt on ACO rising edge. (AIN0 becomes greater than AIN1.)

Figure 2–45 *ACSR Bit Definitions*

The system shown in Figure 2–46 is continuously monitoring the battery without expending any processor time whatsoever, because it is all handled by the analog comparator. When the battery voltage becomes too low, the LED is lighted.

The two analog converter inputs are connected to voltage dividers: one is powered by the regulated +5 V as a reference, and the other is powered directly by the battery. The voltage divider powered by the +5 V is designed to provide approximately 2.2 V at its center point. The other voltage divider is designed so that when the battery discharges to approximately 6 V, the center point of the voltage divider will also measure approximately 2.2 V. The analog comparator is set to detect when the voltage from the battery's voltage divider drops below the 2.2 V provided by the reference divider.

As Figure 2–47 shows, using the analog comparator is relatively simple. In this example, ACSR is loaded with 0x0A to enable the analog comparator, to enable its interrupt, and to

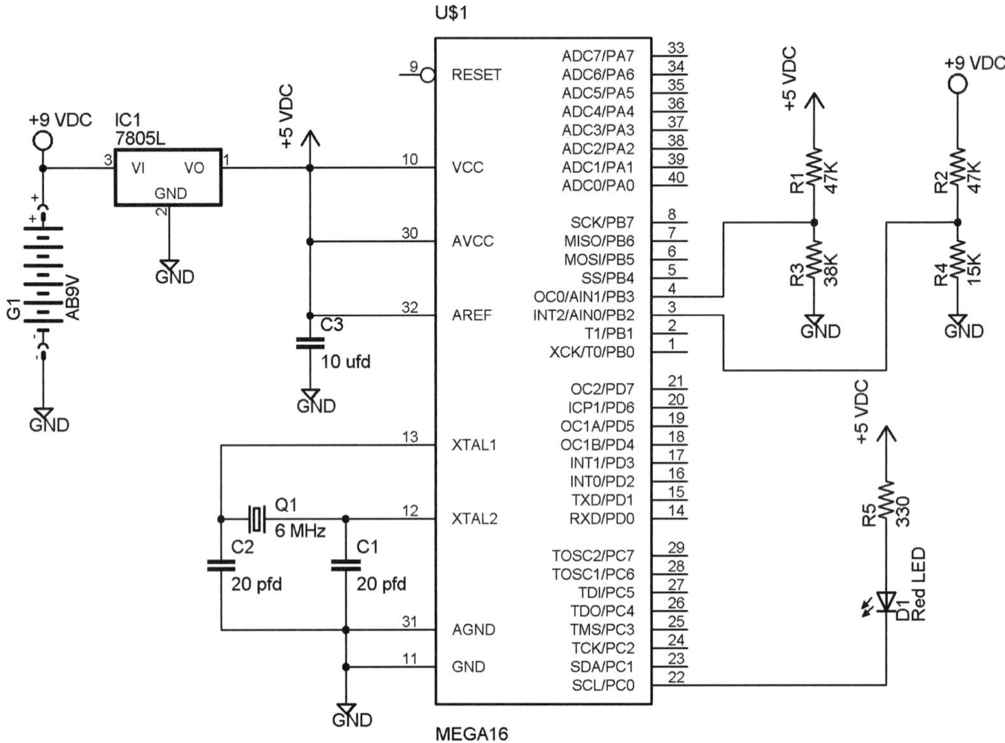

Figure 2–46 *Analog Comparator Example Hardware*

```
#include <mega16.h>
#define LED PORTC.0
unsigned int blink_count = 0;   /*counter for LED blinker*/

/* Analog Comparator interrupt service routine*/
interrupt [ANA_COMP] void ana_comp_isr(void)
{
   ACSR = 0;  /*disable the comparator*/
   TCCR0 = 0x5; /*start the timer*/
}

/* Timer 0 overflow interrupt service routine*/
interrupt [TIM0_OVF] void timer0_ovf_isr(void)
{
   TCNT0 = 256 - 6;  /*set for approx. 1 ms overflow rate*/
   ++blink_count;  /*increment blink counter*/
   switch (blink_count)
   {
```

Figure 2–47 *Analog Comparator Example Software (Continues)*

```
        case 1990:   LED = 0;   /*LED on*/
                     break;
        case 2000:   LED = 1;   /*LED off*/
                     blink_count = 0; /*reset blink counter*/
                     break;
        default:     break;   /*do nothing */
     }
}

void main(void)
{
     PORTC = 0x01;   /*start with LED off*/
     DDRC = 0x01;    /*set bit 1 for output*/
     /*enable analog comp, enable comparator interrupt, falling edge*/
     ACSR=0x0A;
     TIMSK = 0x01;   /*unmask Timer 0 overflow interrupt*/

     #asm("sei")   /*set global interrupt enable bit*/

     while (1)
;     /*do nothing - or, more importantly, put operational code here*/
}
```

Figure 2–47 *Analog Comparator Example Software (Continued)*

set it so that the interrupt occurs on a falling edge when AIN0 drops below AIN1 (AIN1 becomes greater than AIN0). The AC ISR starts Timer0, which is set up to blink the LED on for 10 ms out of every 2 seconds. This blink is designed to show the battery condition while minimizing the additional current draw due to the LED. After all, the battery is low, so we need to protect what is left of the battery power. The AC ISR also disables the analog comparator. This is important to be sure that noise or other instability of the battery voltage will not cause erratic blinking on the LED.

Chapter 2 Example Project: Part D

This is the fourth part of the Chapter 2 example using a stock car data collection system. This portion is concerned with measuring the engine temperature and sending the collected data to the PC.

MEASURING ENGINE TEMPERATURE USING THE ANALOG-TO-DIGITAL CONVERTER (ADC)

Previous investigation showed that the temperature signal is connected to ADC3, the ADC input on Port A, pin 3. Some further investigation using the specification for the RTD thermocouple and the circuitry conditioning the RTD signal shows that the measurement range is, happily, exactly the range your boss wants, 100°F to 250°F. Using the 10-bit measurement mode on the ADC means that the resulting measured values will be as follows:

$$100°F = 0x000 = 0_{10}$$

$$250°F = 0x3FF = 1023_{10}$$

This sets the conversion formula to be

$$Temp = (150°F * ADC\ reading) / 1023 + 100°F$$

The ADC may most conveniently be run in free-running mode for this use. In this way, the temperature will be kept up to date as fast as possible so that when the data is stored at the one-second interval, the most recent value will be recorded. In free-running mode, the ADC interrupt occurs at the end of each conversion and can be used to update the current temperature value.

Part of the ADC initialization is to select an appropriate clock frequency by choosing the prescaler value for the ADC clock. The ADC clock must be between 50 kHz and 200 kHz. In this case, you have a system clock of 8 MHz, and so a prescaler of 64 will give an ADC clock of 125 kHz.

The ADC initialization process involves choosing the channel to be measured, enabling the free-running mode, and starting the first conversion so that an interrupt will occur at the end of the conversion, keeping the process running:

```
/* ADC initialization*/
ADMUX=0x3;  /*select channel 3 and AREF pin as the reference voltage*/
             /*input*/
ADCSRA=0xE9; /*enable ADC, free run, started, clock prescaler of 64*/
```

The ADC ISR has the job of reading the current conversion value and converting it to temperature:

```
/* ADC interrupt service routine*/
interrupt [ADC_INT] void adc_isr(void)
{
    unsigned int ADC_DATA;
    ADC_DATA = ADCW;    /*get data from ADC result register*/
    current_temp = ((long)150 * (long)ADC_DATA)/ (long) 1023    +
            (long)100;
}
```

SENDING COLLECTED DATA TO THE PC

The collected data is already converted to units of rpm and °F, so this function needs only to send the data to a PC. This involves initializing the USART and using built-in functions to format and send the data.

The USART is initialized for 9600 baud, 8 bits, and no parity as follows:

```
/* UART initialization */
UCSRA=0x00;
```

```
UCSRB=0x18;       /* enable transmitter and receiver */
UBRR=0x33;        /*select 9600 baud*/
UBRRHI=0x00;
```

The data sent to the PC is going to be analyzed using a Microsoft Excel spreadsheet. Investigation shows that data can be input directly using a comma-delimited (CSV) format. This means that commas separate the data on each line of the spreadsheet, and CR separates each line. That is, data sent as

data1, data2, data3

data4, data5, data6

will appear in the spreadsheet exactly as shown above occupying a space two cells high by three cells wide. In this case, you decide that the first column in the spreadsheet will contain the engine rpm, the second will contain the shaft rpm, and the third will contain the engine temperature. Each line of the spreadsheet will contain one set of data, so each line will contain data taken one second after the data in the previous line. The following code is the *while(1)* from the *main()* function of the code:

```
while (1)
{
      if (!PINA.0)   /*note: switch must be released before data is*/
                     /*all sent*/
      {
        unsigned char x;   /*temporary counter variable*/

        /*print column titles into the spreadsheet in the first row*/
        printf ("%s , %s , %s \n","Engine RPM", "Shaft RPM",
                   "Temperature");
        for (x = 0; x < 120; x++)
        {
        /* print one set of data into one line on the spreadsheet*/
          printf ("%d , %d , %d \n",e_rpm[x], s_rpm[x], temp[x]);
        }
     }
};
```

This routine sends the 120 sets of data (and a line of column titles) using the built-in *printf()* function. A *newline* character ("\n") is included with each set of data to cause the next set to appear in the following line of the spreadsheet.

2.10 SERIAL COMMUNICATION USING THE SPI

The *serial peripheral interface*, SPI (pronounced "spy"), is another form of serial communication available in the AVR microcontrollers. It is a synchronous serial communication bus, meaning that the transmitter and receiver involved in SPI communication must use the same clock to synchronize the detection of the bits at the receiver. Normally, the SPI bus is used for very short distance communication with peripherals or other microcontrollers that are located on the same circuit board or at least within the same piece of hardware. This is different from the USART, which is used for communication over longer distances, such as between units or between a microcontroller and PC. The SPI bus was developed to provide relatively high-speed, short distance communication using a minimum number of microcontroller pins.

SPI communication involves a master and a slave. Both the master and a slave send and receive data simultaneously, but the master is responsible for providing the clock signal for the data transfer. In this way, the master has control of the speed of data transfer and is, therefore, in control of the data transfer.

Figure 2–48 shows the connections between the master and the slave units for SPI communication. The master supplies the clock and eight bits of data, which are shifted out of the *master-out-slave-in* (MOSI) pin. The same eight bits are shifted into the slave unit, one bit per clock pulse, on its MOSI line. As the eight bits are shifted out of the master and into the slave, eight bits are also shifted out of the slave on its *master-in-slave-out* (MISO) line and into the master on its MISO pin. SPI communication, then, is essentially a circle in which eight bits flow from the master to the slave and a different set of eight bits flows from the slave to the master. In this way, the master and a slave can exchange data in a single communication.

It is entirely possible to connect many different devices together using a SPI bus, because all of the MOSI pins and all of the MISO pins could be hooked together. Any device on the network can become a master by simply deciding to send data. A device on the network becomes a slave when its *slave select* (SS) pin is grounded. Usually, the SS pins from the slaves

Figure 2–48 *SPI Communication Scheme*

Bit 7	Bit 6	Bit 5	Bit 4	Bit 3	Bit 2	Bit 1	Bit 0
SPIE	SPE	DORD	MSTR	CPOL	CPHA	SPR1	SPR0

Bit	**Description**
SPIE	SPI Interrupt Mask bit.
SPE	SPI Enable bit. Set to enable SPI operations.
DORD	Data Order bit. Cleared causes the MSB of the data word to be transmitted first.
MSTR	Master/Slave Select bit. Set for the device to be the master.
CPOL	Clock Polarity bit.
CPHA	Clock Phase bit.
SPR1	SPI Clock Rate bits (see table below).
SPR0	

SPI Clock Rate Bit Definitions:

SPR1	**SPR0**	**SCLK Frequency Select**
0	0	System Clock / 4
0	1	System Clock / 16
1	0	System Clock / 64
1	1	System Clock / 128

Figure 2–49 *SPCR Bit Definitions*

are connected to either a parallel port on the master or to a decoder that determines which device will be the slave.

The *SPI control register*, SPCR, controls the operation of the SPI interface. SPCR bit definitions are shown in Figure 2–49. In a similar manner to the control registers studied previously, there are bits to enable the SPI interface and its associated interrupt, and bits to control the communication speed. There is also a bit to set when a device is acting as the master, MSTR.

In addition, there are bits to control the data order, MSB first or LSB first, and to select the clock polarity and the clock phase. These bits are used when matching the SPI communication peripheral to other SPI devices.

Figures 2–50 through 2–53 show the hardware and software for a simple SPI demonstration system. The system reads the switch data from the dip switch on one system and

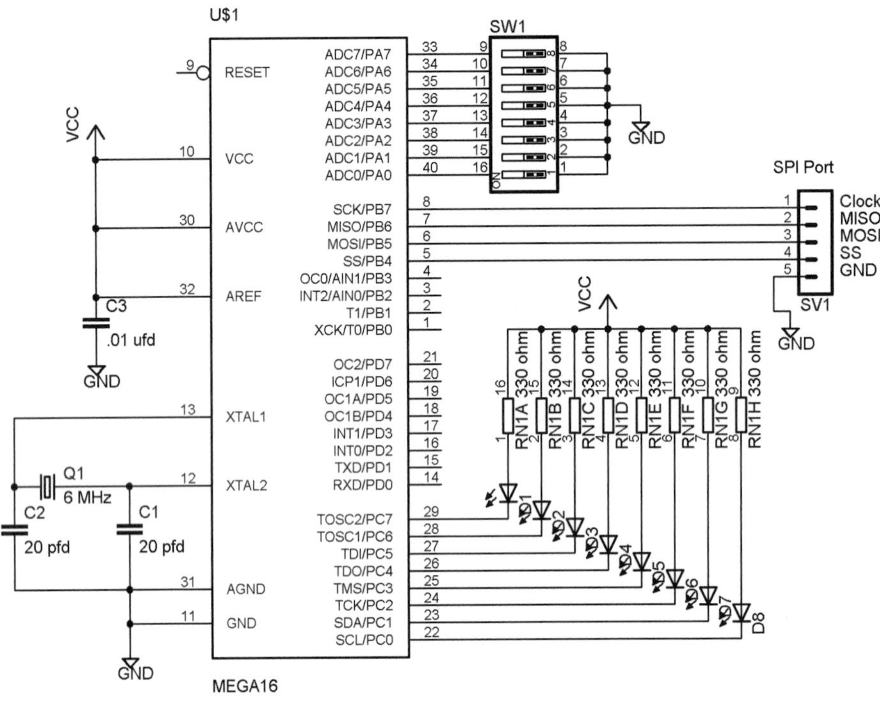

Figure 2–50 *ATMega16 SPI Demonstration System Hardware*

displays the switch data on the other system, and vice versa. All communication takes place via the SPI bus. The Atmega128 is the SPI Master, and the ATMega16 is the SPI slave.

The Atmega128 reads the data to be sent from the dip switches on PORT F and sends it to the ATMega16 slave, where it is displayed on PORT C. At the same time, the data from the ATMega16 (from the dip switch on PORT A) is returned to the ATMega128 and displayed on its PORT A. The system continuously updates, making it appear to be instantaneous.

PORT B (PORT B alternate functions include the SPI pins) on each processor is initialized as shown below:

Pin	*Slave PORT B*	*Master PORT B*
MOSI	Input	Output
MISO	Output	Input
SCK	Input	Output
SS	Input	Output

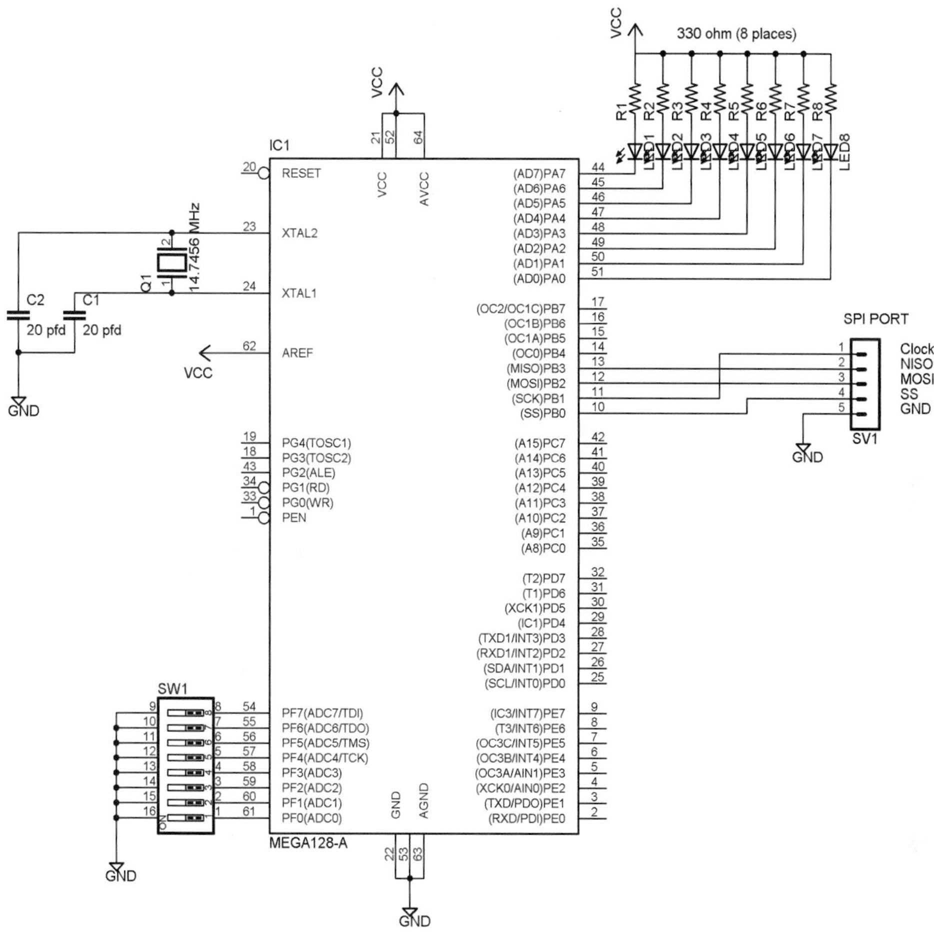

Figure 2–51 *ATMega128 SPI Demonstration Hardware*

The definition of master and slave is reflected in the port initializations. It also allows a direct connection (MOSI-MOSI, MISO-MISO, SCK-SCK, and SS-SS). The ports are initialized, and the SPI system is enabled, along with its interrupt, in the *main()* loop.

Following the initializations, there is a short assembly code routine automatically inserted by CodeVisionAVR that makes sure that the SPI interrupt is cleared. This is an important step to be sure that the SPI interrupt service routine does not execute until you intend it to. Finally, 0x00 is written to SPDR to start the SPI communication process.

Once started, the entire function is handled in the SPI ISR functions. The master's ISR reads the data received and writes it to the output port (the LEDs). Then it reads the input port (the dip switch) and starts a new communication cycle by writing the data from the

```
/*SPI Slave code*/
#include <mega16.h>

#define SPI_output_data PORTC
#define SPI_input_data PINA

/* SPI interrupt service routine*/
interrupt [SPI_STC] void spi_isr(void)
{
   SPI_output_data = ~SPDR;  /*read out new data received*/
   SPDR = SPI_input_data;    /*load new data for NEXT communication cycle*/
}

void main(void)
{
   DDRB = 0x40;      /*sclk, MOSI, SS = input, MISO = output*/
   DDRC = 0xFF;      /*all output*/
   PORTA = 0xFF;     /*pull ups for dip switch*/
   SPCR=0xC1;     /*enable SPI, and its interrupt, slave mode*/
   /* Provided by CodeVisionAVR to clear the SPI interrupt flag*/
   #asm
    in   r30,spsr
    in   r30,spdr
   #endasm
   #asm("sei")// Global enable interrupts

   while (1)
   {
       ;  /*put code here to interpret SPI results if needed*/
   }
}
```

Figure 2–52 *SPI Slave Software Example*

```
#include <mega128.h>
#define SPI_output_data PORTA
#define SPI_input_data PINF

/* SPI interrupt service routine*/
interrupt [SPI_STC] void spi_isr(void)
{
PORTB.0 = 1;  /*ss high*/
SPI_output_data = ~SPDR;  /*read data from last exchange*/
PORTB.0 = 0;   /*lower SS*/
SPDR = SPI_input_data;  /*send new data out*/
}
```

Figure 2–53 *SPI Master Software Example (Continues)*

```
// Declare your global variables here

void main(void)
{
PORTA=0xFF;     /*start with LED's off*/
DDRA=0xFF;      /*all output*/
PORTB = 0x01;   /*start with SS high*/
DDRB=0x07;      /*ss,sck, and MOSI as output*/
PORTF=0xFF;     /*pull-ups for dip switch*/

SPCR=0xD1;      /*375 KHz clock, enable SPI, SPI interrupt*/
SPSR=0x00;

// Clear the SPI interrupt flag
#asm
    in   r30,spsr
    in   r30,spdr
#endasm

// Global enable interrupts
#asm("sei")

SPDR = 0x00;  /*write once to start [process*/

while (1)
      {
         ;  /*place other code here*/
      };
}
```

Figure 2–53 *SPI Master Software Example*

input port to SPDR. The slave ISR then merely accepts the new data and provides its own switch data for the next communication cycle. And the cycle repeats forever.

An important concept to note is that the data going from the slave to the master is actually one cycle *behind* the data going to the slave. This occurs because the slave loads new data *after* it receives data from the master.

The SPI bus is often used to form a high-speed network within a device. The connections may be parallel (all the MISO pins tied together and all of the MOSI pins tied together) or in a large serial loop, where all of the data bits travel through all the devices in turn in a manner that amounts to a large circle of shift registers. The parallel method requires a method of selecting the slave device so that only two devices share the data flow, while in the serial loop scheme all of the data flows through all of the devices. In either method, multiple CPUs and slave devices can exist on a SPI-based network to share data.

2.11 SERIAL COMMUNICATION USING I²C

Inter-Integrated Circuit (I²C) communication is a system developed by Philips Semiconductor to provide communications between IC devices in entertainment devices. It has grown to become a method to communicate with a multitude of peripheral devices.

I²C synchronous communication differs from SPI in two major ways: it is a "protocol-ed" language (it requires a certain protocol), and it uses only two wires for communication. These are typically named *Synchronous Clock* (SCL) and *Synchronous Data* (SDA). Using only two wires does not allow data to flow simultaneously from master to slave and from slave to master as in SPI. I²C is sequential in that data flows first from the master to the slave and then (after the master-to-slave communication is complete) from the slave to the master if needed. Data only flows back to the master when the slave is responding with data to a master's request.

Refer to Figure 2–48 and Figure 2–54 to compare the SPI and I²C schemes. Figure 2–54 shows the clock flowing one way (master to slave) and the data flowing both ways on a single wire.

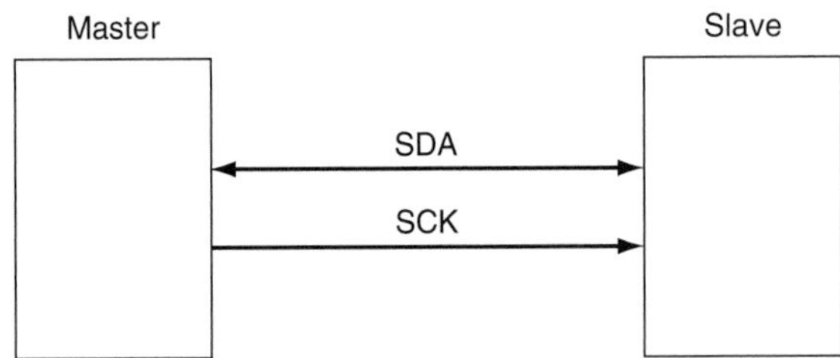

Figure 2–54 *I²C Data Flow*

Using a protocol-based communication scheme is an excellent place to use built-in libraries. Developing the code to implement the protocol is often very time-consuming and a waste of time if the libraries exist.

Figure 2–55 shows the hardware for an I²C example system. The system is composed of an ATMega128 processor and a PCF8563 real time clock (RTC) device. The two components communicate using I²C.

The software looks to be very simple at first glance. However, the protocol is buried in the *i2c.h* and *pcf8563.h* libraries. These libraries are used by putting the *#include* line in the code to include the libraries. As a new programmer, you should open the *i2c.h* header file and look at the function calls listed there. This will tell you how to call and use the library functions. Likewise, inspecting the pcf8563.h file will allow you to see what calls are provided and what data must be passed to and what data may be expected to be returned from the library functions.

The software in Figure 2–56 shows the methods used to poll the seconds register in the RTC. The seconds are then displayed on LEDs attached to PORT B. *Main()* initializes

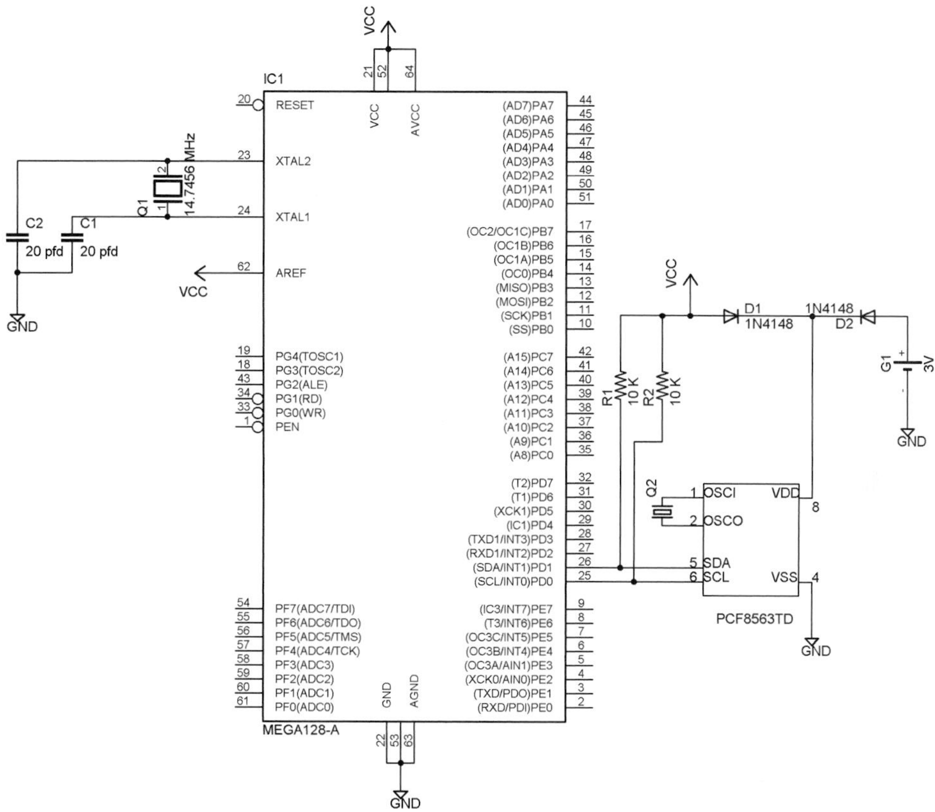

Figure 2–55 *I²C Real Time Clock Example*

```
#include <mega128.h>
/* I2C Bus functions*/
#asm
   .equ __i2c_port=0x12 ;PORTD
   .equ __sda_bit=1
   .equ __scl_bit=0
#endasm
#include <i2c.h>
/* PCF8563 Real Time Clock functions*/
#include <pcf8563.h>

void main(void)
{
/*variables to hold time*/
unsigned char hour, minute, second;
DDRB=0xFF; /*all output*/
/* I2C Bus initialization using library*/
i2c_init();
```

Figure 2–56 *I²C Real Time Clock Software Example (Continues)*

```
/* PCF8563 Real Time Clock initialization using library call*/
rtc_init(1,RTC_CLKOUT_OFF,RTC_TIMER_OFF);

while (1)
      {
      /*read the current time using library functions*/
      rtc_get_time(&hour,&minute,&second);
      PORTB = ~second;   /*display seconds*/
      };
}
```

Figure 2–56 *I²C Real Time Clock Software Example (Continued)*

the ports, the I²C bus (via a library call to *i2c_init()*), and the RTC via a call to *rtc_init()*. The RTC time is then read via a call to rtc_get_time(). Only the seconds are actually used, but the function from the library returns hours, minutes, and seconds, and so all must be stored even if they are not used.

2.12 THE AVR RISC ASSEMBLY LANGUAGE INSTRUCTION SET

The C programming language is defined as a high-level language. A high-level language is one that is easy for humans to read and allows complex operations to be performed in a single line of code. Microcontrollers, on the other hand, can actually only execute what is called machine code. Machine code consists solely of numbers and is a very low-level language.

Assembly language is very close to machine code in that it translates very directly into machine code. Assembly language is also considered to be a low-level language because it is very close to machine language.

C language compilers work by *compiling* the C language statements into the assembly code, which is then *assembled* into machine code. The compilers accomplish this through the use of libraries containing assembly language functions that implement the C language statements. The C language compiler looks at the C language statements and selects those library functions that implement the statements, compiling them into an assembly language program. Finally, the C language compiler uses an assembler to convert the assembly language to machine code.

There are actually two additional steps buried in the compiling and assembling process: linking and locating. *Linking* is the process of integrating library functions and other code into the program being compiled. *Locating* is the process of assigning actual code memory addresses to each of the executable instructions.

CodeVisionAVR adds a further step in that it converts the executable machine code to an Intel-formatted HEX file for downloading into an AVR microcontroller using a chip programmer.

Figure 2–57 shows this process using one line of code from the ADC Example Software (refer to Figure 2–44). The diagram in Figure 2–57 should make the process relatively clear, with two exceptions. The first is that, as you will recall, the program memory spaces are 16 bits wide and the memory addresses shown under step 3 are the word addresses for each instruction word. (Notice that each instruction shown is contained in a single word, which is appropriate in a RISC processor.) The program words, however, are split into bytes in the Intel HEX code line and, as a result, the addresses shown are two times the address shown

STEP 1 Write the C code:

```
if (adc_data >(3*1023)/5) LEDs = red;   //too high (>3V)
```

STEP 2 Compile:

The compiler uses its libraries to compile the C code into following assembly code (comments were added to the assembly code by the author for clarity):

```
LDI   R30,LOW(614)       ;put the low byte of the constant into R30
LDI   R31,HIGH(614)      ;put the high byte of the constant into R31
CP    R30,R16            ;compare the low byte with adc_data low byte
                         ;(must be stored in R16)
CPC   R31,R17            ;now compare the high bytes (with carry)
BRSH _0x2                ;branch if untrue to the 'else' statement
LDI   R30,LOW(3)         ;if true, load the value to light the red LED
                         ;into R30
OUT   0x15,R30           ;and output the value to the port d I/O
                         ;register
```

STEP 3 Assemble:

The assembler produces the machine code shown in the center column. The left-hand column is the code memory address that will hold the instruction word from the center column and the assembly language instructions are shown for clarity.

```
000052 e6e6        LDI   R30,LOW(614)
000053 e0f2        LDI   R31,HIGH(614)
000054 17e0        CP    R30,R16
000055 07f1        CPC   R31,R17
000056 f418        BRSH _0x2
000057 e0e3        LDI   R30,LOW(3)
000058 bbe5        OUT   0x15,R30
```

STEP 4 Convert to Intel HEX for downloading:

This line Intel HEX code contains the machine code (see underlined section):

```
:1000A20015B1E6E6F2E0E017F10718F4E3E0E5BB8C
```

(Underlined in the original: E6E6F2E0E017F10718F4E3E0E5BB)

Figure 2–57 *Compiling Process*

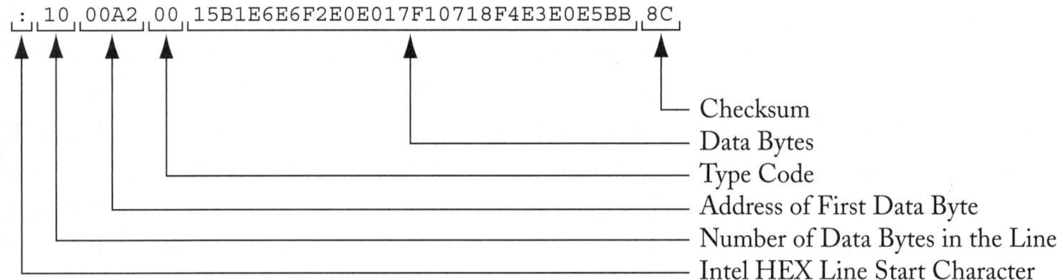

Figure 2–58 *Intel HEX Code Format*

under step 3 (2 * 0x52 = 0xA4). Figure 2–58 shows the format of a line of Intel HEX code to assist you in reading the line.

The second apparent difference between the code in steps 2 and 3 is that the order of the bytes appears to be different. This occurs because under step 3, the words are shown with the most significant byte to the left, as is appropriate when writing numbers, but in the Intel HEX file, the most significant byte is to the right, because it resides in the higher memory location.

The Intel HEX format (refer to Figure 2–58) requires that each line begin with a ":" to denote the beginning of a line of hex code. This is followed by a byte that indicates the number of data bytes contained in a line. (Remember that the numbers are in hexadecimal, so a "10" means there are actually 16 data bytes on the line.) The next two bytes are the address of the first data byte in the line—subsequent data bytes are placed in subsequent addresses. The type code indicates the nature of the data in most older, smaller processors. In the case of AVRs, the type code is actually used as a 64-Kbyte page indication, 0 for page 0, 1 for page 1, and so on, as is necessary when the code is destined for devices with large memory spaces with greater than 16-bit addressing requirements.

Finally, the appropriate number of data bytes is included, followed by a checksum used for error checking.

At this point, it may appear that assembly code is relatively hard to use, especially when compared to a high-level language such as C, and is convoluted and obtuse. You are probably wondering why anyone would be concerned with the assembly code. Assembly code is useful when a programmer needs to do operations directly on the registers of the microprocessor or when the programmer needs to very critically control the speed of execution. Examples of the former would be writing advanced programs such as *real-time operating systems* (RTOS) or other programs that must manipulate the microcontroller memory directly in order to operate. Likewise, it is useful as a debugging tool. Errors in C programs often become apparent when the resulting assembly code is studied to determine what the processor is actually being told to do.

The second reason to use assembly code is to very tightly control the time of execution. C language compilers use a library of functions to implement the C language. Depending on the thoroughness of the compiler, there may be only a few functions that attempt to cover every instance, or there may be many functions that cover every particular instance of a programming command very efficiently. If there are only a few functions that cover all situations, the chances are very good that these functions use much more execution time than is necessary in order to cover all the scenario possibilities. In the case of the better C compilers, such as CodeVisionAVR, they have large libraries of functions to cover most situations very efficiently. In these cases, the need for assembly code programming is reduced.

Should the need for assembly language programming arise, one relatively good way to handle it is to first write the function in C and then, using the assembly code files generated by the compiler, look at the assembly code the compiler proposes to use for your function. You can then analyze the assembly code by looking up each instruction in the function (refer to the instructions in Appendix G) and determine which if any instructions are really not necessary. Experience has shown that even well-optimized code can usually be speeded up somewhat by careful analysis of the assembly code.

2.13 CHAPTER SUMMARY

This chapter has provided you with the basic hardware background necessary to choose and apply a member of the AVR RISC microcontroller family produced by Atmel.

Specifically, the architecture of the microcontrollers, including their memory system design and internal workings, has been presented. Each of the usual peripheral devices has been described, its use discussed, and an example presented showing how to use the peripheral. Should you encounter other peripherals not specifically covered in this chapter, their use should be readily understood because they all work very similarly.

As a student, you should now be ready to meet each of the objectives of this chapter.

Chapter 2 Example Project: Part E

This is the summary of the Chapter 2 example presented in pieces throughout the chapter. The complete program is as follows:

```
#include <mega163.h>
#include <stdio.h>

unsigned char data_set_cntr;   /*counter to track number of data sets*/
                               /*recorded*/
unsigned int e_rpm[120], s_rpm[120];   /*arrays to hold 120 sets of*/
                                       /*data*/
unsigned char temp[120];
unsigned int current_e_rpm, current_s_rpm;    /*variables to hold*/
                                              /*current values*/
unsigned char current_temp;
```

```
unsigned int previous_shaft_capture_time;  /*saved time from previous*/
                                              /*capture*/

/* External Interrupt 0 service routine*/
interrupt [EXT_INT0] void ext_int0_isr(void)
{
   unsigned int current_capture_time, period; /*current time and*/
                                              /*period*/

   current_capture_time = TCNT1;  /*get Timer1 time*/
   /*check for roll-over*/
   if (current_capture_time > previous_shaft_capture_time)
           period = current_capture_time -
           previous_shaft_capture_time;
   else
           period = 0xFFFF - current_capture_time +
             previous_shaft_capture_time;
   current_s_rpm = (unsigned long)60E6 / (unsigned long)period;
   previous_shaft_capture_time = current_capture_time;  /*save for*/
                                                /*next calculation*/
}

/* External Interrupt 1 service routine*/
interrupt [EXT_INT1] void ext_int1_isr(void)
{
   data_set_cntr = 0;    /*clear counter to start counting*/
}

int time_cntr = 0;   /*global variable for number of Timer 09*/
                     /*overflows*/

/* Timer 0 overflow interrupt service routine*/
interrupt [TIM0_OVF] void timer0_ovf_isr(void)
{
   TCNT0 = 6;     /*reload timer 0 for next period*/
   if (time_cntr++ == 500) /*check for one second, increment counter*/
   {
        if (data_set_cntr < 120) /*record data if less than 120 sets*/
        {
            e_rpm[data_set_cntr] = current_e_rpm;    /*record*/
                                                     /*engine rpm*/
            s_rpm[data_set_cntr] = current_s_rpm;    /*record*/
                                                     /*shaft rpm*/
            temp[data_set_cntr++] = current_temp;    /*record*/
                                                     /*engine temp*/
        }
        time_cntr = 0;    /*reset counter for next one-second interval*/
   }
}
```

```
unsigned int previous_capture_time;  /*saved time from previous*/
                                     /*capture*/

/* Timer 1 input capture interrupt service routine*/
interrupt [TIM1_CAPT] void timer1_capt_isr(void)
{
    unsigned int current_capture_time, period;   /*current time and*/
                                                 /*period*/

   current_capture_time = (256* ICR1H) + ICR1L;  /*get captured time*/
   /*check for roll-over*/
   if (current_capture_time > previous_capture_time) period =
     current_capture_time - previous_capture_time;
   else
       period = 0xFFFF - current_capture_time + previous_capture_time;
       current_e_rpm = (unsigned long)120E6 /
       (unsigned long)period; previous_capture_time =
       current_capture_time; /*save for next calculation*/
}

/* ADC interrupt service routine*/
interrupt [ADC_INT] void adc_isr(void)
{
   unsigned int ADC_DATA;

   ADC_DATA = ADCW;   /*get data from ADC result register*/
   current_temp = ((long)150 * (long)ADC_DATA)/ (long) 1023   +
       (long)100;
}

void main(void)
{
  /* Input/Output Ports initialization*/
  PORTA=0x01;    /*enable pull up on bit 0 for upload switch*/
  DDRA=0x00;     /*all input*/
  PORTB=0x00;
  DDRB=0x00;     /*all input*/
  PORTC=0x00;
  DDRC=0x00;     /*all input*/
  PORTD=0x4C;    /*enable pull ups on Port D, Bits 2,3, & 6*/
  DDRD=0x00;     /*all input*/

  /* Timer/Counter 0 initialization*/
  TCCR0=0x03;      /*clock set to clk/64*/
  TCNT0=0x00;      /*start timer0 at 0*/

  /* Timer/Counter 1 initialization*/
  TCCR1A=0x00;
```

```
TCCR1B=0x02;        /*prescaler = 8, capture on falling edge*/
TCNT1H=0x00;        /*start at 0*/
TCNT1L=0x00;

/* External Interrupt(s) initialization*/
GIMSK=0xC0;    /* both external interrupts enabled*/
MCUCR=0x0A;       /* set both for falling edge triggered*/
GIFR=0xC0;     /* clear the interrupt flags in case they are*/
                /*errantly set*/

/* Timer(s)/Counter(s) Interrupt(s) initialization*/
TIMSK=0x21;    /* timer0 overflow, ICP interrupts enabled*/

/* UART initialization*/
UCSRA=0x00;
UCSRB=0x18;     /* enable transmitter and receiver*/
UBRR=0x33;      /*select 9600 baud*/
UBRRHI=0x00;

/* Analog Comparator disabled*/
ACSR=0x80;
SFIOR=0x00;

/* ADC initialization*/
ADMUX=0x3;  /*select channel 3 and AREF pin as the reference*/
            /*voltage input*/
ADCSR=0xE9; /*enable ADC, free run, started, clock prescaler of 64*/

/* Global enable interrupts*/
#asm("sei")

while (1)
{
   if (!PINA.0)   /*note: switch must be released before data*/
                  /* is all sent*/
   {
         unsigned char x;  /*temporary counter variable*/

         /*print column titles in the first spreadsheet row*/
         printf ("%s , %s , %s \n","Engine RPM",
         "Shaft RPM", "Temperature");
         for (x = 0; x < 120; x++)
         {
               /* print one set of data into one line on the*/
               /*spreadsheet*/
               printf ("%d , %d , %d \n",e_rpm[x], s_rpm[x],
                  temp[x]);
            }
      }
   }
}
```

In summary, this program is but one way to accomplish the task; there are many others that would work equally well. And there are some enhancements that would improve system operation. Some of the more likely variations and enhancements are listed here:

- Store the rpm and temperature data as raw data, not converted to rpm or temperature. In this way, time would be saved during each parameter measurement. Each piece of raw data would need to be converted in the routine that sends the data to the PC as the data is sent.
- The rpm measurements could have been acquired by actually counting the pulses for 1 second and multiplying by 60 to get rpm. In this case, the 1-second routine would record the count and clear it for the next 1-second count.
- There is no protection to prevent bumping the start switch after data is recorded and before it is transferred to the PC. Data could be lost in this way. One choice might be to disallow further recording until data is sent to the PC. The problem with this is that if the driver starts the recording process inadvertently, he has no way to recover and transfer the right data. Another choice might be to provide a "data lock" toggle switch. With this switch set one way, data could still be recorded whenever the start switch is pressed. Setting this switch to the other position after the good data is recorded would prevent further recording until the data had been saved to the PC.

There are many more variations and enhancements possible. Having completed this chapter, you should be able to add several yourself.

EXERCISE

2.14 EXERCISES

1. Explain the differences between a microcontroller and a microcomputer (Section 2.3).

*2. Describe the following memory types and delineate their uses (Section 2.4):
   A. FLASH Code memory
   B. Data memory
   C. EEPROM memory

3. Explain why a section of data memory is called I/O (Section 2.4).

4. Explain the function and purpose of the processor *stack* (Section 2.4).

5. Explain how an interrupt finds its way to the interrupt service routine to execute it when the interrupt occurs (Section 2.5).

*6. Write a fragment of C language code to initialize External Interrupt 1 to activate on a falling edge applied to the external interrupt pin (Section 2.5).

7. What is a watchdog timer and why is it used (Section 2.5)?

*8. Write a fragment of C language code to initialize the Port C pins so that the upper nibble can be used for input and the lower nibble can be used for output (Section 2.6).

9. Write a fragment of C code to initialize Port A and set the least significant bit of Port B to be input, and the balance of Port B to be output (Section 2.6).
10. Name and describe at least two distinctly different functions that can be performed by a timer/counter (Section 2.7).

*11. Sketch the waveform appearing at the output of the USART when it transmits an "H" at 9600 baud (Section 2.8). The sketch should show voltage levels and the bit durations, in addition to the waveform.

12. Sketch the waveform appearing at the output of the USART when it transmits an "F" at 1200 baud (Section 2.8). The sketch should show voltage levels and the bit durations, in addition to the waveform.

*13. Compute the missing values to complete the following table relating to analog to digital conversion (Section 2.9):

V_{in}	$V_{fullscale}$	*Digital Out*	*# of bits*
4.2V	10V	8	
1.6V	5V	10	
5V	123	10	
10V	223	8	

14. Describe the differences between the USART and the SPI serial communication (Sections 2.8 and 2.10).
15. Detail the steps in the compiling process (Section 2.11).

"*" preceding an exercise number indicates that the question is answered or partially answered in the appendix.

2.15 LABORATORY ACTIVITIES

1. Create a program that will turn on an LED when a falling edge occurs on external interrupt 0 and will turn it off when a rising edge occurs on external interrupt 1.
2. Create a program that will demonstrate how a watchdog timer resets the processor if the program hangs in an infinite loop.
3. Create a program that will read the data on all eight bits of Port A, swap the nibbles of that data, and output the result to Port B.
4. Create a simulated engine speed monitor that will light a yellow LED if the motor "speed" (a TTL-level square wave) drops below 2000 Hz, a red LED if the speed exceeds 4000 Hz, and a green LED when the speed is between these two limits.

5. Create a program to vary the speed of a small motor (or the brightness of an LED) in sixteen steps from full off to full on using PWM control. Control the speed or brightness through logic levels applied to the lower nibble of Port A.
6. Create a program to output the ASCII character "G" every 50 milliseconds through the USART at 9600 baud. Observe both the TTL-level signal on the microcontroller's TXD line and the RS-232 signal at the output of the media driver using an oscilloscope. On each waveform, identify the start bit, the stop bit, and the data bits of the transmitted signal, as well as the voltage levels of the waveform.
7. Modify the program in Activity 6 so that the "G" is also transmitted from the SPI bus. Use the oscilloscope to display both the transmitted SPI signal and the SPI clock and identify the data bits.
8. Use the analog-to-digital converter to create a simple voltmeter that measures voltage in the range of 0 to 5 V to an accuracy of 0.1 V. Display the results either on the terminal using serial communication or on a port where the upper nibble would be the volts digit and the lower nibble would be the tenths of volts digit.

Standard I/O and Preprocessor Functions

3.1 OBJECTIVES

At the conclusion of this chapter, you should be able to:

- Use standard I/O functions to yield data from your programs for informational as well as debugging purposes
- Redefine the basic I/O operations so that the standard I/O functions can work with peripherals other than a USART
- Use the standard input functions to read user data into your programs
- Understand the use of standard output formatting to simplify program coding and get professional-looking results
- Use the **#define** statement to declare constants as well as redefine functions
- Use the **#include** statement to bring information into your program from external files
- Use the **#pragma** statement to optimize or customize your program during the compilation process

3.2 INTRODUCTION

The standard C language input/output (I/O) functions are a method of sending information to and gaining information from your program as it executes. These functions are built from simple character input and output functions that are easily tailored to meet system hardware requirements. Some of the higher-level functions associated with the standard C language I/O include the ability to format numeric and text output, as well as process and store numeric or text data input.

The standard C language I/O functions outlined here were adapted to work on embedded microcontrollers with limited resources. The prototypes for these functions can be found in the stdio.h file. This file must be brought into the C source code using the **#include** preprocessor command before the functions can be used.

Preprocessor commands are used to add readability and expand the functionality of your program. These commands include **#define, #include,** and **#pragma,** as well as other conditional commands such as **#ifdef, #else,** and **#endif.** These commands are used to guide the compiler and provide information about the program before it is actually compiled. Preprocessor commands are also used to make a program more portable (capable of being moved to another hardware platform). The preprocessor allows a programmer to provide more detail in the definition and construction of a program.

3.3 CHARACTER INPUT/OUTPUT FUNCTIONS – *getchar()* AND *putchar()*

At the lowest level, the input/output functions are *getchar()* and *putchar()*. These functions provide a method of getting a single character into a program or sending a single character out. These functions are generally associated with the *universal synchronous and asynchronous receiver-transmitter* (*USART*) or serial communication port of the microcontroller. The higher-level standard C language I/O functions use these "primitives" as a baseline for their operation. This allows a programmer to easily change the data input and output source requirements for an entire program by simply altering these two functions. The standard form of these functions is as follows:

```
char getchar(void);        // returns a character received
                           // by the USART, using polling.


void putchar(char c);      // transmits the character c using
                           // the USART, using polling.
```

Typically in an embedded system, prior to using these "stock" functions, you must do the following:

- Initialize the USART baud rate.
- Enable the USART transmitter.
- Enable the USART receiver.

For example, in the code below, the program initializes the USART and then sits in a **while** loop. While in this loop, the program calls *getchar()* to wait for a character to be received. Once a character is received, it is echoed back by calling the *putchar()* function:

```
#include <mega8515.h>
            /* include processor specific information */
#include <stdio.h>
            /* include the standard C I/O function definitions */


#define xtal 4000000L          /* quartz crystal frequency [Hz] */
#define baud 9600              /* Baud rate */
```

```
void main(void)
{
   char k;

   /* initialize the USART's baud rate */
   UBRRH=0;
   UBRRL=xtal/16/baud-1;

   /*    Initialize the USART control register
         RX & TX enabled, no interrupts,
         8 data bits */
   UCSRB=0x18;

   /*    Initialize frame format, 8 data bits,
         1 stop bit, asynchronous operation and
         no parity */
   UCSRC=0x86;

   while (1)
   {
        k=getchar(); /* receive the character */
        putchar(k);  /* and echo it back */
   };
}
```

All the standard input/output functions use *getchar()* and *putchar()*. If you intend to use peripherals other than the USART for standard input/output, you must modify the *getchar()* and *putchar()* functions accordingly. The source code for these functions is usually located in the stdio.h file.

Here is a typical version of *getchar()* and *putchar()* for an AVR microcontroller as you might find in the stdio.h file:

```
char getchar(void)
{
   while((UCSRA & RXC)==0)   // wait for character to enter USART
         ;
   return UDR;               // return the character to the caller
}

void putchar(char c)
{
   while((UCSRA & UDRE)==0)       // wait for transmit register
         ;                      // to be empty..
   UDR = c;                     // put c out to the USART
}
```

The *getchar()* and *putchar()* functions can be replaced by customized functions that use a different USART, a serial peripheral interface (SPI), or a parallel port.

When *getchar()* is called in the example above, it forces the microcontroller to wait until a character is received before allowing the program to go on. In some applications, it may not be desirable to sit and wait for a character to arrive. Instead, interrupts can be used to allow serial communication to be handled in the background while the main program continues to run. Interrupt-driven serial communication allow characters to be placed in a buffer or queue as they come in, in the background. Then, when the main program is ready for the received characters, it reads the characters from the buffer and processes them.

The same is true for transmitting characters. To prevent waiting for a character to leave the serial port, interrupt-driven transmitting can be used. As the transmit register of a USART empties, it can generate an interrupt that allows the next character from a transmit buffer to be sent. This process would repeat for as long as there were characters in the buffer.

In the example below, *getchar()* and *putchar()* place characters in, and get characters from, the *Rx_Buffer* and *Tx_Buffer* character arrays, respectively. In this case, the *getchar()* and *putchar()* functions and their support might look like this:

```
#include <mega8515.h>

#define xtal 4000000L     /* quartz crystal frequency [Hz] */
#define baud 9600         /* Baud rate */

// USART Receiver buffer
#define           RX_BUFFER_SIZE        24
char Rx_Buffer[RX_BUFFER_SIZE+1]; // character array (buffer)
char RX_Wr_Index; //index of next char to be put into the buffer
char RX_Rd_Index; //index of next char to be fetched from the buffer
char RX_Counter;  //a total count of characters in the buffer
bit  RX_Buffer_Overflow;    // This flag is set on USART Receiver
                            // buffer overflow
// USART Transmit buffer
#define           TX_BUFFER_SIZE        24
char TX_Buffer [TX_BUFFER_SIZE+1]; // character array (buffer)
char TX_Rd_Index; //index of next char to be put into the buffer
char TX_Wr_Index; //index of next char to be fetched from the buffer
char TX_Counter; //a total count of characters in the buffer
bit  fPrimedIt;  // this flag is used to start the transmit
                 // interrupts, when the buffer is no longer empty

// USART Receiver interrupt service routine
interrupt [USART_RXC] void usart_rx_isr(void)
{
    char c;
```

```
    c = UDR;

    Rx_Buffer[RX_Wr_Index] = c;  /* put received char in buffer */

    if(++RX_Wr_Index > RX_BUFFER_SIZE) /* wrap the pointer */
        RX_Wr_Index = 0;

    if(++RX_Counter > RX_BUFFER_SIZE)  /* keep a character count */
    {                                  /* overflow check.. */
        RX_Counter = RX_BUFFER_SIZE;   /* if too many chars came */
        RX_Buffer_Overflow = 1;  /* in before they could be used */
    }                            /* that could cause an error!! */
}


// Get a character from the USART Receiver buffer
char getchar(void)
{
    char c;

    while(RX_Counter == 0)        /* wait for a character... */
       ;

    c = Rx_Buffer[RX_Rd_Index];   /* get one from the buffer..*/

    if(++RX_Rd_Index > RX_BUFFER_SIZE) /* wrap the pointer */
        RX_Rd_Index = 0;

    if(RX_Counter)
        RX_Counter--;          /* keep a count (buffer size) */

    return c;
}

// USART Transmitter interrupt service routine
interrupt [USART_TXC] void usart_tx_isr(void)
{
    if(TX_Counter != 0)
    {
        if(fPrimedIt == 1)
        {                    /* only send a char if one in buffer */
           fPrimedIt = 0;    /* transmission, then don't send the */

                             /* test and wrap the pointer */
           if(++TX_Rd_Index >  TX_BUFFER_SIZE)
```

```
                    TX_Rd_Index = 0;

                TX_Counter--; /* keep track of the counter */
            }

            if(TX_Counter != 0)
            {
                UDR = TX_Buffer[TX_Rd_Index];
                                      /* otherwise, send char out port */

                                      /* test and wrap the pointer */
                if(++TX_Rd_Index > TX_BUFFER_SIZE)
                     TX_Rd_Index = 0;

                TX_Counter--;    /* keep track of the counter */
            }
        }
        UCSRA |= 0x40;           /* clear TX interrupt flag */
}

// Write a character to the USART Transmitter buffer
void putchar(char c)
{
        char stuffit = 0;

        while(TX_Counter > (TX_BUFFER_SIZE-1))
            ;       /* WAIT!! Buffer is getting full!! */

        if(TX_Counter == 0) /* if buffer empty, setup for interrupt */
            stuffit = 1;

        TX_Buffer[TX_Wr_Index++]=c; /* jam the char in the buffer.. */

        if(TX_Wr_Index > TX_BUFFER_SIZE)       /* wrap the pointer */
            TX_Wr_Index = 0;
                                   /* keep track of buffered chars */
        TX_Counter++;

        if(stuffit == 1)
        {                  /* do we have to "Prime the pump"? */
            fPrimedIt = 1;
            UDR = c;     /* this char starts the TX interrupts.. */
        }
}
```

```
// These defines tell the compiler to replace the stdio.h
// version of getchar() and putchar() with ours..
// That way, all the other stdio.h functions can use them!!
#define         _ALTERNATE_GETCHAR_
#define         _ALTERNATE_PUTCHAR_

// now, we include the library and it will understand our
// replacements

#include <stdio.h>

void main()
{
     char k;

     // initialize the USART's baud rate
     UBRRH=0;
     UBRRL=xtal/16/baud-1;

     //   Initialize the USART control register:
     //   RX & TX enabled,
     //   RX & TX interrupts enabled,
     //   8 data bits
     UCSRB=0xD8;

     //   Initialize the frame format, 8 data bits,
     //   1 stop bit, asynchronous operation and
     //   no parity
     UCSRC=0x86;

     //   Global interrupt enable
     #asm("sei")

     while(1)
     {
         if(RX_Counter)    // are there any received characters??
         {
             k = getchar();     // get the character
             putchar(k);        // and echo it back
         }

         // since there is no waiting on getchar or putchar..
         // other things can be done!!
     }
}
```

In this example, the variable *RX_Counter* contains the number of characters that have been placed in the receiver character array *RX_Buffer* by the interrupt routine. This prevents the program from calling *getchar()* and then sitting idle until a new character arrives. *RX_Counter* is tested in each pass of the **while** loop. If it is not zero, then characters exist in the *RX_Buffer* character array, and a call to *getchar()* will retrieve a character without hesitation.

Since *putchar()* is interrupt driven as well, it adds almost no time to the program execution. *putchar()* places a character in the transmit buffer and starts the transmit interrupts. From that point on, interrupts move the characters from the transmit buffer to the USART each time it becomes empty, until the transmit buffer is empty.

The names "`_ALTERNATE_GETCHAR_`" and "`_ALTERNATE_PUTCHAR_`" are used as a signal to the compiler to use the newly declared *getchar()* and *putchar()* functions, instead of the ones provided in the stdio.h header file.

3.4 STANDARD OUTPUT FUNCTIONS

The output functions of a standard library include the put string, *puts()*, print formatted, *printf()*, and print string formatted, *sprintf()*, functions. In the case of an AVR microcontroller with its RAM and FLASH memory areas, the additional function of put string FLASH, or *putsf()*, was added to allow the printing of a constant string located in FLASH memory space.

3.4.1 PUT STRING—*puts()*

The standard form of *puts()* is

```
void puts(char *str);
```

This function uses *putchar()* to output the null terminated character string *str*, located in SRAM, followed by a new line character ('`\n`'). The *puts()* function is called by the program, and a pointer to a string must be passed to it. The following example calls *puts()* to output the string "Hello" followed by the new line character to the USART:

```
#include <mega8515.h>
            /* include processor specific information */
#include <stdio.h>
            /* include the standard C I/O function definitions */

#define xtal 4000000L      /* quartz crystal frequency [Hz] */
#define baud 9600          /* Baud rate */

char s[6]= "Hello";        /* declare a RAM based string and
                              initialize it */


void main(void)
{
      /* initialize the USART's baud rate */
      UBRRH=0;
```

```
        UBRRL=xtal/16/baud-1;
        /*    Initialize the USART control register
              RX & TX enabled, no interrupts,
              8 data bits */
        UCSRB=0x18;

        /*    Initialize frame format, 8 data bits,
              1 stop bit, asynchronous operation and
              no parity */
        UCSRC=0x86;

        puts(s);      // prints Hello followed by a newline character
        while(1)
            ;
}
```

In the program above, "s", which is the address of the array "`s[]`", is passed to the *puts()* function as a pointer to the array.

3.4.2 PUT STRING FLASH—*putsf()*

The put FLASH (constant) string function has the standard form of

```
void putsf(char flash *str);
```

This function uses *putchar()* to output the null-terminated character string *str*, located in FLASH, followed by a new line character. The following example calls *putsf()* to output the string "Hello", located in FLASH memory, followed by the new line character to the USART:

```
#include <mega8515.h>
            /* include processor specific information */
#include <stdio.h>
            /* include the standard C I/O function definitions */

#define xtal 4000000L     /* quartz crystal frequency [Hz] */
#define baud 9600         /* Baud rate */

flash char s[6]= "Hello"; /* declare a FLASH based string and
                             initialize it */

void main(void)
{
      /* initialize the USART's baud rate */
      UBRRH=0;
      UBRRL=xtal/16/baud-1;
      /*    Initialize the USART control register
            RX & TX enabled, no interrupts,
            8 data bits */
```

```
        UCSRB=0x18;

        /*      Initialize frame format, 8 data bits,
                1 stop bit, asynchronous operation and
                no parity */
        UCSRC=0x86;

        putsf(s);           // prints Hello followed by a line-feed

        putsf("Hello"); // prints Hello followed by a line-feed

        while(1)
            ;
    }
```

3.4.3 PRINT FORMATTED—*printf()*

The print formatted function has the standard form of

```
void printf(char flash *fmtstr [ , arg1, arg2, ...]);
```

This function outputs formatted text, using *putchar()*, according to the format specification in the constant string *fmtstr*. The format specification string *fmtstr* is a constant string and must be located in FLASH program memory. *fmtstr* is processed by the *printf()* function. *fmtstr* can be made up of constant characters to be output directly as well as by special format commands or specifications. As *printf()* is processing the *fmtstr*, it outputs the characters and expands each argument according to the format specifications that may be embedded within the string. A percent sign (%) is used to indicate the beginning of a format specification. Each format specification is then related to an argument, *arg1, arg2,* and so on, in sequence. There should always be an argument for each format specification and vice versa.

The described implementation of the *printf()* format specifications is a reduced version of the standard C function. This is done to meet the minimum requirements of an embedded system. A full ANSI-C implementation would require a large amount of memory space, which in most cases would make the functions useless in an embedded application.

Table 3–1 shows the format specifications available in the CodeVisionAVR library.

When the format specifications are used, there are some rules and modifiers that apply:

- All numeric values are right aligned and left padded with spaces.
- If a **0 (zero)** character is inserted between the % and **d, i, u, x,** or **X,** then the number will be left padded with 0s.
- If a "–" **(minus)** character is inserted between the % and **d, i, u, x,** or **X,** then the number will be left aligned.

- A width specification between 1 and 9 can be inserted between the % and **d, i, u, x,** or **X** to specify the minimum width of the displayed number.
- The format string specifications are used to tell the *printf()* function how many arguments to process.

Specification	*Format of Argument*
%c	outputs the next argument as an ASCII character
%d	outputs the next argument as a decimal integer
%i	outputs the next argument as a decimal integer
%u	outputs the next argument as an unsigned decimal integer
%x	outputs the next argument as an unsigned hexadecimal integer using lowercase letters
%X	outputs the next argument as an unsigned hexadecimal integer using uppercase letters
%s	outputs the next argument as a null terminated character string, located in SRAM
%%	outputs the % character

Table 3–1 *printf—Format Specifications*

Here are some examples of how the format specifications affect the output. Pipe marks (|) are used to help clarify the effect of the formatting by showing where the beginning and end of the text and white space are located. They are not required by the *printf()* function:

```
int i = 1234;
char p[10] = "Hello";

printf("|%d|",i);          // would print:    |1234|
printf("|%5d|",i);         // would print:    | 1234|
printf("|%-5d|",i);        // would print:    |1234 |
printf("|%05d|",i);        // would print:    |01234|
printf("|%x|",i);          // would print:    |4d2|
printf("|%04X|",i);        // would print:    |04D2|
printf("|%s:%04X|",p,i);   // would print:    |Hello:04D2|
```

Here is a program that uses *printf()* to print a string and an integer value from the same statement:

```
#include <mega8515.h>
            /* include processor specific information */
#include <stdio.h>
            /* include the standard C I/O function definitions */

#define xtal 4000000L    /* quartz crystal frequency [Hz] */
#define baud 9600        /* Baud rate */


void main(void)
```

```
{
        int I;

        /* initialize the USART's baud rate */
        UBRRH=0;
        UBRRL=xtal/16/baud-1;
        /*      Initialize the USART control register
                RX & TX enabled, no interrupts,
                8 data bits */
        UCSRB=0x18;

        /*      Initialize frame format, 8 data bits,
                1 stop bit, asynchronous operation and
                no parity */
        UCSRC=0x86;

        for(I=0; I<100; I++)
        {       // print a formatted string and tell about I
                printf("printf example, I = %d\n",I);
        }
        while(1)
          ;
}
```

Running this program would result in the text string "printf example, I = " being printed 100 times. Each time, the number 0 to 99 would be printed following the string. Note that the *fmtstr* contains all the text, plus the "`%d`" indicating where the number *I* should appear, as well as the end of line character ('`\n`'). That one statement does a lot of work!!

You should be aware of the costs associated with the use of *printf()*. The *printf()* function is doing a great deal of work and therefore needs code space to do so. Using the *printf()* function adds flexible, powerful, formatted output, without many lines of code being typed. However, the standard library *printf()* function itself contains several hundred lines of assembly code that are automatically added to your program in order to support *printf()*. Once the function has been added to the code, each call to the function adds but a few lines of assembly.

3.4.4 STRING PRINT FORMATTED—*sprintf()*

The string print formatted function has the standard form of

```
void sprintf(char *str, char flash *fmtstr [ , arg1, arg2, ...]);
```

The operation of this function is identical to *printf()* except that the formatted text is placed in the null-terminated character string *str* instead of calling *putchar()*.

For example:

```
#include <mega8515.h>
                /* include processor specific information */
```

```
#include <stdio.h>
            /* include the standard C I/O function definitions */

#define xtal 4000000L      /* quartz crystal frequency [Hz] */
#define baud 9600          /* Baud rate */

char s[28]; /* declare a RAM based string and
                              initialize it */

void main(void)
{
      int I;

      /* initialize the USART's baud rate */
      UBRRH=0;
      UBRRL=xtal/16/baud-1;
      /*    Initialize the USART control register
            RX & TX enabled, no interrupts,
            8 data bits */
      UCSRB=0x18;

      /*    Initialize frame format, 8 data bits,
            1 stop bit, asynchronous operation and
            no parity */
      UCSRC=0x86;

      for(I=0; I<100; I++)
      {           // load the string s with the formatted data
            sprintf(s,"sprintf example, I = %d",I);
            puts(s);     // then print 's'!!
      }
      while(1)
        ;
}
```

The execution of this program would result in the text string "sprintf example, I = " being printed 100 times. Each time, the number 0 to 99 would be printed following the string. In this example, the *sprintf()* function formats the output and places it in a character array "s" instead of sending it out the USART using *putchar()*. The *puts()* function then sends the string "s" out to the USART using the *putchar()* function and appends the new line character ('\n') to the end of the string.

3.5 STANDARD INPUT FUNCTIONS

The input functions of a standard library include the get string, *gets()*, scan formatted, *scanf()*, and scan string formatted, *sscanf()*, functions. Just as the root output function is *putchar()*, the root input function is *getchar()*. In all cases, *getchar()* is called to gather the necessary data. So by changing the operation of *getchar()*, the data source can easily be altered.

3.5.1 GET STRING—*gets()*

The standard form of *gets()* is

```
char *gets(char *str, unsigned char len);
```

This function uses *getchar()* to input characters into the character string *str*, which is terminated by the new line character. The new line character is replaced with a null ('\0') by the *gets()* function. The maximum length of the string is *len*. If *len* characters were read without encountering the new line character, then the string is terminated with a null and the function ends. The function returns a pointer to *str*.

For example:

```
#include <mega8515.h>
            /* include processor specific information */
#include <stdio.h>
            /* include the standard C I/O function definitions */

#define xtal 4000000L    /* quartz crystal frequency [Hz] */
#define baud 9600        /* Baud rate */

char s[12]; /* declare a RAM based string */

void main(void)
{
      /* initialize the USART's baud rate */
      UBRRH=0;
      UBRRL=xtal/16/baud-1;
      /*    Initialize the USART control register
            RX & TX enabled, no interrupts,
            8 data bits */
      UCSRB=0x18;

      /*    Initialize frame format, 8 data bits,
            1 stop bit, asynchronous operation and
            no parity */
      UCSRC=0x86;

      while(1)
      {
        gets(s,10);       // get a string from the standard input

        puts(s);   // prints string 's' followed by a line feed

      }
}
```

In the program above, characters will be read from the standard input, ten characters at time, and then echoed to the standard output with a line feed appended.

SCAN FORMATTED—*scanf()*

The scan formatted function has the standard form of

```
signed char scanf(char flash *fmtstr [ , arg1 address, arg2
address, ...]);
```

This function performs formatted text input, using *getchar()*, according to the format specifications in the *fmtstr* string. The format specification string *fmtstr* is constant and must be located in FLASH program memory.

Just as in the *printf()* function, *fmtstr* is processed by the *scanf()* function. *fmtstr* can be made up of constant characters to be expected in the input stream, as well as special-format commands or specifications. As *scanf ()* is processing the *fmtstr*, it inputs data using the *getchar()* function and either matches the received character to the one found in *fmtstr* or converts a series of characters for each argument according to the format specifications that may be embedded within the *fmtstr* string. A percent sign (%) is used to indicate the beginning of a format specification. Each format specification is then related to an argument, *arg1, arg2,* and so on, in sequence. There should always be an argument for each format specification and vice versa.

The arguments, *arg1, arg2,* and so on, are the *address of* or *pointers to* variables. (Not using an address or a pointer as an argument in this function can lead to some very disappointing results!) The function returns the number of successful entries or –1 on error.

The described implementation of *scanf()* format specifications is a reduced version of the standard C function. This is done to meet the minimum requirements of an embedded system. A full ANSI-C implementation would require a large amount of memory space, which in most cases would make the functions useless in an embedded application.

Table 3–2 shows the format specifications available.

Specification	*Format of Argument*
%c	inputs the next argument as an ASCII character
%d	inputs the next argument as a decimal integer
%i	inputs the next argument as a decimal integer
%u	inputs the next argument as an unsigned decimal integer
%x	inputs the next argument as an unsigned hexadecimal integer
%s	inputs the next argument as a null terminated character string

Table 3–2 *scanf—Format Specifications*

When the scan format specifications are used, there are some rules that apply:

- The white space characters, blank (space), and tab are ignored.
- Ordinary characters (except for %) are expected to match the next non–white space character from the input.

Here is an example that prompts the user for inputs, separated by a comma, and then calls *scanf()* to wait for the input and store it into variables. Finally, *printf()* is called to format and print the numbers back to the user:

```
#include <mega8515.h>
            /* include processor specific information */
#include <stdio.h>
            /* include the standard C I/O function definitions */

#define xtal 4000000L        /* quartz crystal frequency [Hz] */
#define baud 9600            /* Baud rate */

void main(void)
{
      int i,j;

      /* initialize the USART's baud rate */
      UBRRH=0;
      UBRRL=xtal/16/baud-1;
      /*    Initialize the USART control register
            RX & TX enabled, no interrupts,
            8 data bits */
      UCSRB=0x18;

      /*    Initialize frame format, 8 data bits,
            1 stop bit, asynchronous operation and
            no parity */
      UCSRC=0x86;

      putsf("Enter two integers 'i,j':");
      scanf("%d,%x\n",&i,&j);
                  // get two integers, one decimal
                  // and one hexadecimal with a comma
                  // separating them

      printf("%d, %04X",i,j);  // print the values and stop..

      while(1)
        ;
}
```

3.5.3 SCAN STRING FORMATTED—*sscanf()*

The scan string formatted function has the standard form of

```
signed char sscanf(char *str, char flash *fmtstr [ , arg1 address,
                                              arg2 address, ...]);
```

This function is identical to *scanf()* except that the formatted text is read from the null-terminated character string *str*, located in SRAM.

Modifying the above example to use *sscanf()* to get the data from a string instead of from the standard input (the USART using *getchar()*) produces the following:

```
#include <mega8515.h>
      /* include processor specific information */
#include <stdio.h>
      /* include the standard C I/O function definitions */

#define xtal 4000000L     /* quartz crystal frequency [Hz] */
#define baud 9600         /* Baud rate */

/* declare a RAM based string and initialize it*/
char s[14] = {"1234,abcd\n"};

void main(void)
{
      int i,j;

      /* initialize the USART's baud rate */
      UBRRH=0;
      UBRRL=xtal/16/baud-1;
      /*    Initialize the USART control register
            RX & TX enabled, no interrupts,
            8 data bits */
      UCSRB=0x18;

      /*    Initialize frame format, 8 data bits,
            1 stop bit, asynchronous operation and
            no parity */
      UCSRC=0x86;

                  // get the values from the RAM string 's'
      sscanf(s,"%d,%x\n",&i,&j);
                  // get two integers, one decimal
                  // and one hexadecimal with a comma
                  // separating them

      printf("%d, %04X",i,j);  // print the values and stop..
```

```
        while(1)
          ;
}
```

3.6 PREPROCESSOR DIRECTIVES

Preprocessor directives are not actually part of the C language syntax, but they are accepted as such because of their use and familiarity. The preprocessor is a separate step from the actual compilation of a program and happens before the actual compilation begins. The most common directives are **#define** and **#include.** The preprocessor directives allow you to do the following:

- Include text from other files, such as header files containing library and user function prototypes.
- Define macros that reduce programming effort and improve the legibility of the source code.
- Set up conditional compilation for debugging purposes and to improve program portability.
- Issue compiler-specific directives to generalize or optimize the compilation.

3.6.1 THE #INCLUDE DIRECTIVE

The **#include** directive can be used to include another file in your source. There can be as many files included as needed, or as are allowed by the compiler, and includes can be nested (meaning that an included file can contain another, non-recursive, **#include**). Typically, there is a limit to the depth of the nesting, and the limit will vary from one compiler to another. In the CodeVisionAVR C Compiler a maximum of sixteen nested files are allowed. The standard form of the statement is

```
#include <file_name>

        or

#include "file_name"
```

The less than (<) and greater than (>) sign delimiters indicate that the file to be included is part of a standard library or set of library files. The compiler will typically look for the file in the "\inc" directory. When the file name is delimited within quotation marks, the compiler will first look in the same directory as the C file being compiled. If the file is not found there, then the compiler will look for the file in the default library directory ("\inc").

The **#include** statement can be located anywhere in the code, but it is typically used at the top of a program module to improve the program readability.

3.6.2 THE #DEFINE DIRECTIVE

A **#define** directive should be thought of as a "text substitution" referred to as a "macro." The standard form of a **#define** is

```
#define          NAME          Replacement_Text
```

Typically, **#define** directives are used to declare constants, but the tag "`NAME`" and the "`Replacement_Text`" may also have parameters. The preprocessor will replace the tag "`NAME`" with the expansion "`Replacement_Text`" whenever it is detected during compilation. If "`NAME`" contains parameters, the real ones found in the program will replace the formal parameters written in the text "`Replacement_Text`".

There are a couple of simple rules that apply to macro definitions and the **#define** directive:

- Always use closed comments (`/* . . . */`) when commenting the line of a **#define.**
- Remember that the end of the line is the end of the **#define** and that the text on the left will be replaced by *all* of the text on the right.

For example, given the following macro definitions

```
#define ALPHA 0xff   /* mask off lower bits */
#define SUM(a,b) a+b
```

the following expression

```
int x = 0x3def & ALPHA;
```

would be replaced with

```
int x = 0x3def & 0xff    /* mask off lower bits */ ;
```

and this expression

```
int i=SUM(2,3);
```

would be replaced with

```
int i=2+3;
```

The diversity and capability of the **#define** directive is somewhat dependent on the sophistication of the compiler's preprocessor. Most modern compilers have the ability to redefine functions and operations as well as the abillity to sort out the parameters that go with them (as in the examples above). The CodeVisionAVR compiler is very capable in these areas. Lesser compilers may only allow the simplest operations, such as defining constants and strings. Using a **#define** to alias functions (create alternate calls for functions) can be a powerful and handy thing! For example, an existing function in a library may appear as

```
int _write_SCI1(int c)   // this function sends a char to
{                        // an alternate comm. port
      . . .
    return c;
}
```

Giving the function an alias

```
#define putchar2(c) _write_SCI1(c)
```

allows the function to be referred to in two ways:

```
putchar2('b');
```

or

```
_write_SCI1('b');
```

This is quite useful when reutilizing code from another program or combining several libraries in a large development. Instead of the locating and renaming of every call or reference, the function was simply given an alias, allowing it to be referred to in more than one way.

This use of **#define** can also enhance the readability of a program by allowing a function to be "re-declared." Let's assume a function called *set_row_col()* places the cursor at a position on a four-row by twenty-character LCD display based on two parameters, a *row* and a *column*:

```
void set_row_col(char row, char column)
{
        . . .
}
```

With the use of a macro to form a re-declaration, the positions on the LCD display can be referred to as an *x* and *y* coordinate instead of a row and column:

```
#define  goto_xy(x,y)     set_row_col(y,x)
```

Another possibility is a re-declaration that can use a character position to set the cursor:

```
#define set_char_pos(p)  set_row_col((p/20),(p%20))
```

The row is calculated by dividing the character position by the number of columns, and the column position will be what is left, a modulus function of the width. This is all done as a function of "fancy text substitution." The compiler sees only what is on the right side of the **#define,** because the preprocessor did all the work ahead of time. These changes in reference potentially make the program more understandable not only to the programmer but to others reading the code as well.

The substitution process can be even more finely controlled. When defining macros, you can use the # operator to convert the macro parameter to a character string. For example,

```
#define PRINT_MESSAGE(t)  printf(#t)
```

will substitute this expression

```
PRINT_MESSAGE(Hello);
```

with

```
printf("Hello");
```

It is also possible to concatenate two parameters using the ## operator. In the example below, the formal parameters "a" and "b" are to be concatenated. In the substitution, the left side, "alpha(a,b)", is effectively replaced with the right side, "ab". In the program, the real parameters x and y are used. So the macro definition

```
#define alpha(a,b) a ## b
char alpha(x,y)=1;
```

will result in

```
char xy=1;
```

Typically, a replacement text is defined on a single line along with the tag name. It is possible to extend the definition to a new line by using a backslash (\) at the end of the line:

```
#define MESSAGE "This is a very \
long text..."
```

The backslash character will not appear in the expansion, and the new line character will be eliminated as well. The result is that "This is a very long text..." will be treated as a single, unbroken string.

A macro can be undefined using the **#undef** directive. This feature is primarily used in conditional compilation where a tag name is used as a flag:

```
#undef ALPHA
```

The **#undef** directive allows a previously defined macro to be redefined. For instance, if a macro was defined as

```
#define A_CHAR     'A'
```

you could redefine it to a lower case 'a':

```
#undef A_CHAR
#define A_CHAR     'a'
```

3.6.3 THE #ifdef, #ifndef, #else, AND #endif DIRECTIVES

The **#ifdef**, **#ifndef**, **#else**, and **#endif** directives can be used for conditional compilation.

The syntax is

```
#ifdef macro_name
      [set of statements 1]
#else
      [set of statements 2]
#endif
```

If "`macro_name`" is a defined macro name, then the **#ifdef** expression evaluates to TRUE and the "`set_of_statements_1`" will be compiled. Otherwise, the "`set_of_statements_2`" will be compiled. The "`#else`" and "`set_of_statements_2`" are optional. Another version is

```
#ifndef macro_name
      [set of statements]
#endif
```

If "`macro_name`" is not defined, the **#ifndef** expression evaluates to TRUE. The rest of the syntax is the same as that for **#ifdef**. The **#if**, **#elif**, **#else**, and **#endif** directives can be used for conditional compilation.

```
#if expression1
        [set of statements 1]
#elif expression2
        [set of statements 2]
#else
        [set of statements 3]
#endif
```

If "`expression1`" evaluates to TRUE, the "`set_of_statements_1`" will be compiled.

If "`expression2`" evaluates to TRUE, the "`set_of_statements_2`" will be compiled.

Otherwise, the "`set_of_statements_3`" will be compiled. The "`#else`" and "`set_of_statements_3`" are optional.

Conditional compilation is very useful in creating one program that runs in several configurations. This makes maintaining the program (or the product the program runs in) easier, since one source code can apply to several operating configurations.

Another, more common, use of conditional compilation is for debugging purposes. The combination of conditional compilation and standard I/O library functions can help you get a program up and running faster without equipment such as in-circuit emulators and oscilloscopes.

The example state machine used to control an "imaginary" traffic light (from Chapter 1) could be modified to add some debugging capability. Using conditional compilation, the debugging features added could be turned on and off as needed, trading between providing information and increased performance, as shown in Figure 3–1:

```
#include <mega8535.h>
#include <delay.h>

/*#define    DEBUGGING_ON */

/* Removing the comment will cause the DEBUGGING_ON "flag" to
   be enabled. This will cause the compiler to include the
   additional library and lines of code required allowing the
   programmer to watch the state machine run from the UART. */

#ifdef DEBUGGING_ON
       /* include the standard C I/O function definitions */
       #include <stdio.h>

       #define xtal 4000000L      /* quartz crystal frequency [Hz] */
       #define baud 9600          /* Baud rate */
#endif
```

Figure 3–1 *Expanded "Imaginary Traffic Light" Software (Continues)*

```
#define    EW_RED_LITE PORTC.0    /* definitions to actual outputs */
#define    EW_YEL_LITE PORTC.1    /* used to control the lights */
#define    EW_GRN_LITE PORTC.2
#define    NS_RED_LITE PORTC.3
#define    NS_YEL_LITE PORTC.4
#define    NS_GRN_LITE PORTC.5

#define    PED_XING_EW PINA.0   /* pedestrian crossing push button */
#define    PED_XING_NS PINA.1   /* pedestrian crossing push button */
#define    FOUR_WAY_STOP PINA.3     /* switch input for 4-Way Stop */

char time_left;            /* time in seconds spent in each state */
int current_state;         /* current state of the lights */
char flash_toggle;         /* toggle used for FLASHER state */

/* This enumeration creates a simple way to add states to the machine
by name. Enumerations generate an integer value for each name
automatically, making the code easier to maintain. */

enum { EW_MOVING , EW_WARNING , NS_MOVING , NS_WARNING , FLASHER };

/* The actual state machine is here.. */
void Do_States(void)
{
   switch(current_state)
   {
      case  EW_MOVING:            /* east-west has the green!! */
            EW_GRN_LITE = 1;
            NS_GRN_LITE = 0;
            NS_RED_LITE = 1;    /* north-south has the red!! */
            EW_RED_LITE  = 0;
            EW_YEL_LITE = 0;
            NS_YEL_LITE = 0;

            if(PED_XING_EW || FOUR_WAY_STOP)
            {      /* pedestrian wishes to cross, or
                    a 4-way stop is required */
                   if(time_left > 10)
                          time_left = 10;    /* shorten the time */
            }
            if(time_left != 0)         /* count down the time */
            {
                   --time_left;
                   return;                     /* return to main */
            }                    /* time expired, so.. */
            time_left = 5;      /* give 5 seconds to WARNING */
            current_state = EW_WARNING;
                                       /* time expired, move */
            break;                     /* to the next state */
```

Figure 3–1 *Expanded "Imaginary Traffic Light" Software (Continues)*

```
        case    EW_WARNING:
                EW_GRN_LITE = 0;
                NS_GRN_LITE = 0;
                NS_RED_LITE = 1;   /* north-south has the red.. */
                EW_RED_LITE  = 0;
                EW_YEL_LITE = 1;   /* and east-west has the yellow */
                NS_YEL_LITE = 0;

                if(time_left != 0)       /* count down the time */
                {
                      --time_left;
                      return;      /* return to main */
                }                  /* time expired, so.. */
                if(FOUR_WAY_STOP)  /* if 4-way requested then start */
                      current_state = FLASHER; /* the flasher */
                else
                {                          /* otherwise.. */
                      time_left = 30;      /* give 30 seconds to MOVING */
                      current_state = NS_MOVING;
                }                          /* time expired, move */
                break;                     /* to the next state */

        case    NS_MOVING:
                EW_GRN_LITE = 0;
                NS_GRN_LITE = 1;
                NS_RED_LITE = 0;   /* north-south has the green!! */
                EW_RED_LITE  = 1;  /* east-west has the red!! */
                EW_YEL_LITE = 0;
                NS_YEL_LITE = 0;

                if(PED_XING_NS || FOUR_WAY_STOP)
                {     /* if a pedestrian wishes to cross, or
                       a 4-way stop is required.. */
                      if(time_left > 10)
                            time_left = 10;    /* shorten the time */
                }
                if(time_left != 0)         /* count down the time */
                {
                      --time_left;
                      return;                      /* return to main */
                }                  /* time expired, so.. */
                time_left = 5;     /* give 5 seconds to WARNING */
                current_state = NS_WARNING;    /* time expired, move */
                break;                     /* to the next state */

        case    NS_WARNING:
                EW_GRN_LITE = 0;
```

Figure 3–1 *Expanded "Imaginary Traffic Light" Software (Continues)*

```
        NS_GRN_LITE = 0;
        NS_RED_LITE = 0;   /* north-south has the yellow.. */
        EW_RED_LITE  = 1;
        EW_YEL_LITE = 0;   /* and east-west has the red.. */
        NS_YEL_LITE = 1;

        if(time_left != 0)        /* count down the time */
        {
              --time_left;
              return;                   /* return to main */
        }                  /* time expired, so.. */
        if(FOUR_WAY_STOP)  /* if 4-way requested then start */
              current_state = FLASHER; /* the flasher */
        else
        {                         /* otherwise.. */
              time_left = 30;     /* give 30 seconds to MOVING */
              current_state = EW_MOVING;
        }                         /* time expired, move */
        break;                    /* to the next state */

  case  FLASHER:
        EW_GRN_LITE = 0;   /* all yellow and */
        NS_GRN_LITE = 0;   /* green lites off */
        EW_YEL_LITE = 0;
        NS_YEL_LITE = 0;

        flash_toggle ^= 1; /* toggle LSB.. */
        if(flash_toggle & 1)
        {
              NS_RED_LITE = 1;   /* blink red lights */
              EW_RED_LITE  = 0;
        }
        else
        {
              NS_RED_LITE = 0;   /* alternately */
              EW_RED_LITE  = 1;
        }
        if(!FOUR_WAY_STOP)        /* if no longer a 4-way stop */
              current_state = EW_WARNING;
        break;                    /* then return to normal */

  default:
        current_state = NS_WARNING;
        break; /* set any unknown state to a good one!! */

  }
}
```

Figure 3–1 *Expanded "Imaginary Traffic Light" Software (Continues)*

```
void main(void)
{
    DDRC = 0xFF;    /* portc all out */
    DDRA = 0x00;    /* porta all in */

    #ifdef DEBUGGING_ON
        /* initialize the USART's baud rate */
        UBRRH=0;
        UBRRL=xtal/16/baud-1;

        /* Initialize the USART control register
        RX & TX enabled, no interrupts,
        8 data bits */
        UCSRB=0x18;

          /* Initialize frame format, 8 data bits,
        1 stop bit, asynchronous operation and
        no parity */
        UCSRC=0x86;

        printf("Compiler version is %d\n\r",__CODEVISIONAVR__);
        #warning "Debug printing is on."
    #endif

    current_state = NS_WARNING;/* initialize to a good starting
                                  state (as safe as possible) */
    while(1)
    {
        delay_ms(1000);       /* 1 second delay.. this time could
                             be used for other needed processes */

        Do_States();    /* call the state machine, it knows
                             where it is and what to do next */

        #ifdef DEBUGGING_ON
           /* send state and time data to serial port */
           printf("Current State = %d :", current_state);
           printf("Time Left = %d  \r", (int)time_left);
        #endif
    }
}
```

Figure 3–1 *Expanded "Imaginary Traffic Light" Software (Continued)*

3.6.4 THE #pragma DIRECTIVE

The **#pragma** directive allows for compiler-specific directives or switches. **#pragma** statements are very compiler dependent, so you should always refer to the compiler user's guide when using these controls. The controls described below pertain to the CodeVisionAVR compiler but will be characteristic of what can be found in other compilers.

#pragma warn

The **#pragma warn** directive enables or disables compiler warnings. This would be used to disable warnings that may be viewed as a nuisance by the programmer. These can include warnings such as "variable declared but never referenced" or "possible loss of precision." Disabling the warnings should not be done on a permanent basis; it could cause a useful warning to be missed, leaving you wondering why your program is yielding bad results. The best course of action is to declare, code, and cast until the warnings are all gone.

```
/* Warnings are disabled */
#pragma warn-

/* Write some code here */

/* Warnings are enabled */
#pragma warn+
```

#pragma opt

The compiler's code optimizer can be turned on or off using the **#pragma opt** directive.

The optimizer is responsible for improving the compiler's generated assembly language output. This is accomplished by reducing the assembly output size, arranging it to run faster, or both. The **#pragma opt** directive must be placed at the start of the source file, and the default is optimization turned on:

```
/* Turn optimization off, for testing purposes */
#pragma opt-

        or

/* Turn optimization on */
#pragma opt+
```

#pragma optsize

A program that is optimized for size may actually run a little slower because common code is converted to subroutines, and subroutine calls replace the common code. A program that is more "in-line" (not optimized for size) runs faster because there is no additional overhead from the calls to and returns from subroutines.

If the code optimization is enabled, you can optimize some of the program, or all, for size or speed using the **#pragma optsize** directive:

```
/* The program will be optimized for minimum size */
   #pragma optsize+

/* Place your program functions here */

        or

/* Now the program will be optimized for maximum execution speed */
```

```
#pragma optsize-

/* Place your program functions here */
```

The default state is determined by the **Project|Configure|C Compiler|Compilation|Optimization** menu setting in the CodeVisionAVR Environment.

#pragma savereg

The automatic saving and restoring of registers R0, R1, R22, R23, R24, R25, R26, R27, R30, R31, and SREG during interrupts can be turned on or off using the **#pragma savereg** directive. This control is used to manually reduce any potential excess overhead associated with saving and restoring the machine state in an interrupt service routine. This can be dangerous to the novice, so some careful study of how the compiler generates code should be done before you use this "guru-level" control:

```
/* Turn registers saving off */
#pragma savereg-

/* interrupt handler */
interrupt [1] void my_irq(void) {
/* now save only the registers that are affected by the rou-
tines in the handler, for example R30, R31 and SREG */
#asm
    push r30
    push r31
    in   r30,SREG
    push r30
#endasm

/* place the C code here */

/* now restore SREG, R31 and R30 */
#asm
    pop r30
    out SREG,r30
    pop r31
    pop r30
#endasm
}
/* re-enable register saving for the other interrupts */
#pragma savereg+
```

The default state is automatic saving of registers during interrupts. The default state of this compiler setting is "savereg+", and there is no menu option for this setting. Care must be used to make sure that this setting is "savereg+" when its function is no longer needed, to prevent erratic operation of your program from registers not being saved.

#pragma regalloc

The CodeVisionAVR compiler will try to use register variables to increase performance as often as possible. There may be instances, like debugging with an In-Circuit-Emulator (ICE), where register variables may become difficult to track, so you might wish to direct the compiler to not perform this automatic register allocation.

The automatic allocation of global variables to registers can be turned on or off using the **#pragma regalloc** directive:

```
/* the following global variable will be automatically
allocated to a register */
#pragma regalloc+
unsigned char alpha;

/* the following global variable will not be automatically
allocated to a register and will be placed in normal SRAM */
#pragma regalloc-
unsigned char beta;
```

The default state is determined by the **Project|Configure|C Compiler|Compilation|Automatic Register Allocation** check box setting in the CodeVisionAVR Environment.

#pragma promotechar

The ANSI standard for the C language treats characters as 16-bit values. This is not necessary (or even desired) in a small, embedded system. The default settings for the CodeVisionAVR compiler are to leave a character a character, but if an application requires characters to be treated as integers, there is a method of enforcing the ANSI standard.

The ANSI character to integer operands promotion (or conversion) can be turned on or off using the **#pragma promotechar** directive:

```
/* turn on the ANSI char to int promotion */
#pragma promotechar+

      or

/* turn off the ANSI char to int promotion */
#pragma promotechar-
```

This option can also be specified in the **Project|Configure|C Compiler|Promote char to int** menu setting in the CodeVisionAVR Environment.

#pragma uchar

When character variables are declared, the default format is as follows:

```
char var;      // declares a signed character variable
unsigned char var2; // declares an unsigned character variable
```

The **#pragma uchar** directive allows the default format to be changed as it pertains to characters. If the **uchar** directive is enabled, the following format is now in effect:

```
signed char var;   // this is now a signed character variable
char var2;   // and this is an UNSIGNED character variable
```

uchar can be turned on or off using the **#pragma uchar** directive:

```
/* char will be unsigned by default */
#pragma uchar+

or

/* char will be signed by default */
#pragma uchar-
```

This option can also be specified in the **Project|Configure|C Compiler|char is unsigned** menu in the CodeVisionAVR Environment.

#pragma library

You may wish to create your own function library to be used as commonly as the standard I/O library. Once your library has been created, it can be included in the program at compile time using the **#pragma library** directive:

```
#pragma library mylib.lib
```

The creation of libraries is compiler specific; you should refer to the compiler user's guide for instruction on the required methods.

3.6.5 OTHER MACROS AND DIRECTIVES

Table 3–3 lists some predefined macros. These tag names are specific to CodeVisionAVR but could be found in other compilers. These can be used to get information, such as the compile date and time, into the compiled program. They can also be used to conditionally compile based on memory model or optimization settings.

CodeVisionAVR also supports the directives **#error, #warning,** and **#line.**

The **#error** directive can be used to stop compilation and display an error message.

The syntax is

```
#error error_message
```

For example,

```
#error This is an error
```

Similarly, the **#warning** directive causes a warning message to be displayed, but does not cause compilation to stop. The syntax for the **#warning** directive is

```
#warning warning_message
```

For example,

```
#warning This is a warning to be displayed at completion of
the compilation.
```

Macro	Description
__CODEVISIONAVR__	the version and revision of the compiler represented as an integer, for example V1.24.0 is represented as 1240
__LINE__	the current line number of the compiled file
__FILE__	the current compiled file
__TIME__	the current time in *hh:mm:ss* format
__DATE__	the current date in *mm dd yyyy* format
__BUILD__	the build number of the compiler
_MCU_CLOCK_FREQUENCY_	the clock frequency specified in integer in Hz as set in the compiler options
_MODEL_TINY_	set if compiled using the TINY memory model
_MODEL_SMALL_	set if compiled using the SMALL memory model
_OPTIMIZE_SIZE_	set if compiled with optimization for size
_OPTIMIZE_SPEED_	set if compiled with optimization for speed
_HEAP_START_	the heap starting address
_HEAP_SIZE_	the heap size as set in the compiler options
_UNSIGNED_CHAR_	set if compiler option is enabled or **#pragma uchar+** is used
_8BIT_ENUMS_	set if the 8 bit enums compiler option is enabled or #pragma 8bit_enums+ is used

Table 3–3 *CodeVisionAVR Predefined Macros*

Finally, the **#line** directive modifies the predefined "__LINE__" and "__FILE__" macros as displayed in Table 3–3. The syntax is

```
#line integer_constant ["file_name"]
```

For example,

```
/* this will set __LINE__ to 50 and __FILE__ to "file2.c" */
#line 50 "file2.c"

/* this will set __LINE__ to 100
#line 100
```

3.7 CHAPTER SUMMARY

This chapter has provided the knowledge necessary for you to use standard library functions for printing information, gathering user input, and debugging your programs. You have also learned how to modify the root functions of *getchar()* and *putchar()* such that the standard library can be used on a variety of hardware configurations.

You have learned about the use of the complier directives **#include** and **#define** and how they are used to add reusability and readability to your programs. You have also learned about conditional compilation using **#ifdef, #ifndef, #else,** and **#endif,** providing a method for having a single source code be used for a variety of system configurations or for adding temporary code for debugging purposes.

#pragma statements were also described so that your code could be generalized or optimized during compilation, yielding the best reusability and performance from your program.

3.8 EXERCISES

*1. Write a macro using the **#define** directive to alias the function *putchar()* with a function named *send_the_char()* (Section 3.6).

2. Write a function that prompts the user for two integer values, adds them together, and prints the values and their sum in the format "The values A + B = C", where A, B, and C are decimal integers (Section 3.4).

3. Write a function that uses *scanf()* to receive a date in the format "MM/DD/YYYY" and then uses *printf()* to display the date "DD-MM-YY" (Section 3.5).

*4. Write a function that prints its compile date and time. Use the compiler's internal tag names to get the date and time values (Section 3.6).

5. Write a function that prints its compiled memory model and optimizer settings.

*6. Write a function that inputs a 16-bit hexadecimal value and then prints the binary equivalent. Note: There is no standard output function to print binary—so it is all up to you!

7. Use a **#define** to declare a parallel, 8-bit I/O port, PORTX, at location 0x1200. The declaration should be made such that the port can be accessed by "`PORTX = 34`", or "`i = PORTX`" (Section 3.6).

8. Write a function to initialize the USART on the ATmega8515 for 8 data bits, 1 stop bit, no parity, and asynchronous operation. The baud rate should be set for 9600 baud with a system clock of 4MHz. Transmit and receive should also be enabled. Utilize the _MCU_CLOCK_FREQUENCY macro to automatically get the value for UBRRL (Section 3.6).

3.9 LABORATORY ACTIVITIES

1. Create a program to output the ASCII character "G" every 50 milliseconds through the USART. Observe both the TTL level signal on the microcontroller's TXD line and the RS-232 signal at the output of the media driver using an oscilloscope. On each waveform, identify the start bit, the stop bit, and the data bits of the transmitted signal.

2. Modify the program in Activity 1 so that the "G" is also transmitted from the SPI bus. Use the oscilloscope to display both the transmitted signal and the clock and identify the data bits.

3. Create a program to use *getchar()* to get a serial character (from the PC) and *putchar()* to return a character that is two letters further along in the alphabet.

4. Modify the program from Activity 1 by changing *putchar()* so that the code for the ASCII character being returned appears on port C, where it could be conveniently read by attaching LEDs.

5. Create a program that uses a parallel *getchar()* to get a byte of data in on PORTA and uses the "stock" *putchar()* to echo what it receives to the USART at 9600 baud. Use external interrupt 0, set to trigger either on a rising or a falling edge (whichever is convenient to your hardware), as the strobe that causes PORTB to be read and the data transmitted through the USART.
6. Create a parallel port version of *putchar()* along with the *getchar()* (from Activity 3) to output the data read from port A on port C.
7. Team up with a partner. One of you should create a SPI bus transmitter program, a modification of the program from Activity 5, that uses the SPI bus to transmit the data from port A when the appropriate signal is received on the external interrupt. The other partner should create a SPI bus receiver, a modification of the program from Activity 4, that displays the data received over the SPI bus.
8. For a real challenge, use the programs from Activity 7 and create two-way communication where both microcontrollers read the data on port A when triggered by their own external interrupt and send it through the SPI bus. Both microcontrollers also display data received by means of the SPI bus.
9. Create a version of *getchar()* and *putchar()* using the SPI port of an ATmega8515 or your particular processor.
10. Create a program using Activity 1 that sends "C Rules!!" out of the SPI port using the *puts()* function.
11. Create a program using Activity 1 that sends "C Rules this time *ddd*!!" out of the SPI port 100 times, where *ddd* indicates which time, from 1 to 100.

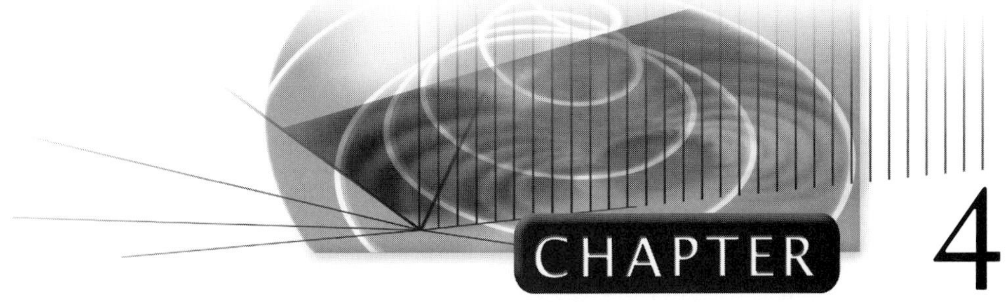

CHAPTER 4

The CodeVisionAVR C Compiler and IDE

4.1 OBJECTIVES

At the conclusion of this chapter, you should be able to

- Operate the CodeVisionAVR C Compiler and integrated development environment
- Correctly utilize the compiler options
- Apply the CodeVisionAVR environment to program a target device
- Apply the CodeVisionAVR code wizard to automatically generate shells of code
- Apply the CodeVisionAVR terminal tool to send and receive RS-232 communications
- Perform basic debugging operations using Atmel's AVR Studio as a software simulator

4.2 INTRODUCTION

The purpose of this chapter is to familiarize you with the CodeVisionAVR C Compiler.

The CodeVisionAVR C Compiler is just one of many C compilers available for the Atmel AVR microcontrollers. Its outstanding development environment and superb compiler determined its use in this text. Unlike a generic compiler modified for the AVR instruction set, CodeVisionAVR was written specifically for the AVR microcontrollers. As a result, it produces very precise code, using the many features of those microcontrollers without waste. In comparison with other compilers, it produces smaller, more efficient code for the Atmel AVR microcontrollers. Finally, CodeVisionAVR has the CodeWizardAVR code generator, an invaluable tool for jump-starting most projects.

This chapter provides a look at the CodeVisionAVR development environment and its many features. You will learn how to create a project, to set specific compiler options for the project, to *make* a project into an output file, and to load that output file into the target device. The CodeWizardAVR code generator will be discussed, along with the terminal tool. Finally, you will be shown some basic features of AVR Studio as a software simulator.

Currently, CodeVisionAVR is designed to run in Windows 95, 98, Me, NT, 2000, and XP environments. The version on the enclosed CD-ROM is the evaluation version. It is limited in the size of the source file that can be compiled and in the number of features supported by the CodeWizardAVR code generator.

4.3 IDE OPERATION

The CodeVisionAVR C Compiler is accessed through its *integrated development environment (IDE)*. The IDE allows the user to build projects, add source code files to the projects, set compiler options for the projects, compile, and *make* projects into executable program files. These executable files are then loaded into the target microprocessor. The goal of this section is to familiarize you with the usual use of the CodeVisionAVR IDE. There are many advanced features in the IDE that are beyond the scope of this text.

As CodeVisionAVR's IDE is discussed, the menu items are referenced by the main menu name followed by "|" and the submenu item name. For example, a reference to the **Save As** menu item within the **File** menu appears as **File|Save As.** The main menu bar for CodeVisionAVR's IDE appears below the application title bar.

4.3.1 PROJECTS

A project is a collection of files and compiler settings that you use to build a particular program. A few projects are provided for you on the CD-ROM included with this book, but it is also important to know how to create and configure a new project. All project files have a *.prj* extension.

An open project appears as in Figure 4–1. Notice the file navigator down the left side of the screen. This lists the files that are part of the project as well as other files that are being used with the project. Only the files listed under the project name are to be compiled as part of the project. In Figure 4–1, the project name is *IOM* and the source files for the project are *IOM.c, cp.c, LCD.c,* and *keypad.c.* The files listed under Other Files are not compiled as source files for the project (*ilcd.h* and *cp.h*). The source code editor occupies the right side of the screen. Many source code files can be opened at once and tiled, stacked, or maximized depending upon the user's needs. Across the top of the screen are the menus and the toolbars, which are referenced throughout this chapter. Finally, across the bottom appears the Messages tab. This is where the compiler errors and warnings are listed upon compilation.

Open Existing Projects

Select the **File|Open** menu command or click the **Open File** button on the toolbar to open an existing project file. When either of these is performed, an Open File dialog box

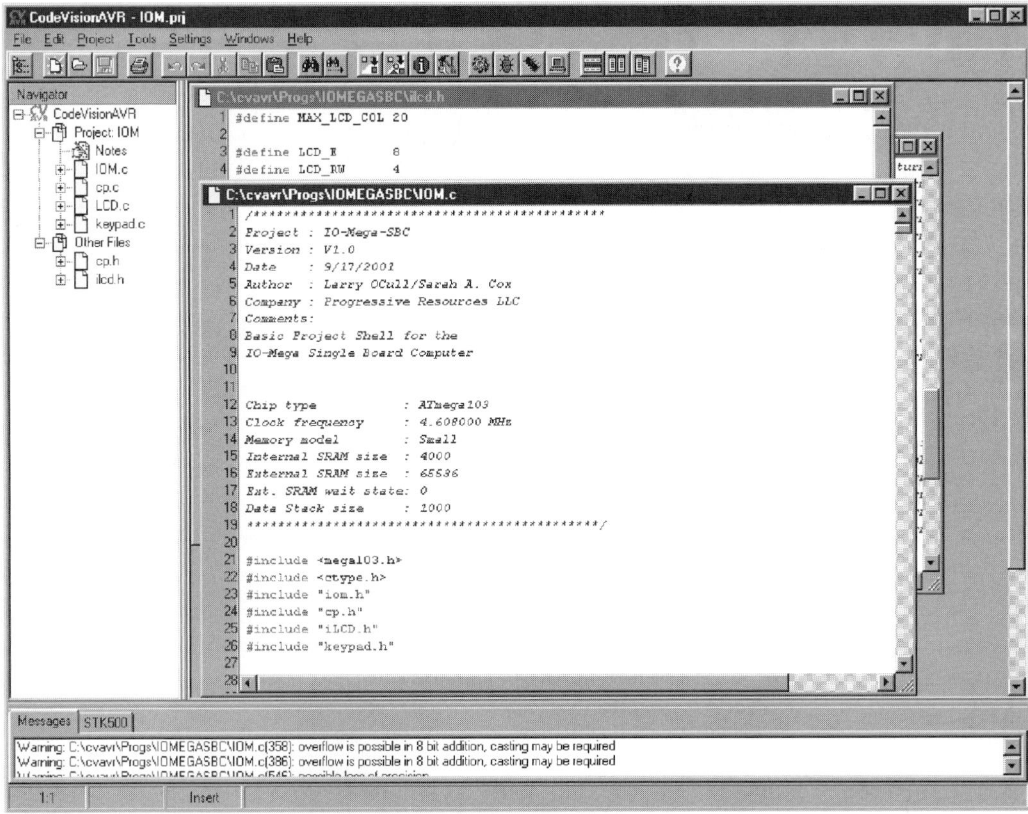

Figure 4–1 *An Open Project in CodeVisionAVR*

appears. Browse to the appropriate directory and select the project file to open. When you open the project file, the project appears as it was last used. Figure 4–1 shows a typical open project. Notice the file navigator down the left side listing the project name and the available files.

Create New Projects

New projects are created with the **File|New** menu command or by clicking the **Create New File** button on the toolbar. These actions open the Create New File dialog box as in Figure 4–2, requesting that you specify the type of file to open. Select the **Project** option and click **OK.** A second dialog box asks you to confirm use of the CodeWizardAVR to create the new project. This dialog box is depicted in Figure 4–3. The code wizard is a tool used to automatically generate part of the source code for your project; it is covered in more depth later in this chapter. To create a project without using the code wizard, select **No** when prompted to use it. A final dialog box prompts you to specify a project file name and location.

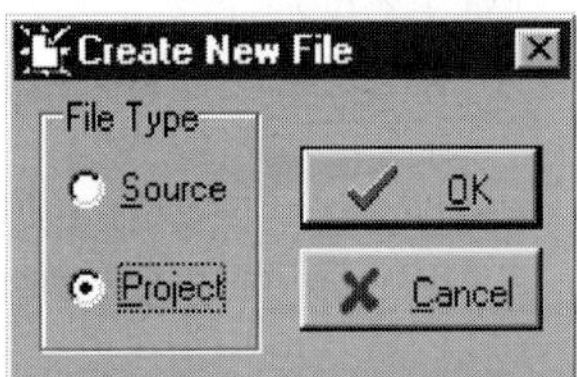

Figure 4–2 *Create New File Dialog Box*

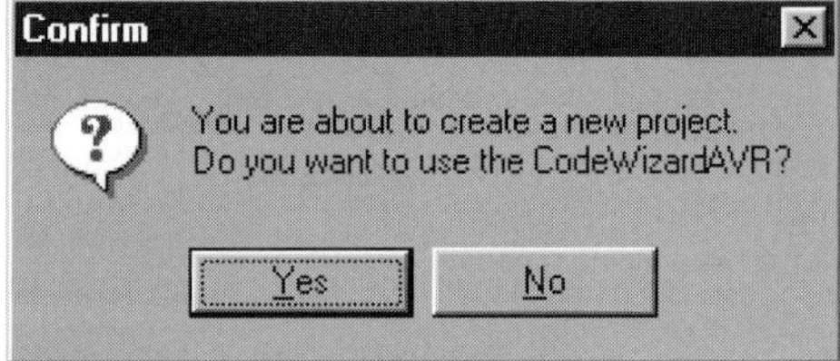

Figure 4–3 *CodeWizardAVR Confirm Dialog Box*

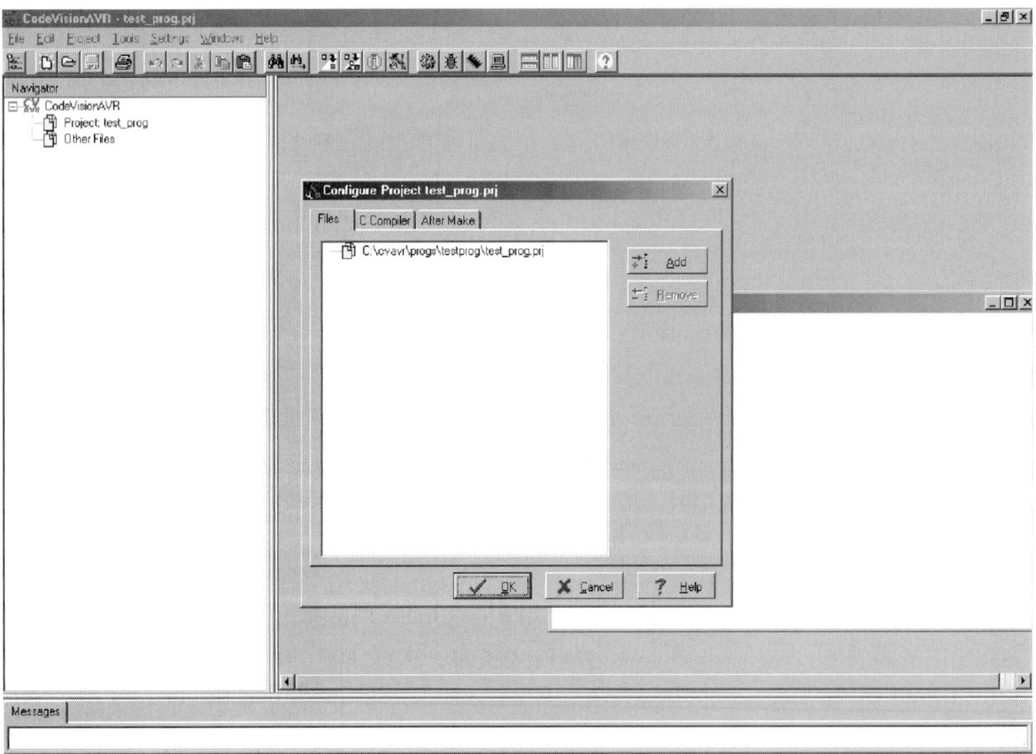

Figure 4–4 *CodeVisionAVR after Creating a New Project*

When these steps are completed, CodeVisionAVR creates a new project and opens the Configure Project dialog box for the new project. This dialog box has several tabbed frames of options that allow you to customize the project for your specific application, as shown in Figure 4–4.

Configure Projects

The project is configured with the **Project|Configure** menu command or the **Project Configure** toolbar button. Selecting either of these opens the Configure Project dialog box.

There are three tabbed frames in this dialog box: Files, C Compiler, and After Make. If you select the Files tab, you are able to add source files to and remove source files from the project. The C Compiler tab allows you to set properties for the project pertaining to the target device and the executable program file. Finally, the After Make tab allows you to select particular programs to run (such as a chip programmer) when the make is complete. The Files and the C Compiler tabs are covered in more detail later in this chapter.

Close Project

To quit working with the current project, use the **File|Close Project** menu command. If the project files were modified and were not yet saved, you are prompted to save the modified files. When you save, the IDE creates a backup file with a *.pr~* extension.

4.3.2 SOURCE FILES

Source files are the files that contain your program source code. They are the files that you painstakingly labor over to make the microprocessor do the right thing. The IDE allows you to add source files to the project and remove them from it. The IDE also has a fairly powerful editor built into it for editing code. There are features of the IDE editor that are very specific to a code editor and are useful in developing and debugging code.

Open an Existing Source File

To open an existing file, use the **File|Open** menu command or click the **Open File** button on the toolbar. An Open File dialog box allows you to browse to the directory to locate the file to open. Select the file and click **OK** to open the file. Alternatively, single-clicking the file name in the navigator also opens the file. The file opens in an editor window within the IDE. Although many files are open in the IDE in Figure 4–5, *cp.c* is the currently selected file. The selected file path and name are listed in the title bar at the top of the editor window.

Create a New Source File

The **File|New** menu command and the **Create New File** button on the toolbar are available to create a new source file. When either is selected, a dialog box appears prompting you to select a file type. Select **Source** and click **OK.** A new editor window opens for the newly created file, as shown in Figure 4–6. The title bar shows the file name as *untitled.c,* and it is listed in the file navigator under Other Files.

To save this file under a new name, use the **File|Save As** menu command. The new file from Figure 4–6 is shown in Figure 4–7 saved under the name *mynewfile.c.* This change shows in the title bar and in the file navigator.

Add an Existing File to the Project

Simply opening a file or creating a file does not automatically add it to the current project. You must open the project and select the **Project|Configure** menu command or click the **Project Configure** toolbar button. This opens the Configure Project dialog box. Select the

Figure 4–5 *The Source File* cp.c *Is Selected*

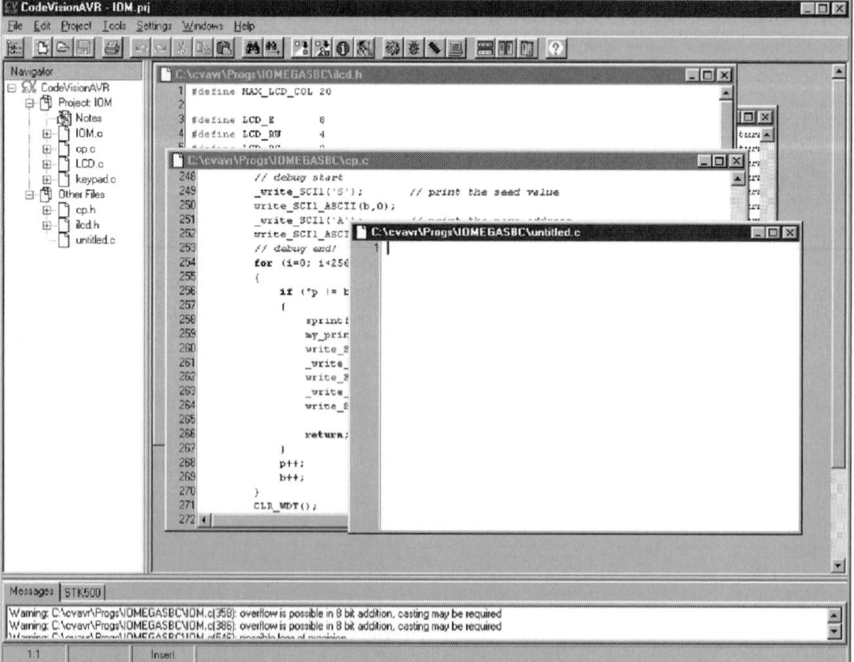

Figure 4–6 *New Source File* untitled.c *Created*

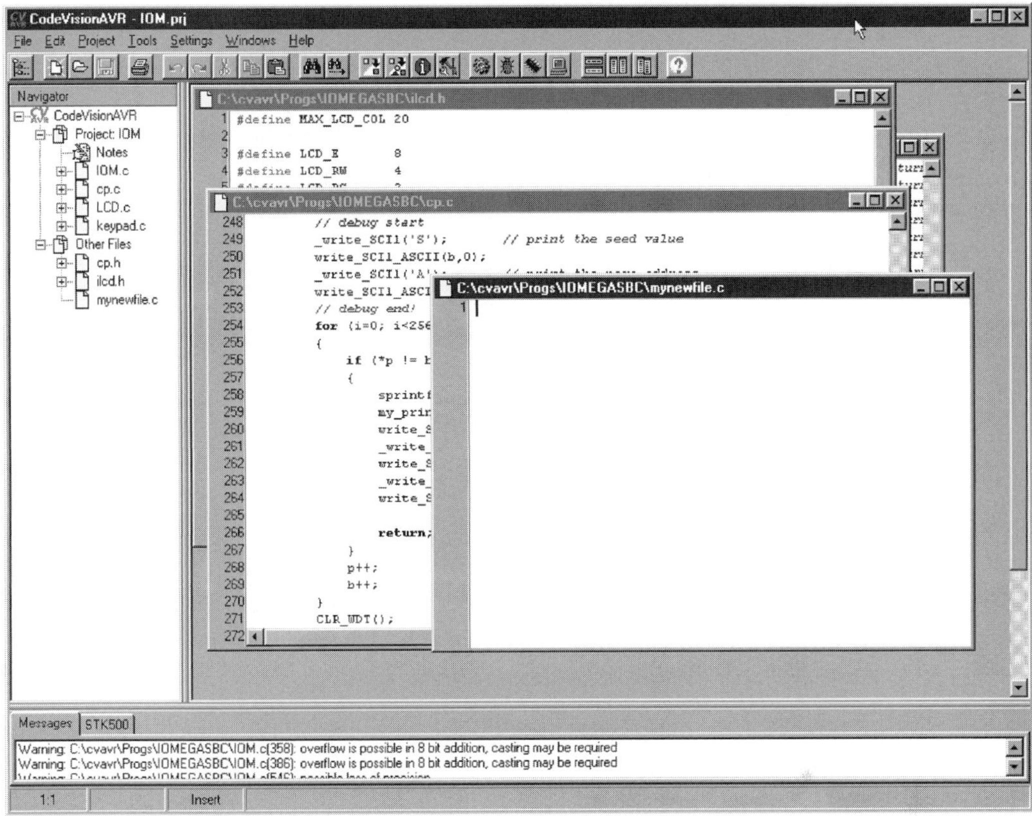

Figure 4–7 *New Source File Renamed as* mynewfile.c

Files tab, which lists the project name and the source file names for the project. Files are added to and removed from the project by clicking the **Add** and **Remove** buttons on the right side of the window.

The Project Configure dialog box is displayed in Figure 4–8, with *iom.c, LCD.c, keypad.c,* and *cp.c* listed as source files for the project.

Clicking the **Add** button and selecting the newly created file *mynewfile.c* adds it to the project as a source file. The new source file list appears in Figure 4–9.

When the Configure Project dialog box is closed, the file navigator shows that the file *mynewfile.c* is now considered part of the source files for the project. In Figure 4–10, the file is no longer listed under Other Files.

4.3.3 EDIT FILES

The editor built into the CodeVisionAVR IDE supports the standard editor functions. The cursor is moved around with the **Home, End,** and arrow keys, as well as the mouse.

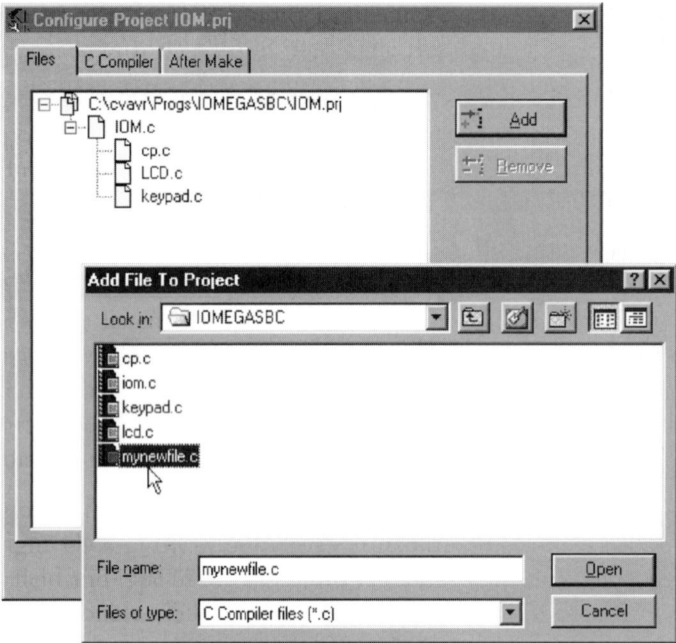

Figure 4–8 *Configure Project Dialog Box, Add Button Clicked*

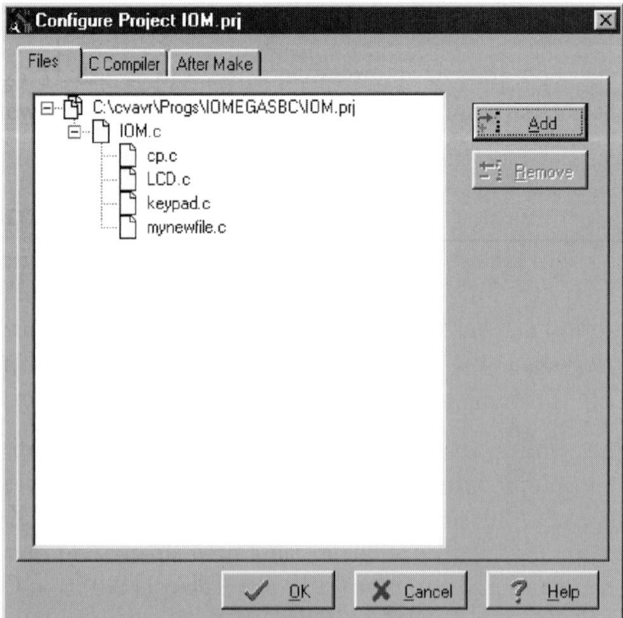

Figure 4–9 mynewfile.c *Added to Project as a Source File*

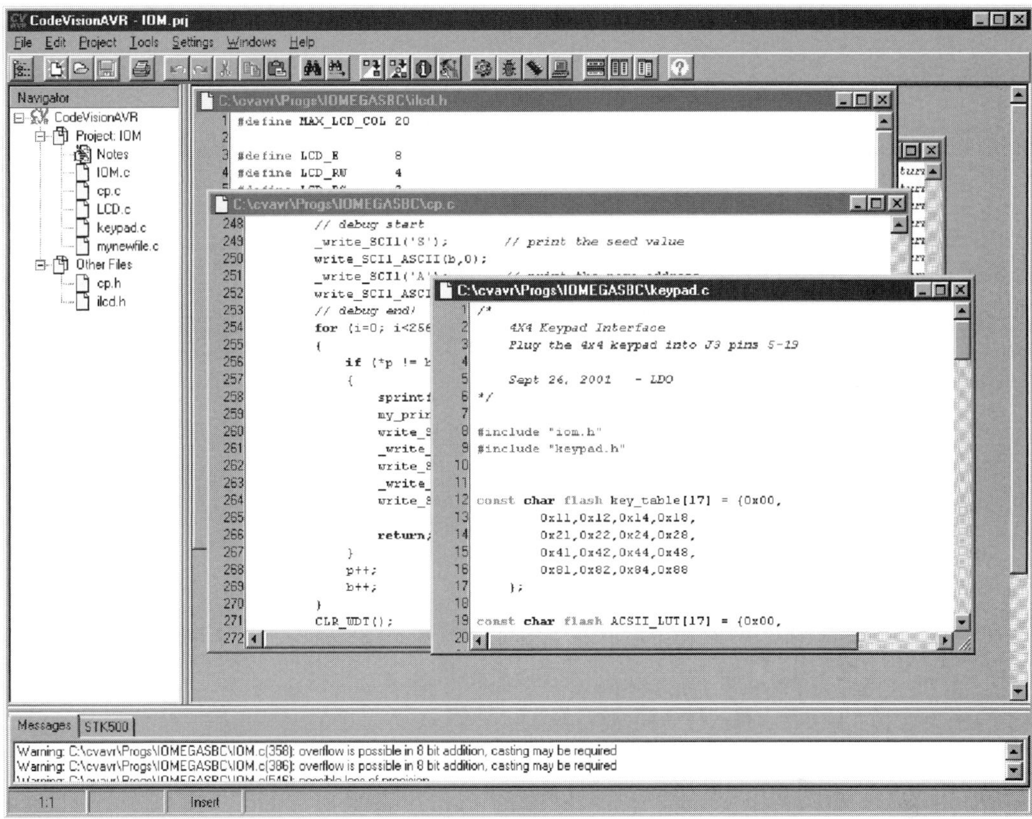

Figure 4–10 *The File Navigator Lists* mynewfile.c *as a Source File*

Clicking and dragging the mouse selects portions of text to be manipulated together. Cut, copy, and paste the selected text with the standard shortcut keys, the **Edit** menu options, or by clicking the right mouse button. Find and replace functions are supported, along with undo and redo editing functions.

The editor has excellent features especially designed to facilitate editing code. Some of these features make it easier to move around in the file. The **Alt+G** key combination or the **Edit|Goto Line** menu command moves the cursor to a specified line number. Bookmarks are toggled on and off at the current cursor position by the **Edit|Toggle Bookmark** menu command or by the **Shift+Ctrl+0** through **9** key combinations. The **Edit|Jump to Bookmark** menu command or the **Ctrl+0** through **9** keys jump the cursor to a previously set bookmark. Bookmarks are useful for moving around within the program without paging through the code or memorizing line numbers.

The editor has two features to simplify formatting code for readability. First, the editor allows you to indent and de-indent blocks of text. Highlight the text by clicking and dragging your mouse, then press the **Ctrl+I** or **Ctrl+U** keys to indent or undo the indent on the block of text by one tab spacing. This is useful for indenting the code in an **if** statement or **while** statement. Second, the editor supports an auto-indenting feature, which moves the cursor directly below the start of the previous line when **Enter** is pressed, instead of moving the cursor to the very beginning of the line. This feature is controlled through the **Settings|Editor** menu command.

The editor also has a brace-matching feature. If the cursor is positioned on an opening or the closing brace, the **Edit|Match Braces** menu command or **Ctrl+M** highlights the portion of the text until the corresponding matching closing or opening brace. This allows the programmer to check for balanced braces around functions, **if** statements, **while** statements, and so on. Pressing any key or clicking the mouse hides the highlighting.

Finally, the editor has a syntax highlighting feature that makes the code easier to read and write. The syntax highlighting uses special colors to display keywords, strings, constants, and comments so that they stand out as you write and review your code. The colors used for highlighting can be modified through the **Settings|Editor** menu command.

4.3.4 PRINT FILES

The **File|Print** menu command and the **Print** button on the toolbar print the currently active file on the default Windows printer. To print only a section of the active file, highlight the portion to be printed and select the **Edit|Print Selection** menu command.

The current print settings are modified by selecting **File|Page Setup.** This opens the Page Setup dialog box, which allows you to select whether or not to print page numbers, page headers, and syntax highlighting. The Page Setup dialog box also establishes the margins for printing and the units they are measured in. Finally, the **Printer** button in the Page Setup dialog box opens a second dialog box for modifying the printer settings or selecting a different printer altogether.

4.3.5 THE FILE NAVIGATOR

The file navigator facilitates displaying and opening source files. Clicking on the file name in the navigator window maximizes or opens the file in an editor window. A typical navigator window appears in Figure 4–11. The project name is listed at the top of the window, with all of the source files for the project listed directly below the project's name. Files that are open but not considered to be part of the project source files are listed under Other Files toward the bottom of the window.

If there is a + or – button next to the file name, more information exists on the file. This information is either expanded or collapsed by clicking on the + or –. The information may include a list of global variables and functions declared in each compiled C source file. (These listings are available only after a compilation has taken place.) Clicking on the variable or function name highlights the variable or function declaration in the appropriate

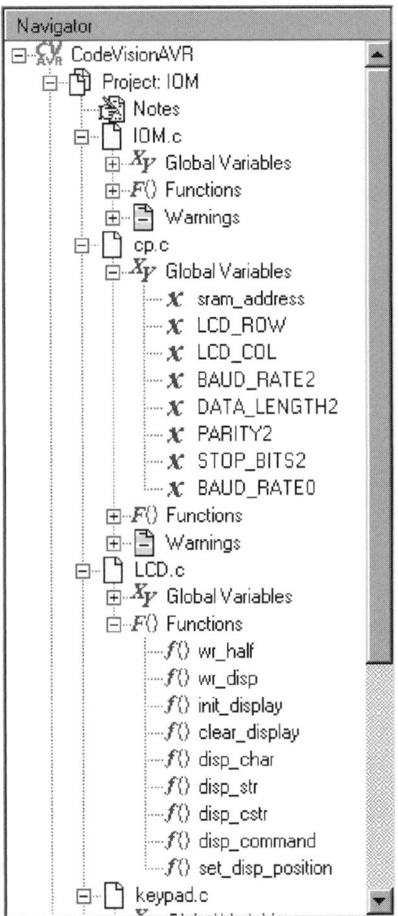

Figure 4–11 *File Navigator Window*

C source file. Errors and warnings generated during compilation are also displayed in the navigator window. Clicking on the error or warning highlights the related line in the source file. Figure 4–11 shows an example with some of the file branches expanded.

The file navigator displays the results of the **Find in Files** command, available under the **Edit** menu. When performed, it searches all of the files listed in the file navigator for the specified search term. The results of the find are listed in the file navigator window for each file in which the term was found. To view the results, click the + button next to the Found line. Figure 4–12 shows the results of an **Edit|Find in Files** command performed on the term "sprintf." At the bottom of the CodeVisionAVR window, a Find in Files tabbed window appears next to the Messages window, providing a second listing of places where the search term was found.

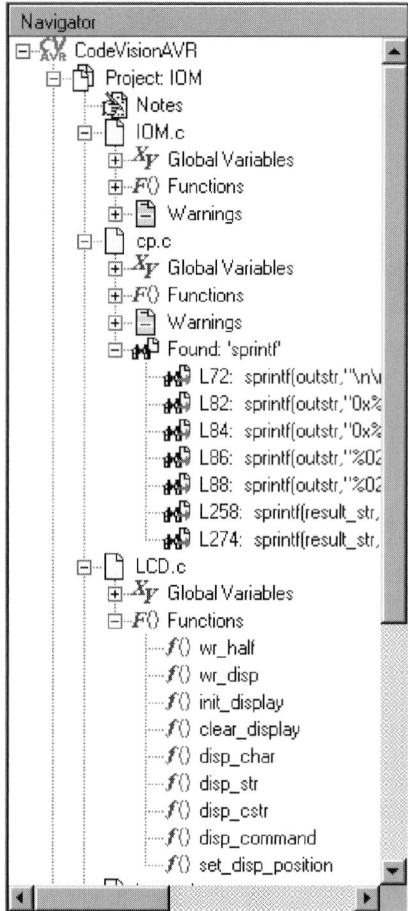

Figure 4–12 *File Navigator Window after* Find in Files

4.4 C COMPILER OPTIONS

The compiler must know some key information about the target device in order to generate the appropriate executable program file. This information includes the chip type, the data stack size of the device, and the size of the internal and external SRAM used in the application. The compiler must also know a few things about how to interpret and optimize the C code in the source files for your application.

To set the C compiler options for the currently opened project, select the **Project|Configure** menu command. The Configure Project dialog box opens, with tabbed frames for different groups of settings. Select the C Compiler tab, as shown in Figure 4–13.

The two most general settings are the chip and clock. The **Chip** list box selects the target AVR microcontroller chip. (Note that the evaluation and lite versions of CodeVisionAVR

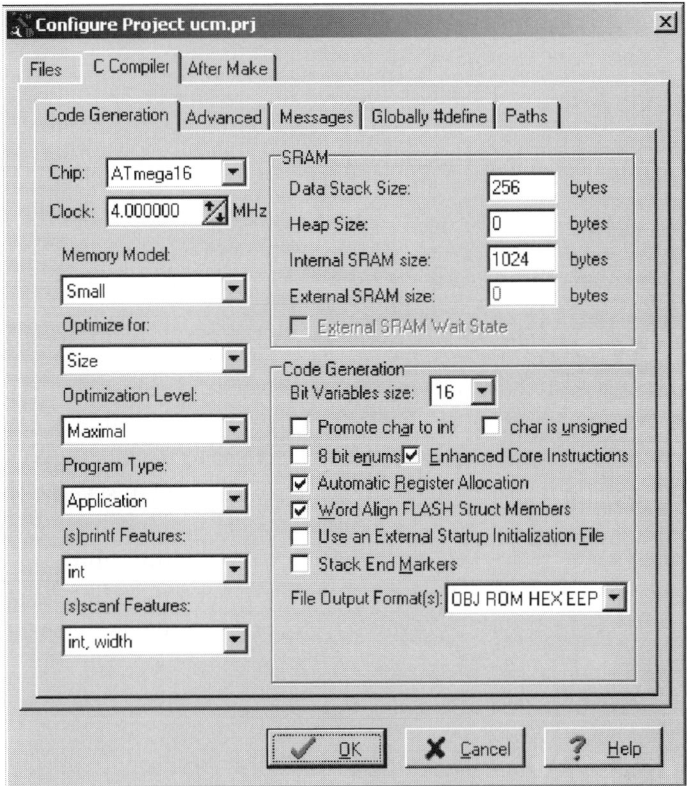

Figure 4–13 *Configure Project Dialog Box, C Compiler Tab*

do not support all of the Atmel AVR processors. Refer to *http://www.hpinfotech.ro* for information on the processors supported by each version.) The **Clock** field refers to the system clock and must also be populated. This is done by either typing in the frequency or using the up and down arrows to scroll to the appropriate value.

4.4.1 MEMORY MODEL

The required memory model is selected under Memory Model. Four memory models are implemented: **Tiny, Small, Medium,** and **Large.**

The **Tiny** option uses 8 bits for storing pointers to the variables placed in SRAM. In this memory model, you have access to only the first 256 bytes of SRAM. The **Small** option uses 16 bits for storing pointers to the variables placed in SRAM. In this memory model, you have access to 65,536 bytes of SRAM. For both the **Tiny** and **Small** options, the pointers to the FLASH and EEPROM memory areas always use 16 bits, so the memory model selection limits the total size of the constant arrays and literal character strings to 64K. However, the total size of the program can be the full amount of FLASH.

The **Medium** memory model is similar to the **Small** memory model except that the **Medium** memory model uses pointers to constants in FLASH that are 32 bits wide. The pointers to functions are 16 bits wide under the **Medium** memory model since they hold the word address of the function. Under word addressing, 16 bits are enough to address a function located in all 128K of FLASH.

The **Large** memory model is similar to the **Medium** memory model except that it uses 32 bits for pointers to the constants in FLASH and for pointers to functions. The **Large** memory model can be used for chips with 256K or more of FLASH.

4.4.2 OPTIMIZE FOR

The compiled program is optimized for minimum size or maximum execution speed by the **Size** or **Speed** option under **Optimize for.** It is usually desirable to optimize the software for size, because the memory on the microcontroller is limited. When optimizing for size, the compiler meticulously pulls out code that repeats in several places and creates a subroutine for it. Each time the code is needed, a call to the subroutine is used instead. If you look at the *.lst* file after compiling, you will notice many subroutines that the compiler created for you to save space. However, each call to a subroutine costs you execution time. So, in some applications, strict timing requirements make it necessary to leave the code in line although it repeats itself later. In these cases, select to optimize for speed, not for size.

4.4.3 OPTIMIZATION LEVEL

The compiled program is optimized according to the **Optimize for** setting as well as the **Optimization Level** setting. In general, you want the smallest, fastest code that the compiler can generate. However, when debugging in an environment such as AVR Studio, you may want to decrease the **Optimization Level** since the maximal optimization can make stepping through the application and debugging more difficult.

4.4.4 PROGRAM TYPE

For devices that allow self-programming, the **Program Type** is selected as either Application or Boot Loader. A boot loader application is one that can reprogram the main program memory while running from a higher memory location. This allows the software in a microcontroller to be updated using serial communication to the processor directly. Once programmed with the boot loader, the microcontroller can be reprogrammed repeatedly without requiring a special programming cable or even resetting the microcontroller.

4.4.5 (s)printf FEATURES AND (s)scanf FEATURES

When *printf, sprintf, scanf,* or *sscanf* are required by an application, the full implementation of these functions can be consuming of the resources. To minimize this consumptive effect, the compiler options **(s)printf Features** and **(s)scanf Features** allows you to select to what level you need the functions implemented. See the compiler documentation for the exact features supported for each function under each option.

4.4.6 SRAM

Default values for the data stack size, heap size, and internal SRAM size are chosen by CodeVisionAVR based on the particular microcontroller selected. The data stack size is the amount of memory to be set aside for variables. For a data-rich application that requires a larger than normal amount of variable space, increase the data stack size. This impacts the amount of memory available for the program stack, so increase the data stack size only by the required amount. If you are using memory allocation functions such as *malloc, calloc,* or *realloc,* then you will need to reserve some heap space as well. The amount of heap space required is the summation of all allocated space plus 4 bytes for each block of memory that is allocated.

If external SRAM memory is connected, specify the size of that memory. If the external memory device is slow, select to use the wait states by selecting the **External SRAM Wait State** check box.

4.4.7 COMPILATION

The Compilation frame contains mostly check boxes that enable or disable particular options. It also contains a list box for determining the type of output file format to be used.

The **Promote char to int** check box enables the ANSI promotion of char operands to int. Promoting char to int leads to larger code size and slower speed for an 8-bit chip microcontroller like the AVR.

The **char is unsigned** check box selection causes the compiler to treat, by default, the char data type as an unsigned 8-bit variable in the range 0 to 255. If the check box is not selected, the char data type defaults to a signed 8-bit variable in the range −128 to 127. Treating char data types as unsigned leads to better code size and speed.

CAUTION: There are some conditional statements that do not operate predictably on unsigned char data types. The "greater than" and "less than" conditional statements are two such statements. If a variable is to be used in this type of statement, it is preferable to use a signed char data type. If the value will exceed a signed char data type, promote the variable to a signed integer. This seems trivial, but it could save you hours of debugging time.

To reduce code space, enumerations can be set up to be 1 byte instead of 2 by selecting the **8 bit enums** option. This should be done with the understanding that any enumerated values under this scheme should be treated in math and all other operations as characters and not integers.

Some of the AVR microcontrollers support an enhanced instruction set. To take advantage of this, you need to select the **Enhanced Core Instrtructions** check box. It allows enabling or disabling the generation of Enhanced Core instructions for microcontrollers such as the Atmega128, Atmega32, or Atmega8. This check box is available only when a microcontroller that supports the enhanced instruction set has been selected.

Word Align FLASH Struct Members is an option that makes compilation backward compatible to previous versions of the compiler where FLASH structures had to be word aligned. This option should not be selected for new projects.

For debugging purposes, the **Stack End Markers** check box causes the compiler to place the strings "DSTACKEND" and "HSTACKEND" at the end of the data stack and the hardware stack areas. When you debug the program with an emulator or simulator like Atmel's AVR Studio, you can see if these strings are overwritten and modify the data stack size accordingly. When your program runs correctly, you can disable the placement of the strings in order to reduce the code size.

The **File Output Format(s)** list box selects the formats for the files generated by the compiler. Most in-system programmers use Intel HEX formatted files. COFF is required by the Atmel AVR Studio debugger; ROM and EEP are required by some in-system programmers.

4.4.8 MESSAGES TAB

The generation of warning messages during compilation is enabled or disabled by the **Enable Warnings** check box found on the Messages tab. Specific messages can be enabled or disabled in the list below the main checkbox. Temporarily disabling the generation of warnings may be useful during debugging to prevent messages about unreachable code or variables being declared and never used. When you work on the final version of a program, it is much better to work through all of the warnings than to simply disable them.

4.5 COMPILE AND MAKE PROJECTS

Obtaining an executable program file requires the following steps:

1. Compile the project's C source files into an assembler source file.
2. Assemble the assembler source file.

Compiling a file executes step 1 only. *Making* a file executes steps 1 and 2 for the main project file.

4.5.1 COMPILE A PROJECT

The CodeVisionAVR C Compiler is called for a project by the **Project|Compile File** menu command, the **F9** key, or the **Compile** button of the toolbar. Execution of the CodeVisionAVR C Compiler produces an assembler source file with the *.asm* extension. After the compilation, the Information window opens, showing the compilation results. See Figure 4–14.

The Information window lists the compiler option settings at the top. Next, the statistics on the number of lines compiled and the number of errors and warnings are listed. The rest of the window details how much memory is used and for what purposes.

If errors or warnings are generated during compilation, they are listed in the Messages window located under the editor window and in the navigator window. Double-clicking the error or warning message highlights the troublesome line of code. In Figure 4–15, the first warning was double-clicked and the associated line of code is highlighted.

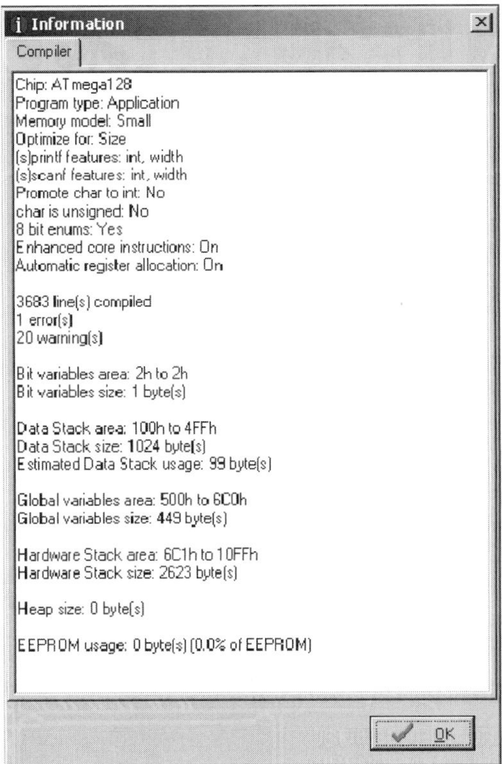

Figure 4–14 *Information Window Showing Compilation Results*

4.5.2 MAKE A PROJECT

An executable file is created by *making* a project. To *make* a project, select the **Project|Make** menu command, press the **Shift+F9** keys, or click the **Make** button of the toolbar. The CodeVisionAVR C Compiler executes, producing an assembler source file with the *.asm* extension. If the compiler generated no errors, the Atmel AVR assembler AVRASM32 is automatically called, and the assembler runs on the newly created assembler source file. The assembler creates an executable program file in the format specified in the C Compiler tab of the Configure Project dialog box (**Project|Configure|C Compiler** command).

After the *make* process is completed, the Information window opens, showing the compilation results. There are two tabs in the window, Compiler and Assembler. These tabs allow you to review the results of both the compiling and the assembling of the program. An example of the compilation results is shown previously in Figure 4–14. Figure 4–16 shows an example of the assembler results.

The top of the Assembler tab displays the version and copyright information for the assembler. The next section in the window is the assembler feedback of how much memory is

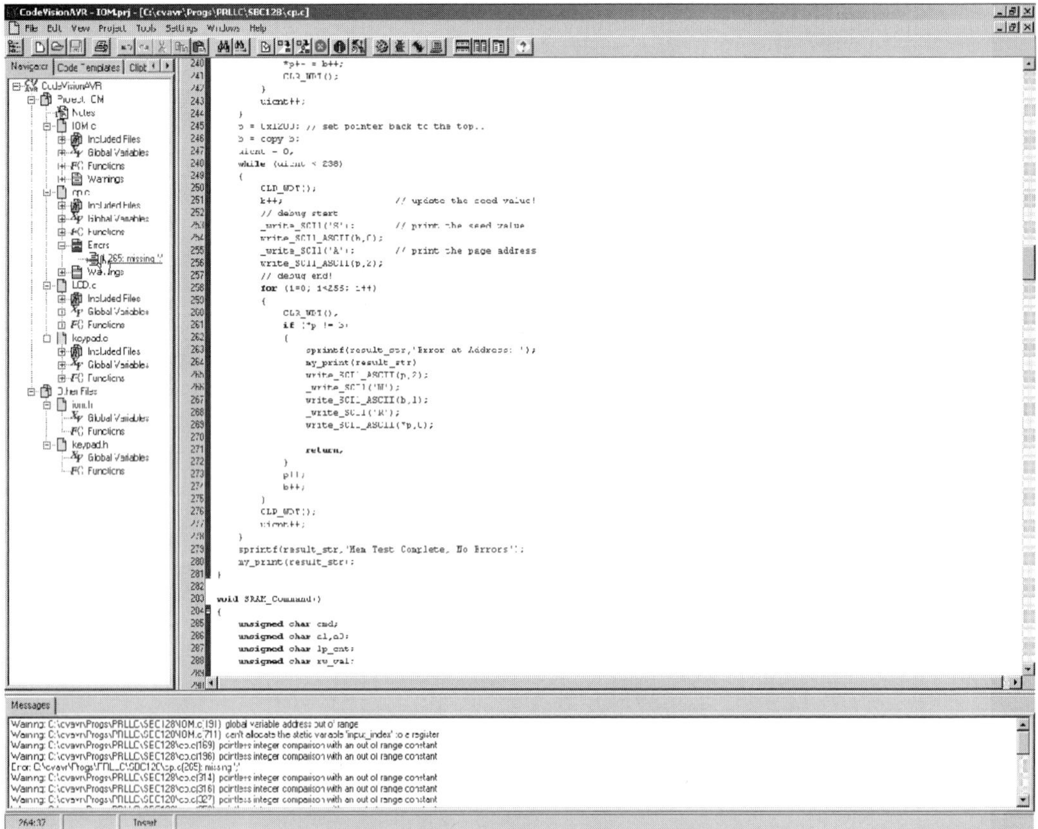

Figure 4–15 *Compilation Errors Displayed in Messages Area and in the Navigator*

used by the program for each area, including the code segment (cseg or FLASH), the data segment (dseg or SRAM), and the EEPROM segment (eseg).

The line below the memory usage table is very important: "Assembly complete with no errors." Occasionally, a program compiles fine but errors occur during the assembly. If this happens, a window opens and displays a message that errors occurred during assembly. This line in the Assembler tab of the Information window tells how many errors occurred. To find the errors, open the *.lst* file and search on the term "Error."

The bottom part of the Assembler tab shows that the *iom.eep* file was deleted. This file was created for EEPROM initializations, but none existed in this program, so the file is not needed.

4.6 PROGRAM THE TARGET DEVICE

The CodeVisionAVR IDE has a built-in In-System AVR Chip Programmer that lets you easily transfer your compiled program to the microcontroller for testing. The programmer

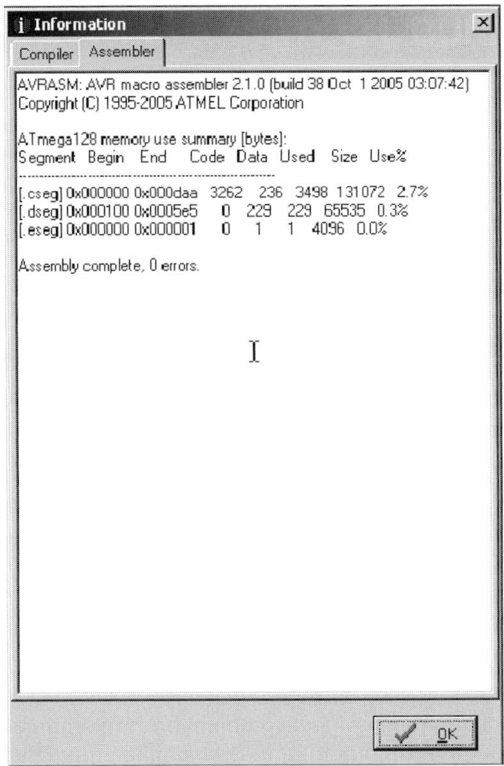

Figure 4–16 *Assembler Results*

is designed to work with development boards such as the Atmel STK500, Kanda Systems STK200+/300, Dontronics DT006, Vogel Elektronik VTEC-ISP, and the MicroTronics ATCPU and Mega2000 development boards, which are connected to your computer's parallel printer port. The **Settings|Programmer** menu item opens a window to select the type of programmer to be used.

The programmer is opened by selecting the **Tools|Chip Programmer** menu command or by clicking the **Chip Programmer** button on the toolbar. Figure 4–17 shows the chip programmer opened with a Kanda Systems programmer selected.

4.6.1 CHIP

The one item on the Chip Programmer dialog box that dictates what is available in the rest of the window is the **Chip** list box, which selects the type of chip you wish to program. The chip selection determines the size of the FLASH and EEPROM buffers. Also, the chip selection determines what is displayed—the Fuse Bit(s), Boot Lock Bit 0, and Boot Lock Bit 1 frames—and whether these frames are displayed at all.

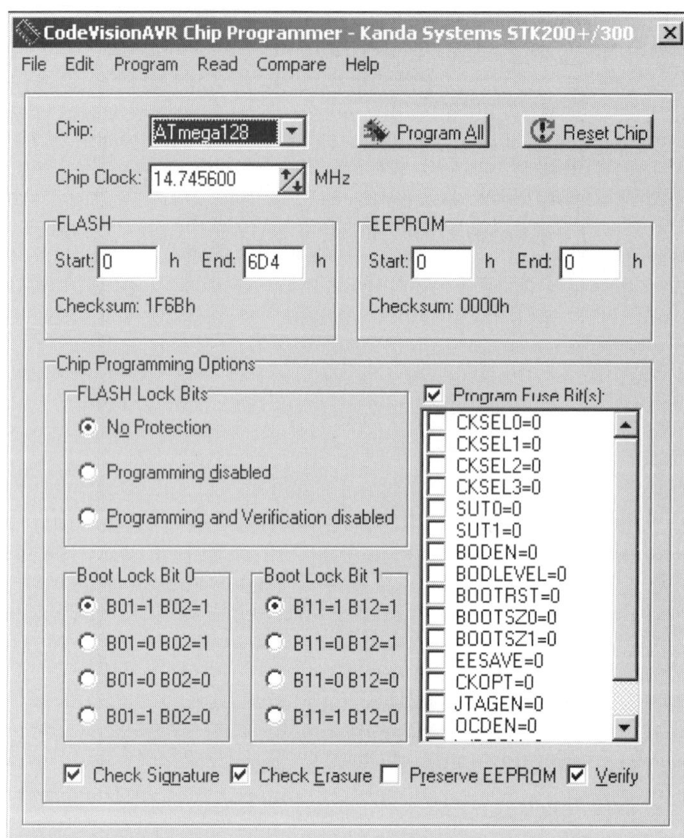

Figure 4–17 *CodeVisionAVR Chip Programmer*

4.6.2 FLASH AND EEPROM

The programmer has two memory buffers: the FLASH memory buffer and the EEPROM memory buffer. The contents of these buffers are loaded and saved with the **File|Load** and **File|Save** menu commands. The supported file formats are as follows:

- Atmel *.rom* and *.eep*
- Intel HEX
- Binary *.bin*

After a file is loaded into the corresponding buffer, the **Start** and **End** addresses are updated accordingly. The **Checksum** field is also updated with the checksum of the loaded file.

It is possible to view and edit the contents of the FLASH and EEPROM buffers. The buffers are displayed and edited by the **Edit|FLASH** and **Edit|EEPROM** menu commands. Upon selection of one of these commands, an Edit window displaying the corresponding buffer contents opens (refer to Figure 4–18).

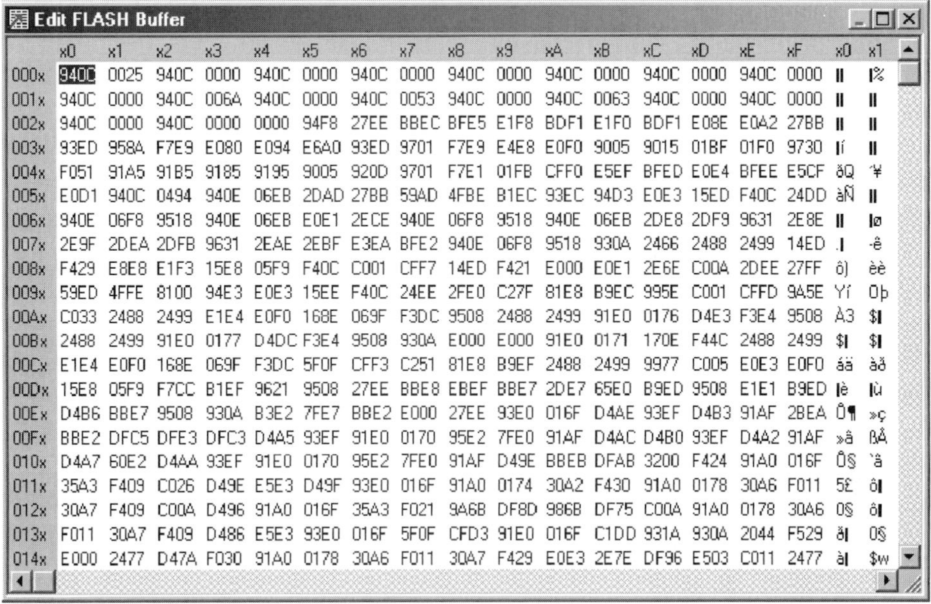

Edit FLASH Buffer

	x0	x1	x2	x3	x4	x5	x6	x7	x8	x9	xA	xB	xC	xD	xE	xF	x0	x1
000x	940C	0025	940C	0000	940C	0000	940C	0000	940C	0000	940C	0000	940C	0000	940C	0000	▌▌	▌%
001x	940C	0000	940C	006A	940C	0000	940C	0053	940C	0000	940C	0063	940C	0000	940C	0000	▌▌	▌▌
002x	940C	0000	940C	0000	0000	94F8	27EE	BBEC	BFE5	E1F8	BDF1	E1F0	BDF1	E08E	E0A2	27BB	▌▌	▌▌
003x	93ED	958A	F7E9	E080	E094	E6A0	93ED	9701	F7E9	E4E8	E0F0	9005	9015	01BF	01F0	9730	▌í	▌▌
004x	F051	91A5	91B5	9185	9195	9005	920D	9701	F7E1	01FB	CFF0	E5EF	BFED	E0E4	BFEE	E5CF	ðQ	´¥
005x	E0D1	940C	0494	940E	06EB	2DAD	27BB	59AD	4FBE	B1EC	93EC	94D3	E0E3	15ED	F40C	24DD	àÑ	▌▌
006x	940E	06F8	9518	940E	06EB	E0E1	2ECE	940E	06F8	9518	940E	06EB	2DE8	2DF9	9631	2E8E	▌▌	▌ø
007x	2E9F	2DEA	2DFB	9631	2EAE	2EBF	E3EA	BFE2	940E	06F8	9518	930A	2466	2488	2499	14ED	.▌	-ê
008x	F429	E8E8	E1F3	15E8	05F9	F40C	C001	CFF7	14ED	F421	E000	E0E1	2E6E	C00A	2DEE	27FF	ô)	èè
009x	59ED	4FFE	8100	94E3	E0E3	15EE	F40C	24EE	2FE0	C27F	81E8	B9EC	995E	C001	CFFD	9A5E	Yí	0þ
00Ax	C033	2488	2499	E1E4	E0F0	168E	069F	F3DC	9508	2488	2499	91E0	0176	D4E3	F3E4	9508	À3	$▌
00Bx	2488	2499	91E0	0177	D4DC	F3E4	9508	930A	E000	E000	91E0	0171	170E	F44C	2488	2499	$▌	$▌
00Cx	E1E4	E0F0	168E	069F	F3DC	5F0F	CFF3	C251	81E8	B9EF	2488	2499	9977	C005	E0E3	E0F0	áä	àð
00Dx	15E8	05F9	F7CC	B1EF	9621	9508	27EE	BBE8	EBEF	BBE7	2DE7	65E0	B9ED	9508	E1E1	B9ED	▌è	▌ù
00Ex	D4B6	BBE7	9508	930A	B3E2	7FE7	BBE2	E000	27EE	93E0	016F	D4AE	93EF	D4B3	91AF	2BEA	Ô¶	»ç
00Fx	BBE2	DFC5	DFE3	DFC3	D4A5	93EF	91E0	0170	95E2	7FE0	91AF	D4AC	D4B0	93EF	D4A2	91AF	»â	ßÅ
010x	D4A7	60E2	D4AA	93EF	91E0	0170	95E2	7FE0	91AF	D49E	BBEB	DFAB	3200	F424	91A0	016F	Ô§	`à
011x	35A3	F409	C026	D49E	E5E3	D49F	93E0	016F	91A0	0174	30A2	F430	91A0	0178	30A6	F011	5£	ô▌
012x	30A7	F409	C00A	D496	91A0	016F	35A3	F021	9A6B	DF8D	986B	DF75	C00A	91A0	0178	30A6	0§	ô▌
013x	F011	30A7	F409	D486	E5E3	93E0	016F	5F0F	CFD3	91E0	016F	C1DD	931A	930A	2044	F529	ð▌	0§
014x	E000	2477	D47A	F030	91A0	0178	30A6	F011	30A7	F429	E0E3	2E7E	DF96	E503	C011	2477	à▌	$w

Figure 4–18 *CodeVisionAVR Edit FLASH Buffer*

The buffer contents, at the highlighted address, are modified by typing in the new value. The highlighted address is modified with the **arrow, Tab, Shift+Tab, PageUp,** and **PageDown** keys.

Occasionally it may be useful to fill a block of FLASH or EEPROM memory with a given value. Right-clicking in the Edit window opens the Fill Memory Block dialog box, as shown in Figure 4–19. This dialog box specifies the **Start Address, End Address,** and **Fill Value** of the memory area to be filled.

The FLASH and EEPROM buffers are programmed to the chip, read from the chip, or compared against the chip through the **Program, Read,** and **Compare** menu items

Fill Memory Block

Start Address: 0 h

End Address: FF h

Fill Value: 0 h

OK Cancel

Figure 4–19 *CodeVisionAVR Chip Programmer Fill Memory Block Dialog Box*

respectively. The chip must always be erased prior to programming the FLASH buffer. If the **Program|All** menu command is used, the chip erase will automatically be performed. If the chip is locked, the read and compare operations on the FLASH and EEPROM from the chip will not work.

4.6.3 FLASH LOCK BITS

The FLASH Lock Bits frame of the Chip Programmer dialog box has options for selecting the security level setting for the chip. This security level must be one of the following:

- **No Protection**—all read and programming operations are allowed
- **Programming Disabled**—reading the FLASH and EEPROM areas is allowed, but writing to them is not allowed
- **Programming and Verification disabled**—no reading to or writing from either area is allowed.

After programming the chip is complete, regardless of the state of the lock bits, the FLASH cannot be written to again until a chip erase is performed.

The lock bits are automatically written to the chip during a **Program|All** operation or can be written independently through the **Program|Lock Bits** menu command. The current value of the lock bits is read with the **Read|Lock Bits** command. A chip erase always sets the lock bits back to no protection.

4.6.4 FUSE BITS

If the chip you have selected has fuse bits that can be programmed, a supplementary Fuse Bit(s) frame appears. In this frame are various check boxes, one for each fuse bit available on the selected chip. The fuses and their respective uses are described in the Atmel datasheets. The fuse bits are programmed to the chip and read from the chip using the **Program|Fuse Bits** and **Read|Fuse Bits** menus. The fuse bits are also programmed as part of the **Program|All** automated programming process.

4.6.5 BOOT LOCK BIT 0 AND BOOT LOCK BIT 1

Chips that support boot-loading applications have additional security bits that lock the boot loader section of memory. The Boot Lock Bit 0 and Boot Lock Bit frames and options support these additional security bits. For information on how these bits are used, refer to the Atmel datasheets.

4.6.6 SIGNATURE

Each different chip type in the Atmel AVR line has a unique signature. The **Read|Chip Signature** menu item reads the signature of the chip to determine the type of chip connected to the programmer. Many programmers use the signature to verify the chip selected against the actual hardware connected before performing any operation. The different chip types have different programming parameters, so unexpected results can occur if the actual chip and the selected chip type do not match.

However, occasionally it is necessary to program or read a chip as another chip type. For example, when the Atmega128 was first released, many programmers supported it only in its Atmega103 mode, so it was necessary to ignore the signature of the Atmega128. It is for this reason that a check box is provided to disable the reading of the signature. If the signature is not correct and **Check Signature** is selected, the programming will not continue. If the **Check Signature** check box is not selected, programming will continue regardless of the signature.

4.6.7 CHIP ERASE

A chip erase is automatically performed whenever the **Program|All** menu command is used. To erase a chip independent of the programming operation, the menu command **Program|Erase Chip** is available. The **Program|Blank Check** menu command verifies that the chip's EEPROM and FLASH memory areas are erased. This verification is also performed automatically by the **Program|All** function unless the **Check Erasure** check box is cleared.

The **Preserve EEPROM** check box does just that: it enables the preservation of the EEPROM data through a chip erase cycle. This is accomplished by reading the EEPROM into a buffer, performing the chip erase, then writing the data back to the EEPROM. If the chip is locked to prevent reading, the EEPROM is not preserved, because the data could not be retrieved.

4.6.8 PROGRAMMING SPEED

To speed up the programming process, clear the **Check Erasure** check box. In this case, there is no verification of the correctness of the chip erase. You can also clear the **Verify** check box. When the **Verify** check box is cleared, there is no verification of the correctness of the FLASH and EEPROM programming. These are useful when you debug code and program the device repeatedly for testing. When you program the chip with the final versions of code, it is recommended that you have the **Check Erasure** and the **Verify** check boxes selected.

4.6.9 PROGRAM ALL

The **Program|All** menu command automates the programming process. This is probably the menu item that is most used in the programmer. The steps taken are as follows:

1. Erase the chip.
2. FLASH and EEPROM blank check (if **Check Erasure** is checked).
3. Program and verify the FLASH.
4. Program and verify the EEPROM.
5. Program the Fuse and Lock Bits.

4.6.10 OTHER PROGRAMMERS

In-system programmers that are not directly supported by CodeVisionAVR can be added to the tools list if they support command-line calls. Our examples show TheCableAVR in-system programmer as a stand-alone programmer, but this can also be called by CodeVisionAVR through the command line. To add a new tool, select the **Tools|Configure** menu item. When the Configure Tools dialog box opens, click the **Add** button. This opens a browser to select the executable for the tool. Figure 4–20 shows TheCableAVR being added in the Configure Tools dialog box.

Once the tool is added, click the **Settings** button to open the Tool Settings dialog box which establishes the settings for the tool: the name of the tool as you want it to appear in the Tools menu, the path and file name for the executable file, any command line parameters for the tool, and the working directory for the tool. Figure 4–21 shows the settings for TheCableAVR.

The command line parameter for TheCableAVR is simply the TheCableAVR project name (*iom.isp*). TheCableAVR allows you to create a project containing the program files for the FLASH and EEPROM memory areas, as well as fuse bit and lock bit settings. Several other settings are part of the project file, but these particular settings depend on the target device.

Once the tool is added and configured, it appears in the **Tools** menu. Figure 4–22 shows *TheCableAVR.exe* listed in the **Tools** menu. Simply selecting this menu item calls TheCableAVR and programs the chip. TheCableAVR window opens, loads the project file, and begins the automatic programming cycle. If an error occurs during the programming

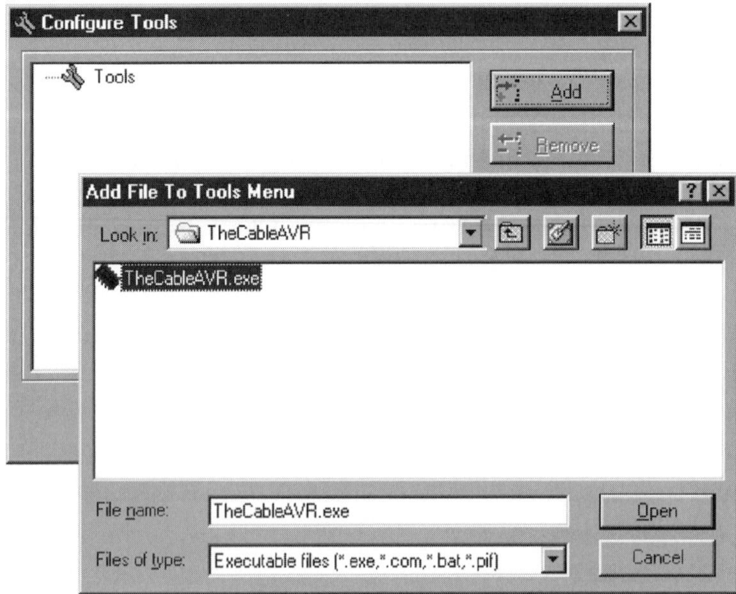

Figure 4–20 *Configure Tools Dialog Box, Add Button Clicked*

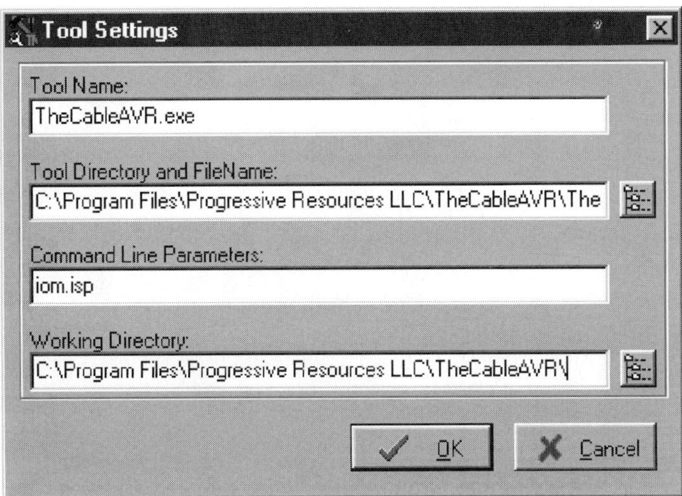

Figure 4–21 *Tool Settings Dialog Box*

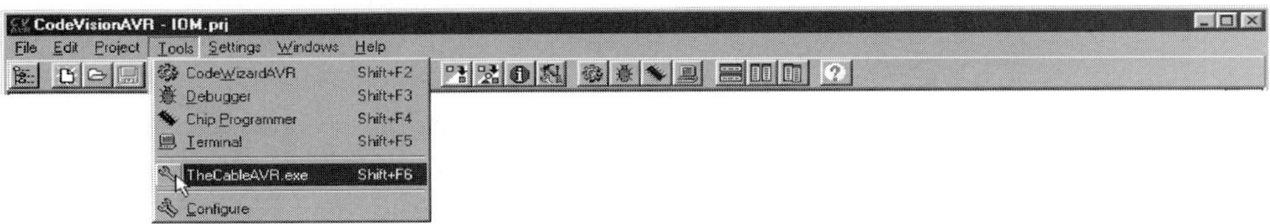

Figure 4–22 *CodeVisionAVR Tools Menu*

cycle, TheCableAVR window remains open for the user to address the error (refer to Figure 4–23). If the programming is successfully completed, TheCableAVR closes.

4.7 CODEWIZARDAVR CODE GENERATOR

When creating a new project, you are given the option of using the CodeWizardAVR to automatically generate a shell of code for your project. The CodeWizardAVR is a great tool for saving time during the startup phase of any project. It automatically generates code to set up timers, communications ports, I/O ports, interrupt sources, and many other features. This saves you from having to dig through the data book for control registers and their required settings.

To effectively use the CodeWizardAVR, you must know how the microcontroller is to be used in your project and you must have a basic knowledge of the hardware of the processor. Chapter 2, **"The Atmel RISC Processors,"** of this book covers many of the architectural features available in the Atmel AVR microcontrollers. Chapter 5, "Project Managment," details good project planning and assists in project planning and preparation for writing software. For now, carefully consider which I/O pins are used as inputs or outputs, what

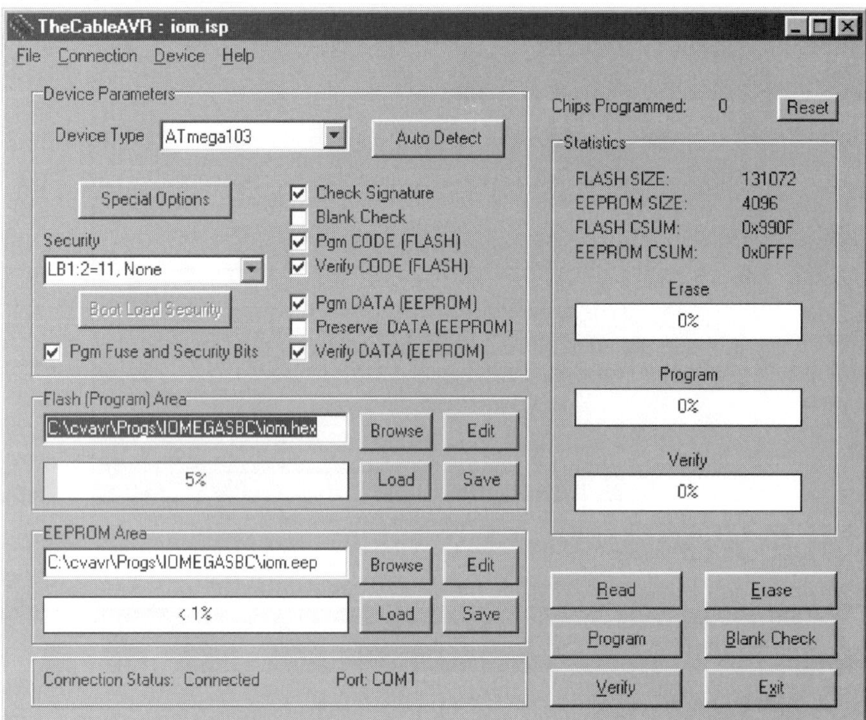

Figure 4–23 *TheCableAVR Software*

baud rates are required by the communication ports, what timers are required, and what interrupt sources are needed.

The CodeWizardAVR places the generated code directly in the source file you specify. This is the starting point for your program. Since you will be editing this file, any of the settings can be modified later, but you will have to look up the new settings in the datasheets for the microprocessor.

The features available through the CodeWizardAVR depend upon the microcontroller selected and the version of CodeVisionAVR in use. (Remember that the evaluation and lite versions of CodeVisionAVR do not support all of the Atmel AVR processors and do not support the same features in the CodeWizardAVR as the standard version. Refer to *http://www.hpinfotech.ro* for information on each version.) Figure 4–24 shows the tabbed frames of options available for the ATTiny22 in the standard version of CodeVisionAVR. The Atmega16 shown in Figure 4–25 has many more tabbed frames available.

This section uses the Atmega16 as the target device and covers some of its basic options available in the standard version of CodeVisionAVR. The example code fragments provided in this section are specific to the Atmega16 device. The initialization code generated by the CodeWizardAVR is placed at the top of *main()*. The interrupt routines are placed above *main()*.

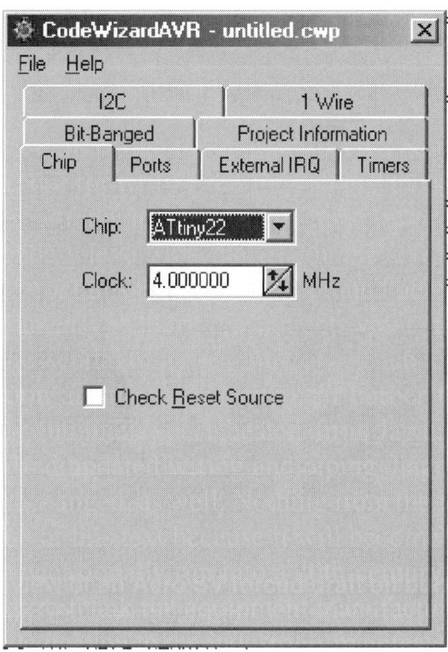

Figure 4–24 *CodeWizardAVR ATTiny22 Options*

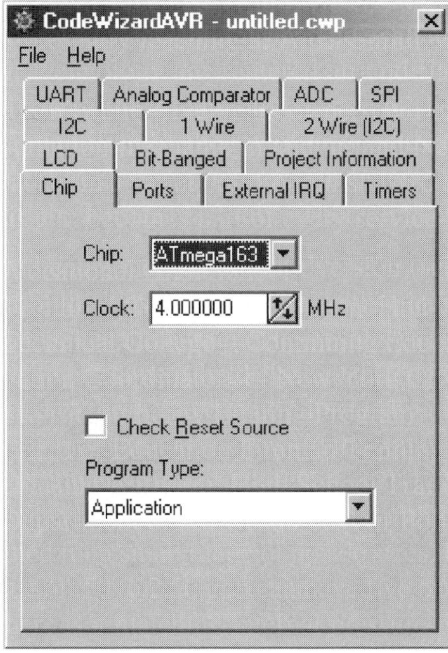

Figure 4–25 *CodeWizardAVR Atmega16 Options*

Note: The CodeWizardAVR is constantly being updated and expanded. This section is not meant to be an exhaustive explanation of its features and possibilities, but rather an overview of the current capabilities of this tool.

4.7.1 CHIP TAB

The Chip tab is the first tab selected as the CodeWizardAVR window opens. The **Chip** list box sets the target device. Changing of the selection in this list box updates the rest of the window. The **Clock** frequency in megahertz is either selected with the up and down arrow buttons or entered directly in the edit field. Selecting the correct frequency is very important for getting the other time-sensitive settings correct.

At the bottom of the Chip tab, the **Check Reset Source** check box enables code to check the reset source. If this is selected, the following code is generated. This code allows the program to take certain actions depending on the source of the last reset. This code is placed at the very beginning of *main()*.

```
// Reset Source checking
if (MCUCSR & 1)
{
   // Power-on Reset
   MCUCSR&=0xE0;
   // Place your code here

}
else if (MCUCSR & 2)
{
   // External Reset
   MCUCSR&=0xE0;
   // Place your code here

}
else if (MCUCSR & 4)
{
   // Brown-Out Reset
   MCUCSR&=0xE0;
   // Place your code here

}
else if (MCUCSR & 8)
{
   // Watchdog Reset
   MCUCSR&=0xE0;
   // Place your code here

}
else if (MCUCSR & 0x10)
```

```
{
    // JTAG Reset
    MCUCSR&=0xE0;
    // Place your code here

};
```

Below the **Check Reset Source** check box is the **Program Type** list box. This selects between generating a boot loader type of program and a normal application. For this example, the application type program is used.

4.7.2 PORTS TAB

The Atmega16 has four I/O ports: A, B, C, and D. In Figure 4–26, Port A is selected. For each pin in Port A, two options are available, data direction and pullup/output value. These settings do two things. First, the Data Direction settings determine the value of the DDRx registers (DDRA for PORTA) and control whether each pin is an input or an output. Second, the Pullup/Output Value setting is used to establish the initial value for the port. If the pins are inputs, the Pullup/Output Value enables or disables the internal pull-up resistors. If the pins are outputs, this setting determines the initial value output on the port.

If, for example, we use the upper half of PORTA for input with pull-up resistors enabled and the lower half of PORTA for output with the value 0x0A as an initial value, the Ports tab settings appear as in Figure 4–26.

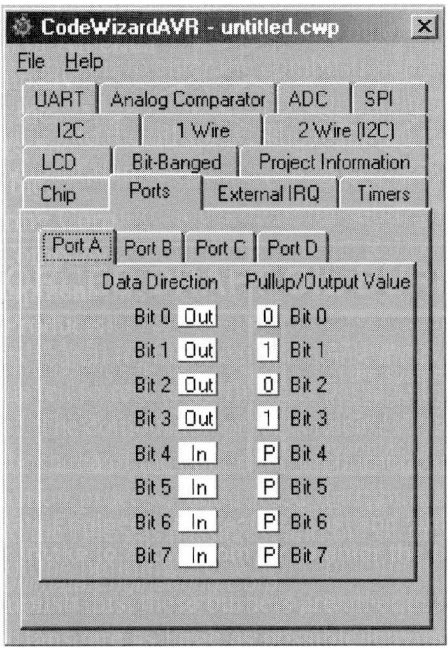

Figure 4–26 *CodeWizardAVR Atmega16 Ports Tab*

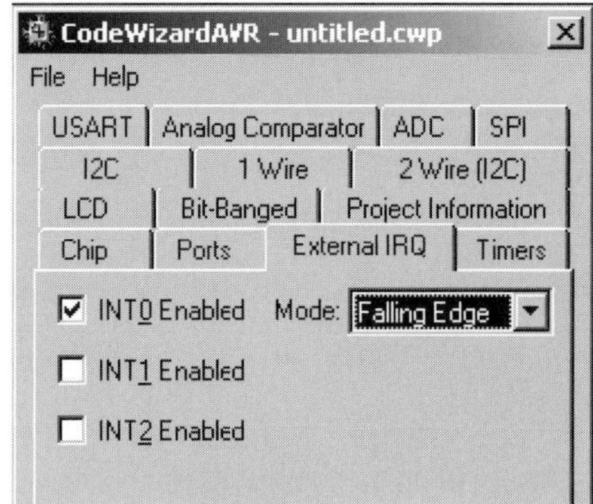

Figure 4–27 *External IRQ Tab*

The generated initialization code for PORTA appears as follows:

```
// Input/Output Ports initialization
// Port A
PORTA=0xFA;
DDRA=0x0F;
```

4.7.3 EXTERNAL IRQ TAB

The External IRQ tab has two check boxes that enable or disable the external interrupts: INT0 and INT1. If the interrupt is enabled, then the **Mode** list box appears and determines what conditions cause the interrupt. For example, INT0 is enabled and is set to interrupt on the falling edge of the signal, as shown in Figure 4–27.

The following interrupt routine is added to the source file:

```
// External Interrupt 0 service routine
interrupt [EXT_INT0] void ext_int0_isr(void)
{
     // Place your code here
}
```

The code generated to enable INT0 interrupt on the falling edge is added to *main()* and appears as

```
// External Interrupt(s) initialization
// INT0: On
// INT0 Mode: Falling Edge
// INT1: Off
// INT2: Off
GICR|=0x40;
MCUCR=0x02;
MCUCSR=0x00;
GIFR=0x40;
```

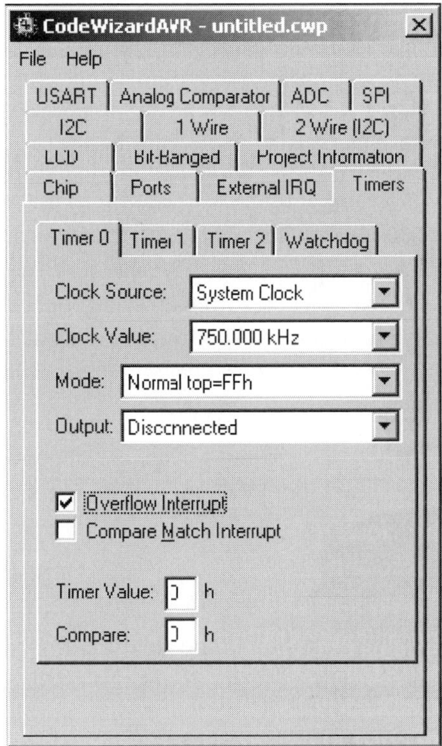

Figure 4–28 *Timers Tab*

4.7.4 TIMERS TAB

The Timers tab, shown in Figure 4–28, is a very versatile tool. It covers Timer 0 through Timer 2 setups and the Watchdog timer setup. For this example, Timer 0 is set up to use the System Clock as its clock source and to run at 750 kHz. When the timer rolls over, an IRQ is generated. The code wizard outputs both the interrupt routine and the code to initialize Timer 0.

The generated initialization code follows:

```
// Timer/Counter 0 initialization
// Clock source: System Clock
// Clock value: 750.000 kHz
// Mode: Normal top=FFh
// OC0 output: Disconnected
TCCR0=0x02;
TCNT0=0x00;
OCR0=0x00;
```

A shell for the interrupt routine called when the timer overflows is placed above *main()*:

```
// Timer 0 overflow interrupt service routine
interrupt [TIM0_OVF] void timer0_ovf_isr(void)
```

```
{
        // Place your code here
}
```

4.7.5 USART TAB

The USART tab has many options once the **Receiver** and/or **Transmitter** check boxes are selected. For each of these check boxes, there is an option to make the receiver or transmitter interrupt driven and a place to specify the length of the associated data buffer. When the CodeWizardAVR generates interrupt-driven USART routines, the standard I/O library routines *putchar()* and *getchar()* are replaced with the code generated in the source file. This allows the *putchar()* and *getchar()* functions to be interrupt driven. This in turn makes all of the functions that use *putchar()* and *getchar()*, such as *printf()* and *scanf()*, interrupt driven as well.

Below the transmitter setup fields is the **USART Baud rate** list box, with many of the common baud rates available. The actual baud rates available depend upon the frequency of the clock used to drive the device. The CodeWizardAVR displays the error percentage associated with the selected baud rate and the system clock. In this example, shown in Figure 4–29, a baud rate of 9600 is chosen, and the baud rate error is 0.2%. This error percentage is small enough not to affect communication with other devices.

The **Communications Parameters** list box is the final field on the USART tab. Select the desired setting, and the CodeWizardAVR generates the needed initialization code for the

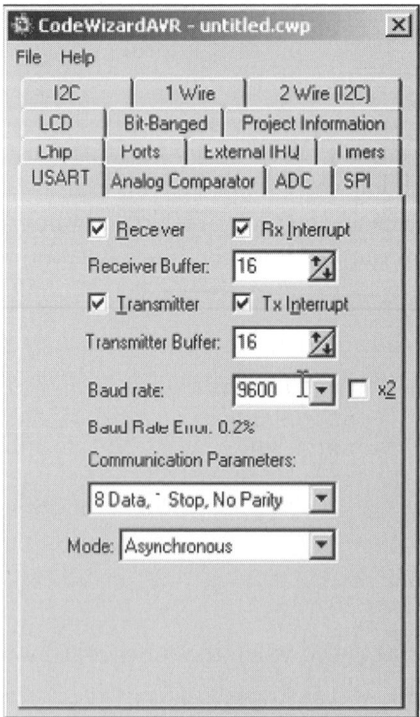

Figure 4–29 *USART Tab*

USART to achieve those settings. In Figure 4–29, the settings are chosen as 8 data bits, 1 stop bit, and no parity.

Interrupt-driven code generated for the USART is placed at the top of the source file. No modifications to this code should be necessary for the use of the standard I/O library routines. (This code is not listed here because it is lengthy. See Figure 4–34.)

The initialization code placed in *main()* for the USART settings in Figure 4–29 is as follows:

```
// USART initialization
// Communication Parameters: 8 Data, 1 Stop, No Parity
// USART Receiver: On
// USART Transmitter: On
// USART Mode: Asynchronous
// USART Baud rate: 9600
UCSRA=0x00;
UCSRB=0xD8;
UCSRC=0x86;
UBRRH=0x00;
UBRRL=0x26;
```

4.7.6 ADC TAB

The ADC (analog-to-digital converter) tab allows you to generate code to initialize the ADC. If the **ADC Interrupt** check box is selected, a shell is generated for the interrupt routine called at the completion of the conversion. The **Volt. Ref.** list box selects the reference voltage for the converter. The final field is the **ADC Clock** list box, which determines the prescaler setting for the converter. The prescaler is set up to divide down the system clock to reach the selected frequency. Figure 4–30 shows a simple setup for the ADC.

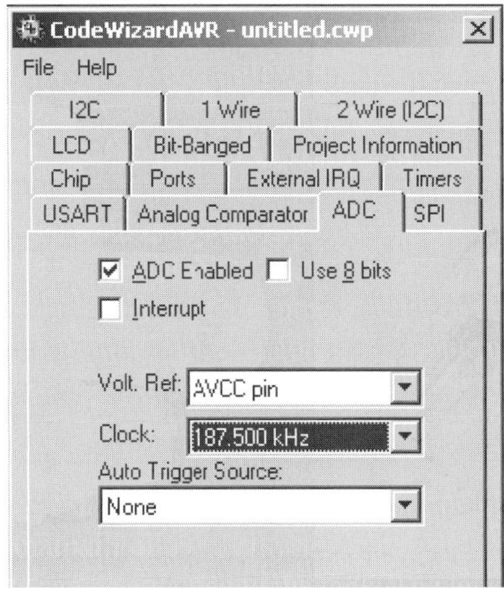

Figure 4–30 *ADC Tab*

The code wizard generates the following code to initialize the ADC and to read values from it according to the settings in Figure 4–30:

```
// ADC initialization
// ADC Clock frequency: 187.500 kHz
// ADC Voltage Reference: AVCC pin
// ADC Auto Trigger Source: None
ADMUX=ADC_VREF_TYPE;
ADCSRA=0x85;

// Read the AD conversion result
unsigned int read_adc(unsigned char adc_input)
{
   ADMUX=adc_input|ADC_VREF_TYPE;
   // Start the AD conversion
   ADCSRA|=0x40;
   // Wait for the AD conversion to complete
   while ((ADCSRA & 0x10)==0);
   ADCSRA|=0x10;
   return ADCW;
}
```

4.7.7 PROJECT INFORMATION TAB

The Project Information tab is just that; it allows you to enter project information and then formats it into the generated source file along with information about the setup of the project. Figure 4–31 shows the project information entered for the example project.

An example of the addition to the source file for the project information tab follows:

```
/*****************************************************
This program was produced by the
CodeWizardAVR V1.24.7d Professional
Automatic Program Generator
© Copyright 1998-2005 Pavel Haiduc, HP InfoTech s.r.l.
http://www.hpinfotech.com
e-mail:officehpinfotech.com

Project : Chapter 4 Example Project
Version : 1.0
Date    : 12/14/2005
Author  : Sally
Company : Sally's Source Code
Comments:
ATmega16 Example Project

Chip type           : ATmega16
Program type        : Application
```

```
Clock frequency     : 6.000000 MHz
Memory model        : Small
External SRAM size  : 0
Data Stack size     : 256
*********************************************************/
```

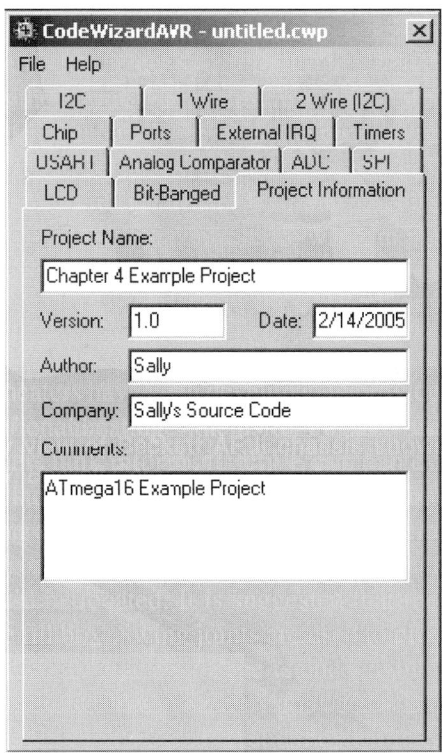

Figure 4–31 *Project Information Tab*

4.7.8 GENERATE SOURCE CODE

Once the CodeWizardAVR options are set for the project, the CodeWizardAVR is ready to generate a source file. Select the **File|Generate, Save and Exit** menu item (refer to Figure 4–32).

This opens several browser windows to select the name of the source file, the name of the project file, and the name of the code wizard file. Once the appropriate file names are entered, the project is created and opened. CodeVisionAVR then appears somewhat as in Figure 4–33. Figure 4–34 lists the source code file that is created with the settings in the previous parts of this section.

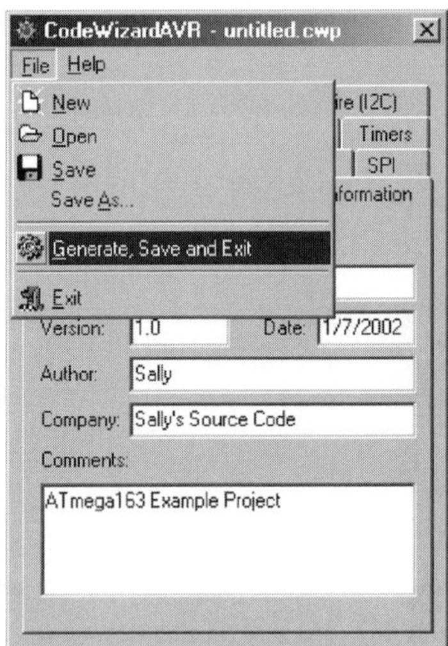

Figure 4–32 *File|Generate, Save and Exit Menu Item*

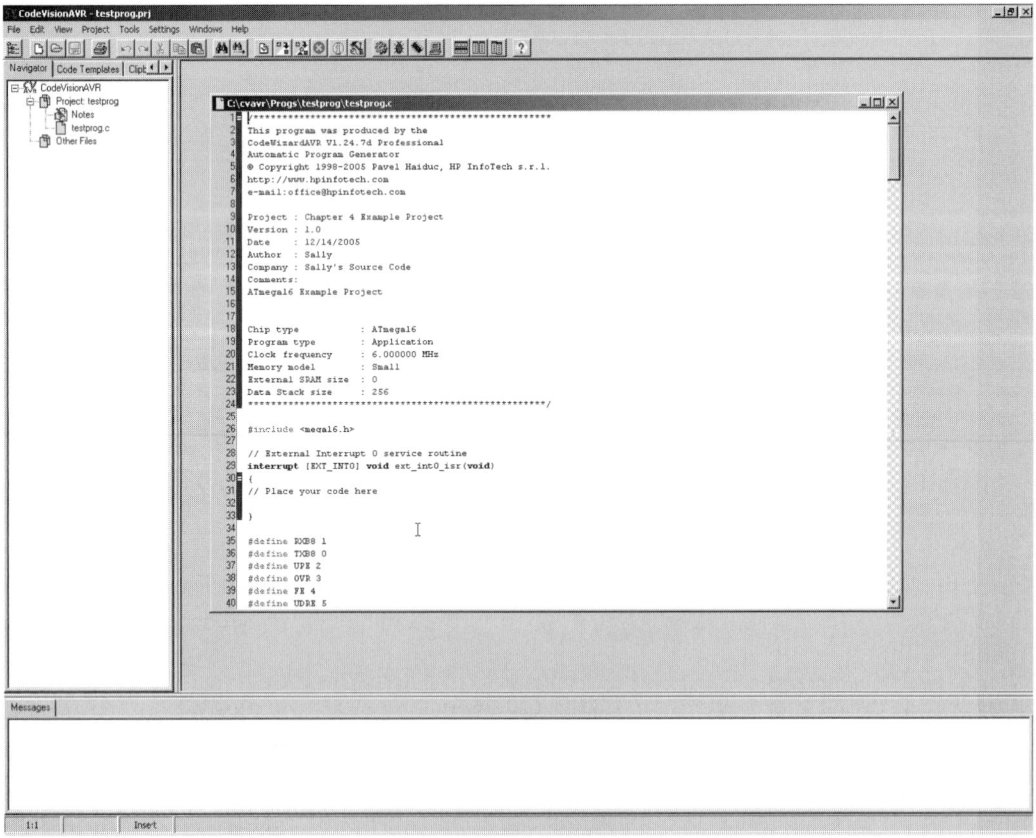

```
/*****************************************************
This program was produced by the
CodeWizardAVR V1.24.7d Professional
Automatic Program Generator
© Copyright 1998-2005 Pavel Haiduc, HP InfoTech s.r.l.
http://www.hpinfotech.com
e-mail:office@hpinfotech.com

Project : Chapter 4 Example Project
Version : 1.0
Date    : 12/14/2005
Author  : Sally
Company : Sally's Source Code
Comments:
ATmega16 Example Project


Chip type           : ATmega16
Program type        : Application
Clock frequency     : 6.000000 MHz
Memory model        : Small
External SRAM size  : 0
Data Stack size     : 256
*****************************************************/

#include <mega16.h>

// External Interrupt 0 service routine
interrupt [EXT_INT0] void ext_int0_isr(void)
{
// Place your code here

}

#define RXB8 1
#define TXB8 0
#define UPE 2
#define OVR 3
#define FE 4
#define UDRE 5
```

Figure 4–33 *New Project Created by the CodeWizardAVR*

```
/*****************************************************
This program was produced by the
CodeWizardAVR V1.24.7d Professional
Automatic Program Generator
© Copyright 1998-2005 Pavel Haiduc, HP InfoTech s.r.l.
http://www.hpinfotech.com
e-mail:office@hpinfotech.com

Project : Chapter 4 Example Project
Version : 1.0
Date    : 12/14/2005
Author  : Sally
Company : Sally's Source Code
Comments:
ATmega16 Example Project


Chip type           : ATmega16
Program type        : Application
Clock frequency     : 6.000000 MHz
Memory model        : Small
External SRAM size  : 0
Data Stack size     : 256
*****************************************************/

#include <mega16.h>

// External Interrupt 0 service routine
interrupt [EXT_INT0] void ext_int0_isr(void)
{
// Place your code here

}

#define RXB8 1
#define TXB8 0
#define UPE 2
#define OVR 3
#define FE 4
#define UDRE 5
#define RXC 7

#define FRAMING_ERROR (1<<FE)
#define PARITY_ERROR (1<<UPE)
#define DATA_OVERRUN (1<<OVR)
#define DATA_REGISTER_EMPTY (1<<UDRE)
#define RX_COMPLETE (1<<RXC)
```

Figure 4–34 *Source Code Created by the CodeWizardAVR (Continues)*

```
// USART Receiver buffer
#define RX_BUFFER_SIZE 16
char rx_buffer[RX_BUFFER_SIZE];

#if RX_BUFFER_SIZE<256
unsigned char rx_wr_index,rx_rd_index,rx_counter;
#else
unsigned int rx_wr_index,rx_rd_index,rx_counter;
#endif

// This flag is set on USART Receiver buffer overflow
bit rx_buffer_overflow;

// USART Receiver interrupt service routine
interrupt [USART_RXC] void usart_rx_isr(void)
{
char status,data;
status=UCSRA;
data=UDR;
if ((status & (FRAMING_ERROR | PARITY_ERROR | DATA_OVERRUN))==0)
   {
   rx_buffer[rx_wr_index]=data;
   if (++rx_wr_index == RX_BUFFER_SIZE) rx_wr_index=0;
   if (++rx_counter == RX_BUFFER_SIZE)
      {
      rx_counter=0;
      rx_buffer_overflow=1;
      };
   };
}

#ifndef _DEBUG_TERMINAL_IO_
// Get a character from the USART Receiver buffer
#define _ALTERNATE_GETCHAR_
#pragma used+
char getchar(void)
{
char data;
while (rx_counter==0);
data=rx_buffer[rx_rd_index];
if (++rx_rd_index == RX_BUFFER_SIZE) rx_rd_index=0;
#asm("cli")
--rx_counter;
#asm("sei")
return data;
}
#pragma used-
#endif
```

Figure 4–34 *Source Code Created by the CodeWizardAVR (Continues)*

```
// USART Transmitter buffer
#define TX_BUFFER_SIZE 16
char tx_buffer[TX_BUFFER_SIZE];

#if TX_BUFFER_SIZE<256
unsigned char tx_wr_index,tx_rd_index,tx_counter;
#else
unsigned int tx_wr_index,tx_rd_index,tx_counter;
#endif

// USART Transmitter interrupt service routine
interrupt [USART_TXC] void usart_tx_isr(void)
{
if (tx_counter)
   {
   --tx_counter;
   UDR=tx_buffer[tx_rd_index];
   if (++tx_rd_index == TX_BUFFER_SIZE) tx_rd_index=0;
   };
}

#ifndef _DEBUG_TERMINAL_IO_
// Write a character to the USART Transmitter buffer
#define _ALTERNATE_PUTCHAR_
#pragma used+
void putchar(char c)
{
while (tx_counter == TX_BUFFER_SIZE);
#asm("cli")
if (tx_counter || ((UCSRA & DATA_REGISTER_EMPTY)==0))
   {
   tx_buffer[tx_wr_index]=c;
   if (++tx_wr_index == TX_BUFFER_SIZE) tx_wr_index=0;
   ++tx_counter;
   }
else
   UDR=c;
#asm("sei")
}
#pragma used-
#endif

// Standard Input/Output functions
#include <stdio.h>

// Timer 0 overflow interrupt service routine
interrupt [TIM0_OVF] void timer0_ovf_isr(void)
```

Figure 4–34 *Source Code Created by the CodeWizardAVR (Continues)*

```
{
// Place your code here

}

#define ADC_VREF_TYPE 0x40

// Read the AD conversion result
unsigned int read_adc(unsigned char adc_input)
{
ADMUX=adc_input|ADC_VREF_TYPE;
// Start the AD conversion
ADCSRA|=0x40;
// Wait for the AD conversion to complete
while ((ADCSRA & 0x10)==0);
ADCSRA|=0x10;
return ADCW;
}

// Declare your global variables here

void main(void)
{
// Declare your local variables here

// Reset Source checking
if (MCUCSR & 1)
   {
   // Power-on Reset
   MCUCSR&=0xE0;
   // Place your code here

   }
else if (MCUCSR & 2)
   {
   // External Reset
   MCUCSR&=0xE0;
   // Place your code here

   }
else if (MCUCSR & 4)
   {
   // Brown-Out Reset
   MCUCSR&=0xE0;
   // Place your code here

   }
```

Figure 4–34 *Source Code Created by the CodeWizardAVR (Continues)*

```
else if (MCUCSR & 8)
   {
   // Watchdog Reset
   MCUCSR&=0xE0;
   // Place your code here

   }
else if (MCUCSR & 0x10)
   {
   // JTAG Reset
   MCUCSR&=0xE0;
   // Place your code here

   };

// Input/Output Ports initialization
// Port A initialization
// Func7=In Func6=In Func5=In Func4=In Func3=Out Func2=Out Func1=Out
// Func0=Out
// State7=P State6=P State5=P State4=P State3=1 State2=0 State1=1
// State0=0
PORTA=0xFA;
DDRA=0x0F;

// Port B initialization
// Func7=In Func6=In Func5=In Func4=In Func3=In Func2=In Func1=In
// Func0=In
// State7=T State6=T State5=T State4=T State3=T State2=T State1=T
// State0=T
PORTB=0x00;
DDRB=0x00;

// Port C initialization
// Func7=In Func6=In Func5=In Func4=In Func3=In Func2=In Func1=In
// Func0=In
// State7=T State6=T State5=T State4=T State3=T State2=T State1=T
// State0=T
PORTC=0x00;
DDRC=0x00;

// Port D initialization
// Func7=In Func6=In Func5=In Func4=In Func3=In Func2=In Func1=In
// Func0=In
// State7=T State6=T State5=T State4=T State3=T State2=T State1=T
// State0=T
PORTD=0x00;
DDRD=0x00;
```

Figure 4–34 *Source Code Created by the CodeWizardAVR (Continues)*

```
// Timer/Counter 0 initialization
// Clock source: System Clock
// Clock value: 750.000 kHz
// Mode: Normal top=FFh
// OC0 output: Disconnected
TCCR0=0x02;
TCNT0=0x00;
OCR0=0x00;

// Timer/Counter 1 initialization
// Clock source: System Clock
// Clock value: Timer 1 Stopped
// Mode: Normal top=FFFFh
// OC1A output: Discon.
// OC1B output: Discon.
// Noise Canceler: Off
// Input Capture on Falling Edge
// Timer 1 Overflow Interrupt: Off
// Input Capture Interrupt: Off
// Compare A Match Interrupt: Off
// Compare B Match Interrupt: Off
TCCR1A=0x00;
TCCR1B=0x00;
TCNT1H=0x00;
TCNT1L=0x00;
ICR1H=0x00;
ICR1L=0x00;
OCR1AH=0x00;
OCR1AL=0x00;
OCR1BH=0x00;
OCR1BL=0x00;

// Timer/Counter 2 initialization
// Clock source: System Clock
// Clock value: Timer 2 Stopped
// Mode: Normal top=FFh
// OC2 output: Disconnected
ASSR=0x00;
TCCR2=0x00;
TCNT2=0x00;
OCR2=0x00;

// External Interrupt(s) initialization
// INT0: On
// INT0 Mode: Falling Edge
// INT1: Off
// INT2: Off
GICR|=0x40;
```

Figure 4–34 *Source Code Created by the CodeWizardAVR (Continues)*

```
MCUCR=0x02;
MCUCSR=0x00;
GIFR=0x40;

// Timer(s)/Counter(s) Interrupt(s) initialization
TIMSK=0x01;

// USART initialization
// Communication Parameters: 8 Data, 1 Stop, No Parity
// USART Receiver: On
// USART Transmitter: On
// USART Mode: Asynchronous
// USART Baud rate: 9600
UCSRA=0x00;
UCSRB=0xD8;
UCSRC=0x86;
UBRRH=0x00;
UBRRL=0x26;

// Analog Comparator initialization
// Analog Comparator: Off
// Analog Comparator Input Capture by Timer/Counter 1: Off
ACSR=0x80;
SFIOR=0x00;

// ADC initialization
// ADC Clock frequency: 187.500 kHz
// ADC Voltage Reference: AVCC pin
// ADC Auto Trigger Source: None
ADMUX=ADC_VREF_TYPE;
ADCSRA=0x85;

// Global enable interrupts
#asm("sei")

while (1)
      {
      // Place your code here

      };
}
```

Figure 4–34 *Source Code Created by the CodeWizardAVR (Continued)*

4.8 TERMINAL TOOL

The terminal tool is intended for debugging embedded systems, which employ serial communication (RS-232, RS-422, RS-485). The **Tools|Terminal** menu command or the **Terminal** button on the toolbar opens the Terminal window, as shown in Figure 4–35.

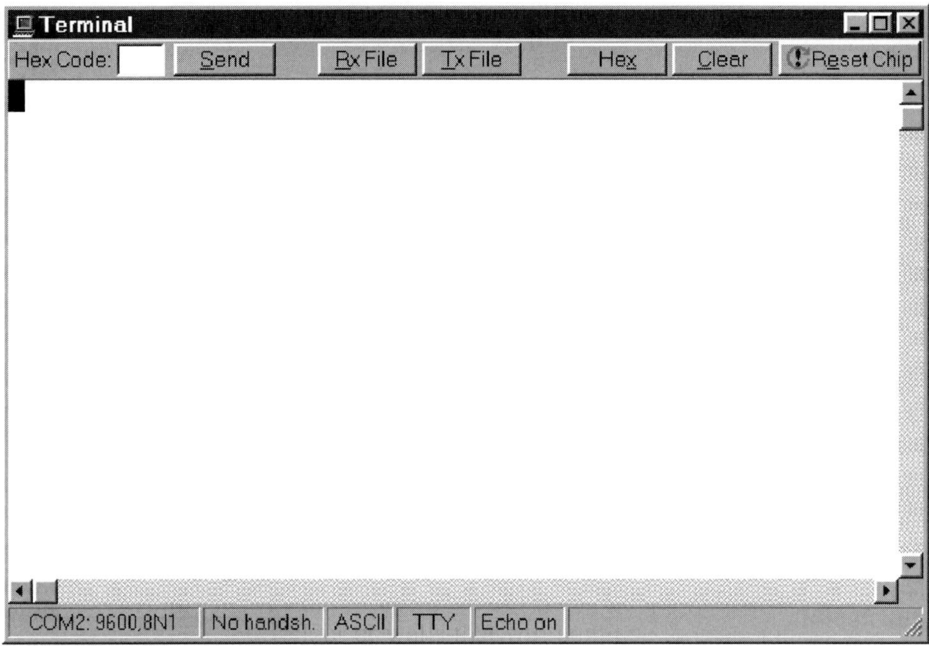

Figure 4–35 *CodeVisionAVR Terminal Window*

The received characters are displayed in the main body of the Terminal window in ASCII or hexadecimal format. The display mode of these characters is toggled by the **Hex/ASCII** button. The received characters are saved to a file with the **Rx File** button.

Any characters typed in the Terminal window are transmitted through the PC serial port. The characters entered can be deleted by the **Backspace** key. Clicking the **Send** button causes the Terminal to transmit a character whose hexadecimal ASCII code value is specified in the **Hex Code** edit box. A file is transmitted by clicking the **Tx File** button and selecting the appropriate file.

Clicking the **Reset Chip** button resets the AVR chip on a STK200+/300, VTEC-ISP, DT006, ATCPU, Mega2000, or other similar development board.

At the bottom of the Terminal window, a status bar displays the following:

- Selected PC communication port
- Communication parameters
- Handshaking mode
- Received characters display mode
- Type of emulated terminal
- State of the transmitted characters echo setting

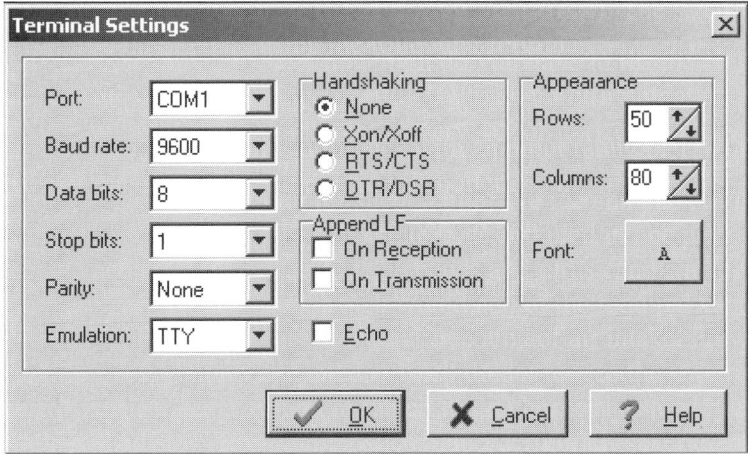

Figure 4–36 *CodeVisionAVR Terminal Settings Dialog Box*

The settings for the Terminal window are modified by the **Settings|Terminal** menu item. The Terminal Settings dialog box is shown in Figure 4–36. The Terminal window must be closed before you modify the settings. The current settings are shown in the bottom status bar on the Terminal window.

4.9 THE ATMEL AVR STUDIO DEBUGGER

Both software simulators and in-circuit-emulators are useful for debugging software and are sometimes referred to as *debuggers*. Most include some method of starting, stopping, and stepping through a program. Most also allow the user to peek at and even modify registers and variable values. They are invaluable in helping you find out why your program is not operating as you expect it to.

CodeVisionAVR is designed to work in conjunction with a particular debugger, the Atmel AVR Studio debugger version 4.06 or later. AVR Studio is a software simulator for Atmel's AVR parts. AVR Studio also supports in-circuit emulators such as Atmel's ICE 200, ICE 50, and JTAG ICE. AVR Studio is available free from Atmel's Web site at *http://www.atmel.com/*.

While it is beyond the scope of this book to go into great detail about the operation of AVR Studio, there are some basic operations that are useful to every programmer: loading a C file for debugging; starting, stopping, and stepping through a program; setting and clearing breakpoints; and viewing and modifying variables and machine state. You are encouraged to utilize the help files included with AVR Studio for more advanced operations.

Debuggers also usually require specially formatted files. These files contain information about the format of the C file, variable names, function names, and other such information. The files used by AVR Studio are denoted with a *.cof* extension. CodeVisionAVR is designed to create this type of file and to launch AVR Studio from a toolbar button.

4.9.1 CREATE A COFF FILE FOR AVR STUDIO

A *COFF* file is required to perform C source-level debugging in AVR Studio. To create this type of file in CodeVisionAVR, select the **Project|Configure** menu item to open the Configure Project dialog box. Select the C Compiler tab; in the lower right corner is the **File Output Format(s)** list box. Select the COFF, ROM, HEX, EEP list item. Click **OK** to close the window and save the setting.

4.9.2 LAUNCH AVR STUDIO FROM CODEVISIONAVR

CodeVisionAVR has two methods for launching a debugger application, a menu item and a toolbar command button. Before using either method, however, you need to specify the path to the debugger. To do this, select the **Settings|Debugger** menu item and click the **Browse** command button in the window that appears. Browse to the debugger and select it. The default path for AVR Studio is

```
C:\Program Files\Atmel\AVR Tools\AvrStudio4\AVRStudio.exe
```

However, it might be placed differently in your system. You can perform a search for *AVRStudio.exe* to locate the file if necessary. Once the file is selected, simply use the **Tools|Debugger** menu item, the command button in the toolbar with the bug icon, or the keys **Shift+F3** to launch the application from CodeVisionAVR.

4.9.3 OPEN A FILE FOR DEBUG

AVR Studio refers to the debugger source files as *Object Files*. To open the *.cof* file created by CodeVisionAVR for debugging, select the **File|Open Objectfile** menu item and browse to your file. Select your *.cof* file and click **OK.** This opens the Select Device and Debug Platform dialog box. In the left column, select the platform you are using as your debugging platform (this example uses the simulator). In the right column, select the device you are simulating or emulating (this example uses the Atmega16). Click **Finish** to close the dialog box and open the debugger window.

As the program opens, it is displayed as C code with the debugger active. The **Run** button is in the upper right part of the tool bars and appears as a page of text with an arrow running down the right side of it. The **Stop** button is located to the right of the **Run** button and appears as a "pause" button (two vertical bars). The **Debug** menu can also be used to run and break the program. Yellow arrows appear in the left margin of the code windows when the program breaks; these denote where the code execution halted when **Break** was clicked.

When the source file changes, because you found what was wrong and fixed it, you need to reload the debugger file. Usually, AVR Studio detects when this happens and prompts you to reload the file. Follow the prompts, or simply re-open the file using the **File** menu.

4.9.4 START, STOP, AND STEP

Before we dive into this very important debugger topic, we need to review one point about AVR Studio: It has two modes, editing and debugging. Under the **Debug** menu are two items that tell which mode you are in, **Start Debugging** and **Stop Debugging.** Only one of these items is available at a time. If **Start Debugging** is available (not disabled, grayed out), select it to enter debugging mode. If **Start Debugging** is disabled, then you are already in debugging mode. None of the debugging menu options are available unless you are in debugging mode.

The buttons to start (run), stop, and step the debugger are located in the upper right corner of AVR Studio's toolbars. The buttons show their function when your mouse is held over them for a moment. Equivalent commands are also available from the **Debug** menu. The available commands are run, stop, reset, show next statement, step into, step over, step out of, run to cursor, and autostep. Many of these are intuitive, but we should review their functionality. Table 4–1 provides a quick guide as to their respective uses.

Command	***Description***	***Function Key***
Run	Starts or resumes execution of the program. The program will be executed until the user stops it or a break point is encountered.	F5
Break	Stops or breaks execution of the program. When execution is halted, all information in all windows is updated.	Ctrl+F5
Reset	Resets the execution target. If currently executing, the program is halted. If executing in source code mode, the program runs until the first source statement is encountered. (The startup code is executed.)	Shift+F5
Show Next Statement	Sets the yellow marker at the actual program counter location and focusses the window.	(None)
Step Into	If in disassembly mode, one line of assembler is executed. If in source code mode, one line of source is executed.	F11
Step Over	Executes one instruction. If the instruction is a function call, the function is executed as well before halting.	F10
Step Out Of	Executes instructions until the current function is complete or a breakpoint is encountered.	Shift+F11
Run to Cursor	Executes instructions until the program has reached the instruction indicated by the cursor in the source file.	Ctrl+F10
AutoStep	Executes instructions one at a time until a breakpoint is encountered or the user issues a break command. The delay between instructions is determined in the **Debug Options** menu. All information in all windows is updated after each instruction.	(None)

Table 4–1 *Debugger Command Buttons*

4.9.5 SET AND CLEAR BREAKPOINTS

A very useful tool in most debuggers is the ability to set and clear breakpoints. These are points in your program where you want execution to halt. Once it is halted, you can review variable or register values, or you can choose to start single-stepping the program to determine exactly what it is doing.

To set or clear breakpoints in AVR Studio, right-click on the desired line of source code. A shortcut menu appears with the option of **Toggle Breakpoint**. Select **Toggle Breakpoint,** and a large red dot either appears or disappears in the left margin of the code window.

Once the breakpoints are set, run your program as usual. The program halts before executing the selected line. After the program is halted, all of the information in all of the windows is updated.

4.9.6 VIEW AND MODIFY REGISTERS AND VARIABLES

In the **View** menu are listed the items **Watch, Memory Window,** and **Register.** These menu items bring up windows to display variables, various parts of memory, and registers, respectively. The information in these windows is not updated while the program is running. Once program execution is halted, by either a breakpoint or the user commanding a break, the information in all of the windows is updated.

The Watch window defaults to appearing in the lower right corner. To add or remove items, you can either right-click in the window and utilize the shortcut menu or double-click on an empty Name field and type in the name of the variable you want to watch. To change the value of the variable, double-click the Value column and type in the new value.

The Memory window overlays the other active windows and allows you to view various memory spaces, including EEPROM, Data, I/O, Program, and Register. Here, you can directly watch memory being modified. Double-clicking a value allows you to enter a new value, and the Address box at the top allows you to specify the address of the space you want to view.

Registers 0 through 31 can be viewed in many ways. To simply see them all displayed at once, select the **View|Registers** menu item. Double-clicking a particular register allows you to modify its value.

4.9.7 VIEW AND MODIFY THE MACHINE STATE

Sometimes it is very useful to look at the exact state of the microprocessor. What are the counters doing? Where is the stack pointer? What is going on with the I/O? Along the left side of the AVR Studio screen is a Workspace frame. Along the bottom of this frame are three tabs: Project, I/O, and Info. The I/O tab gives us the capability of looking at the machine state and modifying it.

There are several items listed with a + to the left. Clicking the + expands the list. Expanding the Processor item presents us with the program counter, the stack pointer, the cycle counter, the X, Y, and Z registers, and some other useful tidbits of information. Expanding the I/O item presents us with a list of the features of the microprocessor, such as the state of the I/O port, the USART, the SPI, the timer, and so on. For each of these, the state is shown as well as the associated control registers. Now you can really see what is happening. The port C, pin 2 is low even though you are driving it high? Did you remember to set the pin as an output? Your timer is not generating interrupts—is it even running?

As with the other views into the processor's state and memory, these windows are ***not*** updated while the debugger is executing the program. Once the debugger halts, the information in these windows is updated.

4.10 CHAPTER SUMMARY

This chapter has covered many of the features of the CodeVisionAVR C Compiler and integrated development environment (IDE). The IDE controls the projects and has a powerful code editor. The IDE gives you access to the compiler, the assembler, and their associated options. It also provides tools such as the chip programmer, CodeWizardAVR, and the terminal tool.

You should now be able to create projects in CodeVisionAVR, add source files to them, and make them into executable files. With the proper programming cable, you should also be able to program a target device with the built-in chip programmer. You should understand the basic operation of the CodeWizardAVR code generator and the terminal tool available in CodeVisionAVR.

Finally, we covered a few basics of the AVR Studio debugger, but we barely scratched the surface of the capabilities of this tool. You should now be familiar with loading, starting, stopping, and stepping a program in AVR Studio. Also, you should be familiar with setting and clearing breakpoints and viewing and modifying variables and registers. You are encouraged to download this free program from Atmel, use the help files provided with it, and experiment.

4.11 EXERCISES

1. For the project shown in Figure 4–37, what is the project name? Also list the source file names for the project (Section 4.3).
2. In the compiler options, what type of memory model should be used for optimum code speed and size? When should the other memory model be used (Section 4.4)?
3. Name the two places error and warning messages are displayed after a compile (Section 4.5).
4. Describe the difference between compiling a project and making a project (Section 4.5).
5. Fill in the blanks for the initialization code generated for PORTB if the CodeWizardAVR Ports tab appears as in Figure 4–38 (Section 4.7).

```
// Input/Output Ports Initialization
// PORT B
PORTB = 0x__
______ = 0x3E
```

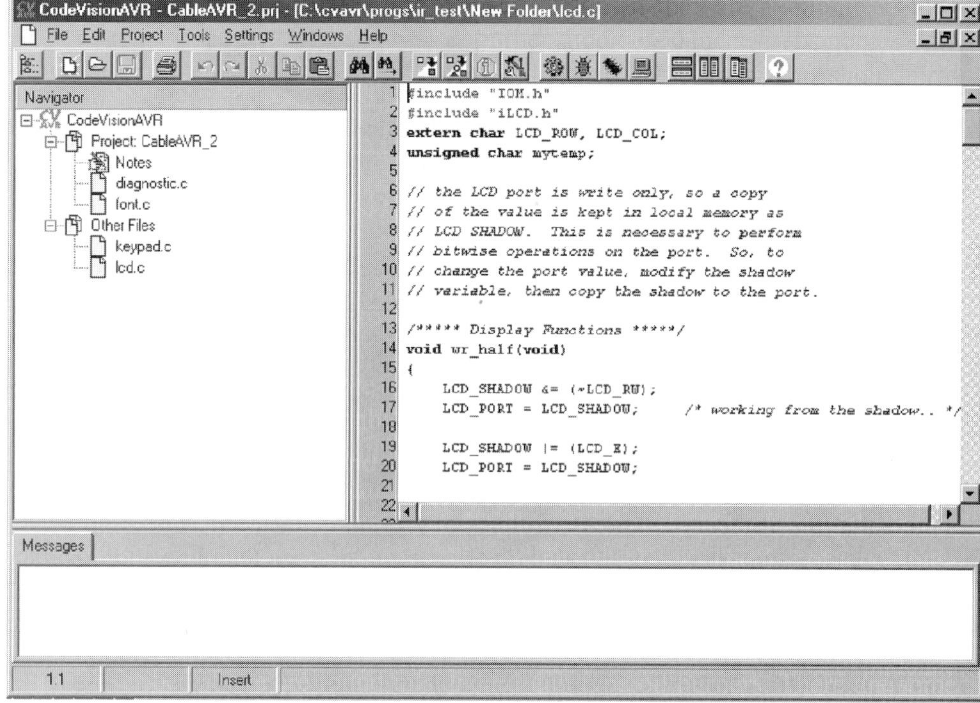

Figure 4–37 *Project Name?*

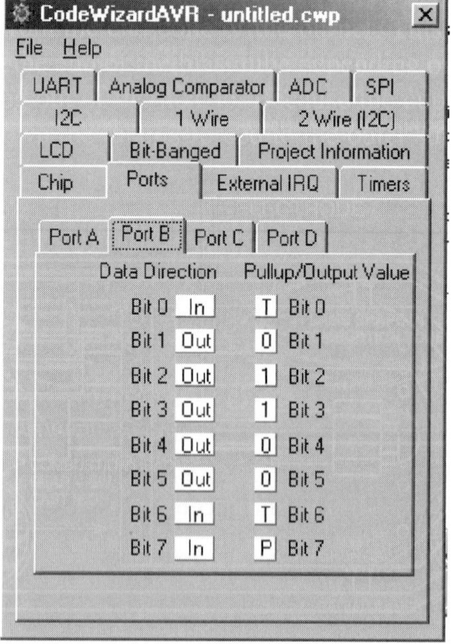

Figure 4–38 *CodeWizardAVR Ports Tab*

EXERCISE

4.12 LABORATORY ACTIVITIES

1. Create a new project by using the CodeWizardAVR to generate the source code shell. The project should be configured as follows:
   - USART: 9600 baud, 8 data bits, 1 stop bit, no parity; transmit and receive enabled
   - PORTA: All inputs with pull-ups turned on
   - PORTC: All outputs
   - Timer 1: Cause an interrupt to occur every 5 ms
2. Modify the project generated in Activity 1 to send the value present at PORTA to the USART once per second. Do not send the value from an interrupt routine.
3. Use the terminal tool available in CodeVisionAVR to read the data read from PORTA in the program in Activity 2 in hexadecimal notation.
4. Send data, in hexadecimal notation, from the terminal tool to the target device. Place the value received by the USART on the target device on PORTC. Verify that the correct value is present at PORTC.
5. Use the CodeWizardAVR to generate a shell to turn on an LED when a falling edge occurs on external interrupt 0 and to turn it off when a rising edge occurs on external interrupt 1. Complete and test the program.

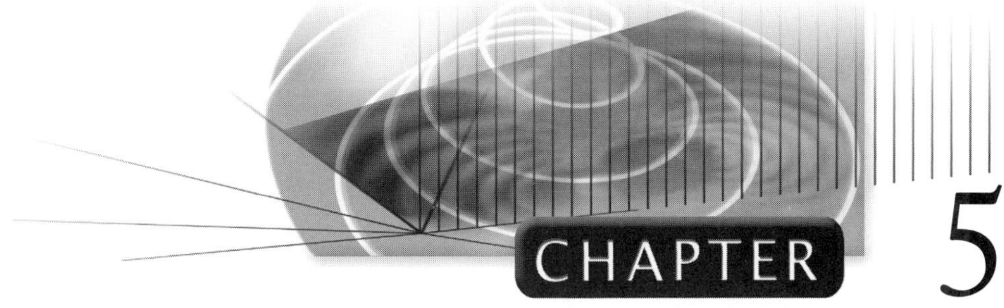

Project Development

5.1 OBJECTIVES

The sole objective of this chapter is for the student (you) to be able to develop electronic projects that involve microcontrollers for either commercial or personal purposes.

5.2 INTRODUCTION

Electronic products involving microcontrollers are most efficiently developed using an orderly approach to the *process,* using a progression of steps, from conception through accomplishment, that virtually always results in success. In this chapter, the process will be described and then applied to a real project to further demonstrate how the process works.

5.3 CONCEPT DEVELOPMENT PHASE

Every project is based on an idea or concept, which comes from any need—somebody "wants one." The need may be to fill a gap in a product market, to improve a production process, to meet a course requirement, or simply to create something that has not been created before. Because projects are often started to satisfy a need, the original description of the project is sometimes called a problem statement or a need statement.

5.4 PROJECT DEVELOPMENT PROCESS STEPS

The steps you should follow in the process of developing a project are as follows:

1. Definition phase
2. Design phase
3. Test definition phase
4. Build and test the prototype hardware phase

5. System integration and software development phase
6. System test phase
7. Celebration phase

5.4.1 DEFINITION PHASE

Defining the project has the objective of clearly stating what the project is to accomplish. This step involves specifying what the device is to do, researching to ascertain that the project is, in fact, feasible, developing a list of specifications that fully describe the function of the project, and, in a commercial environment, providing a formal proposal to go ahead with the project. This step is sometimes called the *feasibility study.*

The purpose of the research conducted during the definition phase is to ensure that the project can, in fact, be accomplished. The early portion of this research will result in a coarse or macro-level block diagram; an example is shown in Figure 5–1. This block diagram will show, in somewhat general terms, the circuits that will make up the final project. If the research shows that certain portions of the project are very sophisticated or are a new use of an existing technology, some of the circuitry can be simulated or built and tested as a "proof of concept" that the circuitry can be used in this project. The goal of the research is for the designer to be reasonably certain that the project will work.

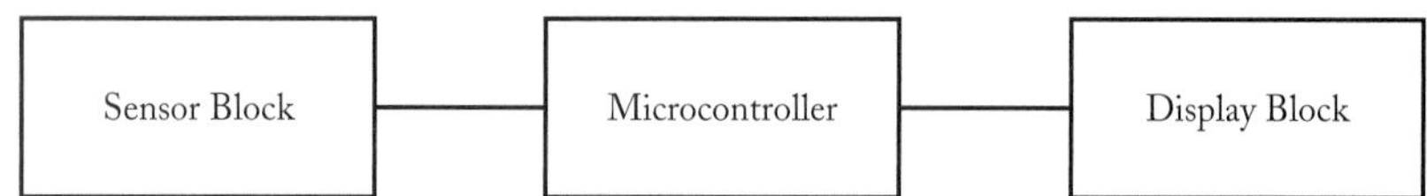

Figure 5–1 *Basic Block Diagram*

At the end of the definition phase, a complete set of specifications for the project is developed. These include electrical specifications as well operating specifications that detail the operation and human interface to the project. Although spending time on the human-factors aspect of the project is usually against the engineer's instincts and inclinations, it is important to ensure that the finished project not only works but also fulfills its purpose in terms of operation. In many systems, a lot more effort goes into creating an operations specification than goes into creating an electrical or functional specification.

In a commercial environment, a Project Proposal is written to summarize the definition phase of the project. In addition to summarizing the research and feasibility testing and providing a written list of the project specifications, the proposal will include an anticipated budget for the project and a schedule for completion of the project. The purpose of this proposal is to give the customer (or upper management) confidence that investing money in this project will result in a successful outcome. Figure 5–2 presents one possible outline for a project proposal.

Project Title

A. Description of project.

This section will fully describe the project, its overall function, the reason or rationale to complete the project, and an indication of the benefits expected to arise from completion of the project. The purpose of this section is for the reader to understand what the project is and why it is important.

B. Summary of Research.

The project research completed during this phase of the project is summarized, including research done on the market and/or competing products, on similar or related projects or products, as well as the results of any proof-of-concept testing completed. The purpose of this section is to show that the project has been thoroughly researched to assure that the relationship of this project to other, existing, projects is clear.

C. Block Diagram.

The overall, macro-level, block diagram is presented in this section. Following the block diagram is a discussion of the *function* of each block *as it relates to the project's overall function*. The purpose of this section is to give the reader confidence that the project can be completed successfully.

D. Project Budget.

A detailed budget is presented here to show all the expected expenditures, including labor and materials, associated with this project.

E. Project Schedule.

The project schedule shows the milestones for starting and completing each block, for starting and completing each phase of the development process, for review or reporting points in the project, and for the final completion and presentation of the project. The purpose of this section is to show, realistically, the time it will take to complete the project and to give the reader confidence that the project has been thoroughly planned out.

F. Appendix.

The appendix includes the bibliography developed during research and any other information that is too large to be included in the text.

Figure 5–2 *Proposal Outline*

The definition phase is a relatively short but very critical step that consumes approximately 10 to 15 percent of the total project time. This step is crucial in guaranteeing that the completed project will do what it is designed to do.

5.4.2 DESIGN PHASE

The major goal of the design phase of the project is to fill in the macro blocks developed during the definition phase with actual circuitry and to plan out, in flowchart form, the project software. Because the hardware and software of the project are intimately related, the hardware design and the software plan must be accomplished concurrently. When thoroughly done, this step will consume 40 to 50 percent of the total project time. Unfortunately, the following two characteristics seem to reflect the approach of many engineers to the definition phase of the project:

- Designing the project is the step that, more than any other step, determines the success of the project, both in terms of function and efficient development—for *efficient*, read "on time and under budget."
- Designing the project is the step that is most neglected, disliked, and even hated by engineers, who vastly prefer hooking up hardware or writing programs to doing research and paperwork.

Designing the project involves going from the somewhat hazy notion of how the project might work, which was developed during the definition phase of the project, all the way to the completed schematics and software flowcharts. The design phase of the project ***does not*** involve any prototyping of hardware or writing of any software. The process of designing the project is shown separately for hardware and software below.

Hardware development steps:

1. Start with the basic block diagram developed during the definition phase. This block diagram will very likely have blocks that encompass more than one function. For example, if you were developing a device to record automobile parameters such as speed of the vehicle, gasoline consumption, and the deflection of the springs, you would probably have one macro block that says "sensors" with a direct line to the microcontroller. Or you might have a single block for each sensor with a line to the microcontroller. And you might have one more block for a display, as shown in Figure 5–1. The objective of this step is to provide a starting point in the form of blocks that you know will provide the desired results if you can fill in the blocks with appropriate circuitry.
2. Thoroughly research both relevant components and circuits to determine what sort of circuitry and components could be used to fill each block. As you determine what components and circuits you are going to use, you will naturally be breaking up the project blocks into smaller blocks. The objective is to break each block into the smallest pieces that you plan to *individually test*. When complete, each block should be labeled with its input and output voltage levels and/or signals. These signal definitions are then used to develop tests for the individual blocks. Figure 5–3 shows only the speed sensor (one-third of the sensor block in Figure 5–1) broken up into reasonable, testable, blocks.

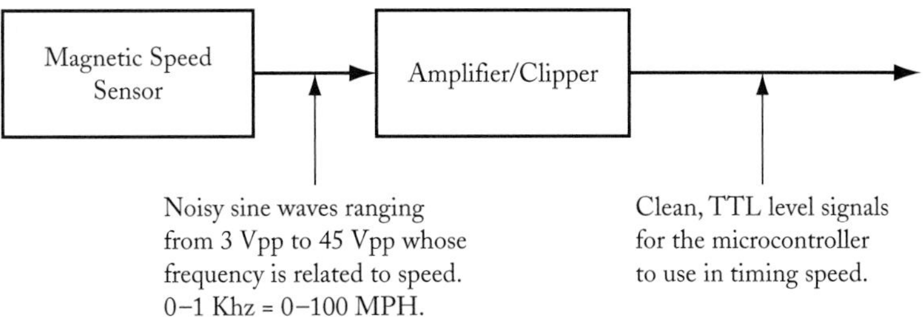

Figure 5–3 *Expanded Sensor Block*

3. Using your research as a basis, develop a tentative circuit schematic for each of the blocks that contain electronic circuitry. Use a circuit simulator such as Spice or Electronic Workbench® to simulate the operation of your circuitry. Modify your circuitry as required until the simulator says it works correctly to fulfill the purpose of the block.
4. Use the individual block circuits to create an overall schematic for the project. Use "cut and paste" techniques or, alternatively, combine the circuit schematics so that errors are not introduced by redrawing the block circuits into a single schematic.

Software development steps:

1. List the tasks to be completed by the software.
2. Prioritize the tasks and determine which are critical and will need to be handled on an interrupt basis and which are less critical. Using our example, timing the speed pulses would probably be considered critical and, hence, a candidate for an interrupt, whereas updating the display would probably be considered a low-priority task that could be completed whenever the processor is not doing more critical tasks.
3. Create a software flowchart or outline for each interrupt function.
4. Create a software flowchart or outline of the program code to complete the project's entire task.

When the project has been thoroughly designed, you will have a detailed block diagram (showing all blocks and signals), a complete set of schematics (as yet untried in hardware), and a set of software flowcharts or outlines.

5.4.3 TEST DEFINITION PHASE

The test definition phase of the project has the goal of ensuring that the project meets all of its specifications and goals (determined in the project definition phase) through the development of a test specification for the project. The test specification is divided into two parts: intermediate tests and final, or system, tests.

The test specification for intermediate tests is a list of functional tests to be performed on each block or group of blocks in the project. The tests are designed to ensure that the section of the project being tested will fulfill its function. For instance, if you are testing a Butterworth filter, you will need to specify the frequency range, the number of data points, and expected results for the block test.

The final test specification is a document that includes all of the test procedures which will be used to verify that the project prototype meets all of its intended specifications. In the commercial environment, the final test specification would very likely need to be approved by the customer, and the results of the final testing presented to the customer upon completion.

5.4.4 BUILD AND TEST THE PROTOTYPE HARDWARE PHASE

Building and testing the hardware should be relatively self-explanatory. In this step, you should construct and thoroughly test the hardware for the prototype project. Use the actual input and output devices for testing wherever possible. If using the real hardware is impractical (such as the speed sensor on a vehicle), then use electronic test equipment and other circuitry as necessary so that you will be able to supply simulated inputs from each of the sensors to test the blocks or sections of the project as specified in intermediate test specifications. Keep any signal simulators developed handy so they can be used during system integration and software development.

At the completion of this step, you should be absolutely sure that the correct input signals are being applied to the microcontroller from each sensor and that the correct output signals from the microcontroller will drive the output circuitry in the expected manner. The purpose of this step is to remove the quandary that occurs when some portion of the program does not seem to be interacting with the hardware correctly. Usually this leads to uncertainty on the developer's part as to whether the problem lies in the circuitry or in the software, but in this case, because the circuitry has all been tested, the developer will be relatively sure that the problem lies with the software.

In some cases, you will find it is easier to test those portions of the hardware that are driven by the microcontroller through the use of the microcontroller itself. If you elect to do this, combine the effort with the next step, the System Integration and Software Development Phase (also known as Write and Test Software), as you develop the functions to drive the output devices. In this way, you will be spending your time developing useful functions rather than simply developing test code that must be discarded later.

5.4.5 SYSTEM INTEGRATION AND SOFTWARE DEVELOPMENT PHASE

This step is the software corollary to the hardware step above. Writing and testing the software means to develop and *individually* test each of the functions that you flowcharted or outlined when you designed the project. Use a simple *main()* function to exercise each of the individual input device functions and output device functions so that, as with the hardware steps above, you are absolutely sure that each of the individual input and output functions works correctly. Again, this step is to remove the uncertainty that will occur later when things do not work as you expect them to.

When the individual functions are working correctly, write the overall software in a stepwise fashion—adding one additional function to the code and debugging it before adding more—until the entire project is functioning correctly using the real or simulated inputs.

It is often useful to develop the code to run the output devices first so that you can use output devices such as displays to show results when testing the input devices. It is also useful to use the serial terminal program in CodeVisionAVR to display intermediate results as a means of debugging and testing your software.

The goal of this step in the process is to have a fully working project using the real or simulated inputs and outputs.

5.4.6 SYSTEM TEST PHASE

The system test is the activity that actually puts the project into use. During the system test phase, the project should be tested in accordance with the final test specification developed earlier to ensure that all of the specifications defined in the definition phase of the process are met.

For a commercial project, the system test phase will also include a demonstration for the customer, and it will often include transfer of the *intellectual property* (IP) relating to the project. The IP includes complete documentation and records relating to the project, and it may be the property of the customer, depending on contract and business stipulations.

5.4.7 CELEBRATION PHASE

Successful completion of a project is always a good reason to celebrate. It is beyond the scope of this text to suggest the ways, means, and extent of your celebration. Just enjoy it.

5.5 PROJECT DEVELOPMENT PROCESS SUMMARY

The project development process is summarized in Figure 5–4. This figure shows each of the steps, the expected intermediate results, and the final results or deliverables from each step.

5.6 EXAMPLE PROJECT: A WEATHER MONITOR

5.6.1 CONCEPT PHASE

As an example, suppose we are engineers at the Wind Vanes R-Us Company, which has recently begun looking for ways to expand its product offerings. At lunch one day with our fellow WVRU employees, someone comments on how they are underdressed for the current weather. This leads to a conversation on the dubious accuracy of the local weather forecaster and how it would be nice to watch the weather indicators ourselves to know how to dress.

Further discussion leads to some speculation on what weather parameters would be useful to measure, and eventually a napkin is used (projects often start with a cocktail napkin and a pen) to start sketching out the attributes of such a system.

Process Step	**Intermediate Expectations/Results**	**Deliverables**
Definition Phase	Research Basic block diagram	Complete project specifications Schedule Budget Proposal
Design Phase	More research Part selection Circuit simulation results	Final, detailed block diagram Block schematics Flowcharts for each interrupt function Flowchart for the project
Test Definition Phase		Test specifications for each block or section Final test specifications
Build and Test the Prototype Hardware Phase	Individual block test results	Working hardware
System Integration and Software Development Phase	Individual function test results	Working software
System Test Phase		A completed, fully functional project Project documentation
Celebration Phase	Up to you	Up to you

Figure 5–4 *Project Development Process Summary*

In the design of the weather monitor, our list might look like this:

- Easy installation
- Collect as many types of weather data as possible
- Low cost
- Simple construction

Expanding on our list of attributes, we speculate that the weather monitor would consist of two units, an indoor unit and an outdoor unit. The outdoor unit should collect as few parameters as possible in order to make it as simple as possible. Temperature, wind, rainfall, and humidity must be collected outside, since that's where the action is. But barometric pressure is basically the same inside or out, so why not measure it indoors?

Indoor temperature and relative humidity must, of course, be measured indoors. There is also a factor of time when looking at weather. We might want to know when something happened, perhaps log some data, such as when there was an occurrence of high winds or very low temperatures. A real-time clock should be incorporated into the indoor unit.

5.6.2 DEFINITION PHASE

We can fill in our list of wants from the concept phase with more details, embellishing or defining each concept that has been put on paper. Notice that no attempt has been made yet to determine what is feasible or possible at this point. Remember that the goal of this phase of the project is to develop a very basic block diagram of the system and a complete set of specifications for the project.

We start by expanding our list of concepts:

1. Easy installation
   a. Wireless—900 MHz, one-way link
   b. No external power required on the outside unit; solar/battery powered
   c. Outside unit battery to last through a rainy week
   d. Simple setup (we would like to be able to just put a post up in the yard)
   e. LCD on inside unit (graphical would be cool, if not too expensive)
   f. Buttons for the user to view different parameters and set the time on the inside unit
   g. Off-the-shelf power supply for the inside unit, with battery backup for the clock
2. Collect as many types of weather data as possible
   a. Temperature, inside and outside
   b. Wind speed
   c. Wind chill would be good—it helps in dressing for the weather
   d. Wind direction
   e. Rain gauge
   f. Barometric pressure
   g. Relative humidity inside and outside (would be neat, if not too expensive)
   h. Dew point (Should I park my convertible inside?)
   i. RS-232 port on the inside unit to monitor weather with a PC?
3. Low cost
   a. Evaluate the costs of each type of measurement
   b. Keep the overall construction inexpensive (as few parts as possible)
4. Simple construction
   a. Parts should be available from easy-to-buy sources
   b. Simple feedback systems, nothing too exotic
   c. Wind and rain gauges should be simple to construct or off-the-shelf

This expanded list provides a better "mental picture" of the envelope that we are trying to work within. As we expanded the list, the wishes formed. At the same time, complexity and cost were also noted.

This is the same process whether you are designing something for yourself as a hobby or you are sitting in an Engineering/Marketing meeting discussing the development of the next product at your company, as we are doing in this case.

5.6.2.1 Electrical Specification

Now that we have a wish list, we can expand the definition to build a performance specification. This is where a set of parameters is assigned to each function or feature, indicating its range of performance and tolerance. Performance specifications do not always need to be created. They can be found, in many cases, by looking at similar or competitive products that already exist. The preliminary functional specifications for a weather monitor might look like those in Table 5–1.

Parameter	***Range***	***Tolerance***
Temperature	–40° to 140°F	+/– 1°F
Humidity	10 to 90% RH	+/– 5% RH
Barometric Pressure	28 to 32 inHg	+/– 0.05 inHg
Rainfall	0.00 to 99.9 inches daily	+/– 4%
Rainfall monthly/yearly	0.00 to 199.99 inches	+/– 4%
Wind Speed	2 to 120 mph	+/– 2 mph
Wind Direction	0–360° @ 1° resolution	+/– 7°
Dew Point	–6°F to 117°F	+/– 2°F
Wind Chill	–102°F to 48°F	+/– 2°F
Time of Day	24 Hr, 1 Sec resolution	+/– 1 Minute/Month
Date (Calendar)	MM-DD-YYYY	N/A

Table 5–1 *Weather Measurement Preliminary Specifications*

5.6.2.2 Operational Specification

Because there is a human-to-product interface involved in our weather station, an operations specification should also be developed.

A good way to begin is to make some sketches of what the display might indicate and write a list or even an Operator's Manual about how the button selections affect the operation of the device. Taking these steps early will greatly reduce the software development time in many cases, simply because of the psychological effects of having a list. As items on the list are completed, check them off. The operational specifications for our weather monitor might look like this:

Outdoor unit Operational Specification:

1. Keep units basic, ADC and timer counts is okay.
2. Measure temperature.

3. Measure humidity.
4. Measure wind speed.
5. Measure wind direction.
6. Measure rainfall.
7. Measure battery/solar-panel voltage.
8. Transmit the information to indoor unit, approximately once per second.
9. Enable the radio frequency (RF) module power only during transmissions to save energy.

Indoor unit Operational Specification:

1. Measure temperature.
2. Measure humidity.
3. Measure barometric pressure.
4. Measure battery/wall-power voltage.
5. Keep a real-time clock.
6. Gather packets of information from outdoor unit, as they come in.
7. Using 4X20 LCD, represent as much information as possible on a single screen, indicating indoor or outdoor and relative units, as much as possible.
8. A "Units" button will allow the user to change display units, °F to °C, mph to kph, and so on.
9. A "Select" button will allow the user to select rainfall readings from inches-per-hour, to inches-per-day, to inches-per-month, to inches-per-year.
10. A "Set" button will allow the user to set the time and date.
11. Enable LCD backlight only when wall power is available.
12. Include a "low battery" indication LED:
    a. The LED flashes when indoor unit battery is low.
    b. The LED is on solid when outdoor unit battery is low.
13. Include RF communications activity LED to indicate outdoor unit transmissions.

5.6.2.3 Basic Block Diagrams

Using the data that we have put together so far—the wish list, the expanded definition, and preliminary performance and operational specifications—we can draw some basic block diagrams of the system. The diagrams shown in Figures 5–5 and 5–6 show the basis for the hardware architecture. The power supply, what parameters are measured where, and the method of outputting the processed data, whether it is by RF or LCD.

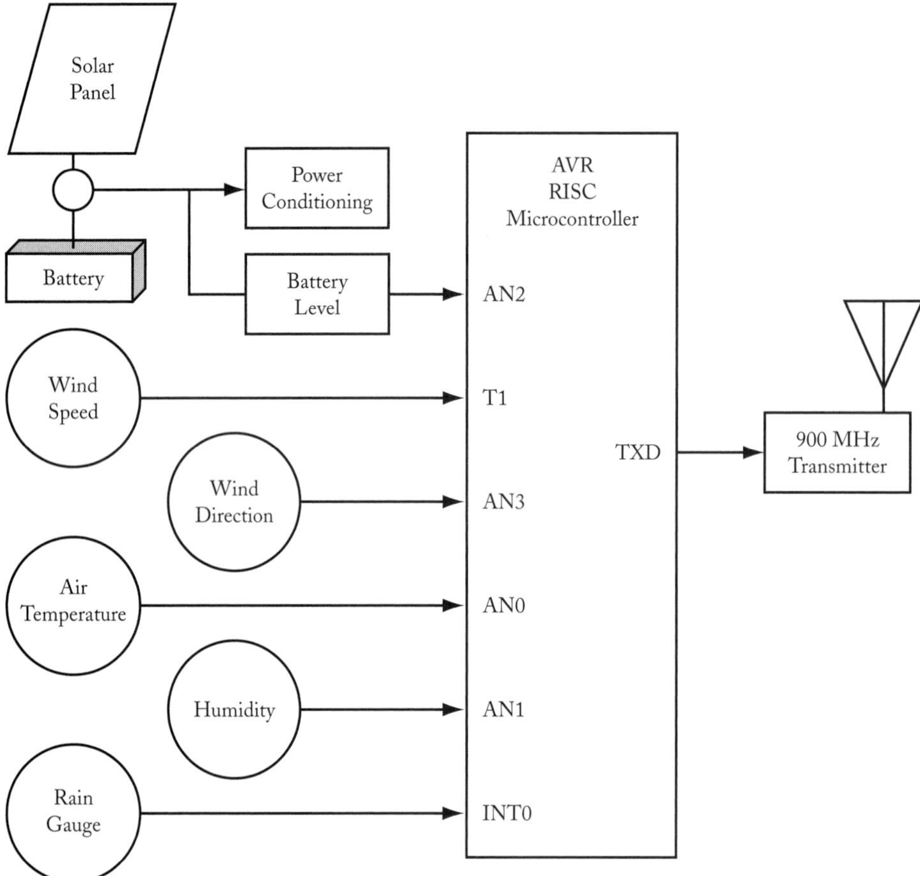

Figure 5–5 *Block Diagram, Outdoor Unit*

The block diagrams show a detail that is determined in the feasibility study of the definition phase. On the block diagram, the inputs to the microcontroller are named by the hardware function of the input. The following section describes how each of the measurements is taken and what kind of input to the microcontroller is required. It is important to do this in this phase of the project because the number of special inputs, such as analog input or timer/counter input, is limited on the microcontroller.

Using the block diagram as a guide, we can choose an appropriate microcontroller with sufficient I/O and special features. As we go on through the design process, we can then check our chosen microcontroller against the project needs to ensure that we have made a good choice with the micro.

The ATMega16 was chosen for the prototype development. The Mega16 is pin-for-pin compatible with the smaller-memory model ATMega8535 as well as with the larger-memory

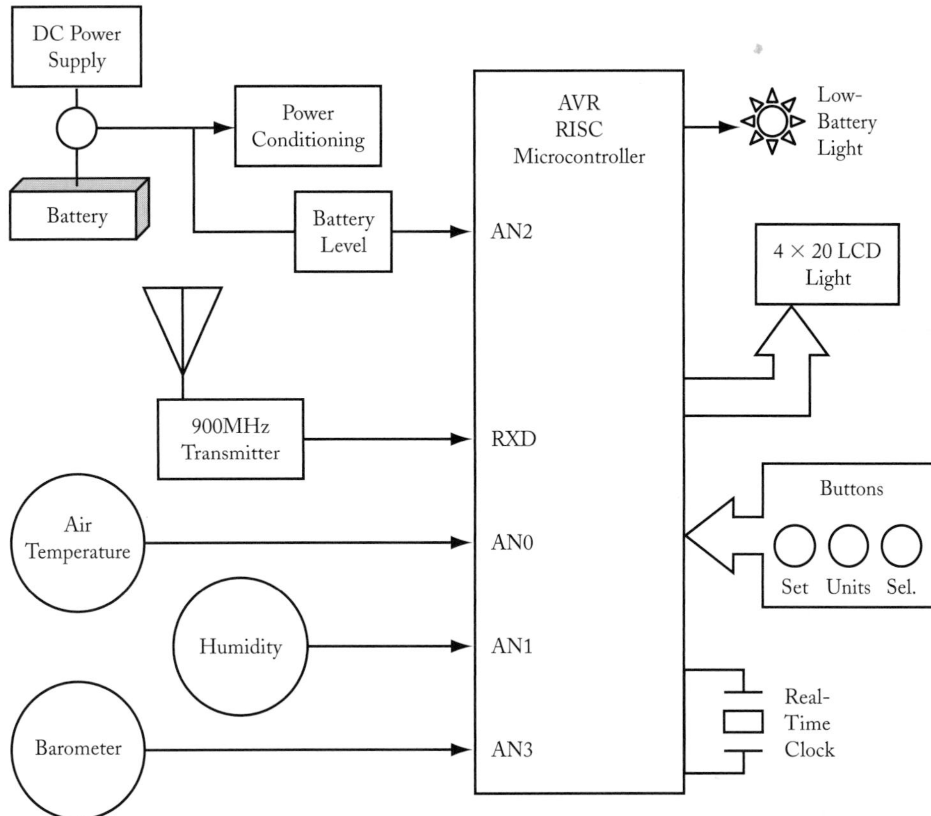

Figure 5–6 *Block Diagram, Indoor Unit*

model ATMega32. This allows some freedom during the design phase. The option of increasing the memory model may be needed, since we do not know exactly how much code space is required in either unit. At the end of the project, and after the features have ceased to creep, if the code would fit into a potentially less expensive component with a smaller memory size, that choice is available as well.

The other, perhaps more obvious points of selection are the availability of ADCs (analog-to-digital converters), a USART, and an oscillator driver for the 32 kHz crystal for the real-time clock.

5.6.3 MEASUREMENT CONSIDERATIONS FOR THE DESIGN

The final step in the definition phase is a feasibility study or "sanity" check of the system. Now that we have specified what we *want* the system to do, can we meet those expectations, or do they need to be revised based on physical or electrical limitations? This section describes the steps taken to ensure that our project is feasible before we move on to the design phase.

It is hard, however, to completely separate this final step of the definition phase from the design phase. In many cases, it is necessary to develop very specific details about the design to ensure we can measure to our specifications. So this section crosses into the design phase in many areas as we explore the physical and electrical requirements of measuring the weather.

Examining the basic block diagrams, you may note that there are common elements. Temperature and humidity are required both indoors and outdoors, but they are fundamentally the same measurement because analog voltages represent them both. Wind speed, wind direction, and rainfall are outdoor-only items. And because barometric pressure is basically the same indoors as it is outdoors, why not measure it in the best possible weather conditions, indoors?

Each parameter in a system may require a different method for determining its value. In some cases, the parameter can be measured in several ways, and it becomes more of a matter of selecting what is appropriate based on performance, cost, or both. The design considerations for each parameter being measured are discussed below.

Once the method of measurement is selected, it is checked for total range and resolution. When you deal with voltages and ADCs, the resolution of the ADC is the pacing factor. When pulses are brought into a counter, the size of the counter and the rate at which the counts are checked determine the range and resolution of the readings.

In this weather monitor project, temperature, humidity, battery voltage, and barometric pressure are all voltages that are processed by the ADCs. These are calculated out to verify that everything is within range and that the specifications can be met as the sections are designed. The wind speed and rainfall are not quite as straightforward.

5.6.3.1 Temperature

Temperature can be detected inexpensively and simply using a PTC (*positive temperature coefficient*) or NTC (*negative temperature coefficient*) resistive device, or thermistor. These devices come in a wide range of values, and with proper selection considerations, no special amplification or conditioning might be necessary.

There are many other temperature sensors available, including bi-metal thermocouples, temperature compensation diodes, and *resistive temperature devices* (RTD). There are also integrated circuits, such as the National Semiconductor LM35, that are calibrated, stable measuring devices which deliver a specific amount of voltage per °C. For example, the LM35 outputs 10 mV/°C. Each of these, however, is either more costly or requires additional circuitry that makes its use more costly.

A glass-encapsulated NTC thermistor was chosen for our design to do the temperature sensing. Figure 5–7 illustrates the resistance versus temperature of a typical NTC thermistor.

Since thermistors move in value as a percentage of resistance per degree C, using a resistor divider allows for a reasonably linear measurement of temperature. The divider's voltage

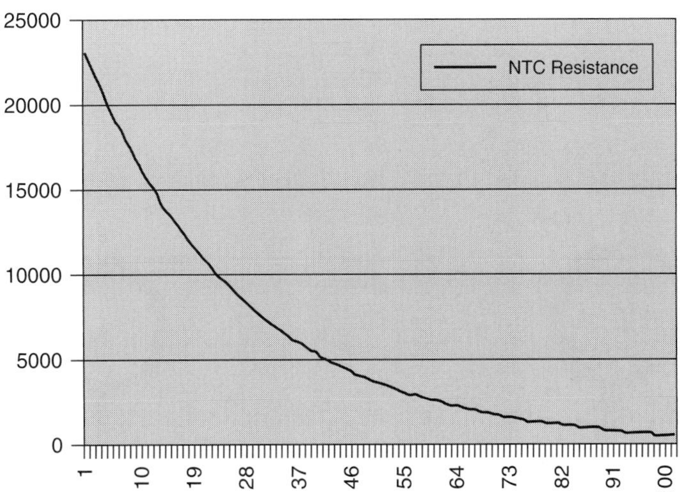

Figure 5–7 *NTC Resistance Versus Temperature*

output range can also be tailored to match the range of the ADC, providing the best possible resolution for desired range of measurement. This is where the NTC device has an advantage over an integrated circuit like the LM35. It would take additional circuitry, plus the added cost of the LM35, to get a result comparable to the NTC solution. The output result of the NTC/Resistor divider may not be perfectly linear, but we can use software savvy and look-up-tables to fix anything that is not deemed acceptable.

The Mega16 processor, as with the entire AVR family, prefers a circuit with a fairly low output impedance (about 5K ohms or less) for input into the ADC. We selected a 4.7K resistor to form the divider with a 2K thermistor (approximate resistance 25°C or 77°F). The selected thermistor changes 3.83 percent in value for every 1°C of temperature change (from the thermistor specification). This provides a range of voltage to the ADC from approximately 0.84 V at –40°F to 4.5 V at 140°F. The thermistor was placed in the top of the divider because of its negative impedance change with temperature. This allows the ADC readings to increase with temperature. Figure 5–8 illustrates the relationship of the output voltage at the resistor divider and the temperature of the thermistor device.

The ADC resolution is 10 bits (or 1024 counts), so that means that each °F is approximately 4.16 counts on the ADC. We arrive at this number by

$$\frac{\text{4.5 V - 0.84 V range}}{\text{5 V (full scale)}} \cdot \text{1024 counts} = \text{749.56 counts for the range}$$

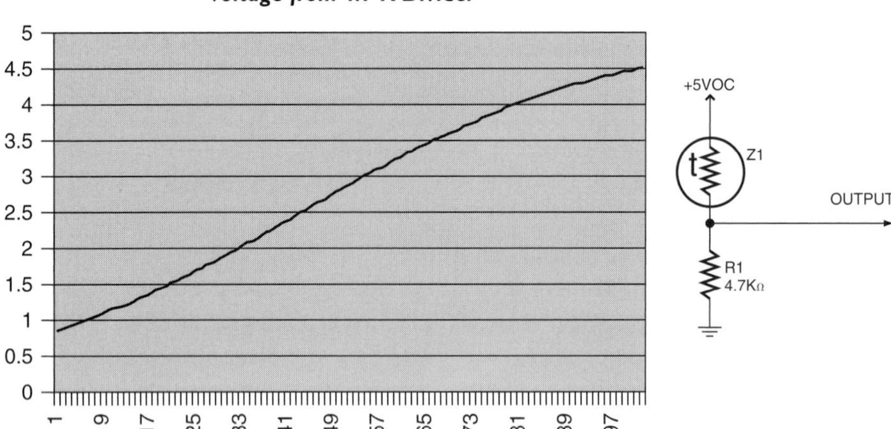

Figure 5–8 *NTC/Resistor Divider Output Versus Temperature*

where the range is –40°F to 140°F. This means we have a resolution of approximately 0.240°F for each count of the ADC. The calculation for this is

$$\frac{140°F - (-40°F)}{749.56} = 0.240°F$$

This resolution shows that the measurement accuracy is sufficient to meet the project temperature accuracy specification of +/– 1°F.

5.6.3.2 Barometric Pressure

Pressure can be detected using off-the-shelf transducers. These devices come in a wide range of values, and with proper selection considerations, no special amplification or conditioning might be necessary. Barometric pressure is measured in units of *inches of mercury* (inHg). In fact, a total range of four inches of mercury is typical, as suggested by research on similar weather stations. Therefore, a fairly sensitive detection is required.

Pressure transducers usually come in three forms: gauge, differential, and absolute. The gauge (PSIG) and differential (PSID) transducers, in both cases, measure pressure differentially (the difference between two pressures). A gauge pressure transducer measures against ambient air—specifically designed to ensure that ***no*** barometric pressure is figured into the reading. A differential transducer measures against a reference pressure, so that the barometric pressure that is available at both inputs is cancelled out. An absolute transducer measures differentially against a vacuum (0 PSIA), so the pressure being read includes the effects of barometric pressure, making it appropriate for this project.

The SenSym ASCX30AN is a 0 to 30 PSIA (*pounds per square inch, absolute*) pressure transducer. In the weather monitor, we simply open the sensor to the outside air. This means that the only pressure detected is barometric. The range of barometric pressure is very narrow. It is generally measured as 28 to 32 inches of mercury. This translates into a pressure range of 13.75 to 15.72 PSIA. According to its specification, the ASCX30AN has a sensitivity of 0.15 V/PSI. That means the actual output voltage from the 13.75 to 15.72 PSIA range would be 2.06 to 2.36 V DC.

This is not much signal! Consequently, it is necessary to provide some amplification and DC offset to the transducer signal before it is applied to the microcontroller's ADC. Providing an amplifier with a gain of 16 boosts the voltage range from the sensor to a useful level:

```
(2.36 V - 2.06 V) • 16 = (0.3 V x 16 ) = 4.8 V
```

A range of 4.8 V uses most of the 5 V input range on the ADC, maximizing the ability to accurately measure the signal. For this to work, however, the 2.36 V DC offset must be subtracted from the sensor voltage before it is applied to the ADC.

A differential amplifier is used to provide a gain of about 16 to the signal as well as an offset of −2.06 V. This allows the ADC to acquire a signal that ranges from 0.0 to 4.8 V DC for the 28 to 32 inches of mercury. The total range of ADC counts for the measurement range is

$$\frac{\texttt{4.8 V - 0.0 V range}}{\texttt{5 V (full scale)}} \bullet \texttt{1024 counts = 983.04 counts for the range}$$

where the range is 28.00 to 32.00 inches of mercury. This means we have a resolution of approximately

$$\frac{\texttt{(32 inHg - 28 inHg)}}{\texttt{983.04}} \texttt{ = 0.0041 inHg}$$

0.0041 inches of mercury for each count of the ADC, which more than meets the desired accuracy specification of +/− 0.05 inHg.

5.6.3.3 Humidity

Humidity can be detected using off-the-shelf transducers as well. These devices are thermally compensated for accuracy and are fairly expensive. Using a humidity-sensing capacitor could create a less expensive solution, but a large effort would be spent on compensating and processing the device input to form an accurate humidity measurement.

In this weather monitor, humidity is measured by an off-the-shelf temperature compensated humidistat. It was a decision based on ease of implementation alone. This is not the least expensive solution, but it is certainly simple to integrate and very accurate. The device used in this design is a Honeywell (Micro Switch) HIH-3602-L. The device outputs a processed DC voltage from 0.8 to 4.0 V for 0 to 100% humidity. The output of this device may be connected directly to the ADC input of the Mega16 processor. The total range of output from the ADC for the humidity range is

```
     4.0 V - 0.8 V range
  ————————————————————————— • 1024 counts = 655.36 counts for the range
      5 V (full scale)
```

Using this number, we determine the resolution for humidity measurement as follows:

```
     100% - 0%
     ————————— = 0.152%
     655.36
```

An accuracy of measurement of 0.156% is way more than adequate to measure the RH +/– 5% as specified in the electrical specifications.

5.6.3.4 Wind Speed

Wind speed is typically measured using a set of rotating cups or a fan blade arrangement called an anemometer. Basically, as the wind blows, the anemometer turns at a speed that is directly related to the wind speed. A device that measures the rate at which something is turning is called a tachometer. Several possible tachometer types are used for wind speed measurement.

One often-used form of a tachometer is like a small electric motor or generator. As it turns, it generates a DC voltage that is related to the rate at which it is spinning. In this case, the resulting DC voltage could be measured using the microcontroller's ADC and the wind speed interpreted from the DC voltage.

Another version of tachometer is the *digital tachometer.* A digital tachometer typically generates a pulse train, associated with the rate of spin, which can be measured as a frequency and converted to *revolutions per minute* (rpm). A digital tachometer is, in most cases, a less expensive alternative to an analog tachometer. It can be easily constructed from a slotted disk and a photo-interrupter or opto-coupler. It can also be an off-the-shelf solution in the form of an optical encoder. Since there are no magnets and much less inertia due to the weight of the armature, the digital tachometer also offers an advantage over the analog tachometer because it usually imposes less drag on the system.

For our weather monitor, an off-the-shelf anemometer was selected. This selection was made for the sake of mechanics alone (see Figure 5–9). If you are a particularly handy individual, you might be able to construct an assembly that would provide similar results at a greatly reduced cost.

This particular unit uses a magnetic reed-switch for encoding the speed. Basically, a small metallic switch is mounted in the base, and a magnet is mounted on the anemometer. As the anemometer turns, the magnet makes and breaks the switch closure. This is a simple solution mechanically, but it lacks some electrical quality. Since the switching is not electronic in nature, there is a high potential for switch "bounce." Switch bounce is just that; the contacts literally bounce like a rubber ball as they are slapped together, creating several transitions as the switch is made. If this signal is brought directly into the timer input of the

WSD-PB Dual Wind Speed/Direction Sensor

The Model 034A Wind Sensor combines wind speed and direction measurements into a single remote sensing unit. The 034A Wind Sensor installs in minutes and will provide accurate, long term, continuous monitoring in hostile environments.

- Wind speed and direction in a single sensor.
- Durable aluminum and stainless steel construction.
- Low starting threshold.
- Stainless steel bearings.
- Low power.

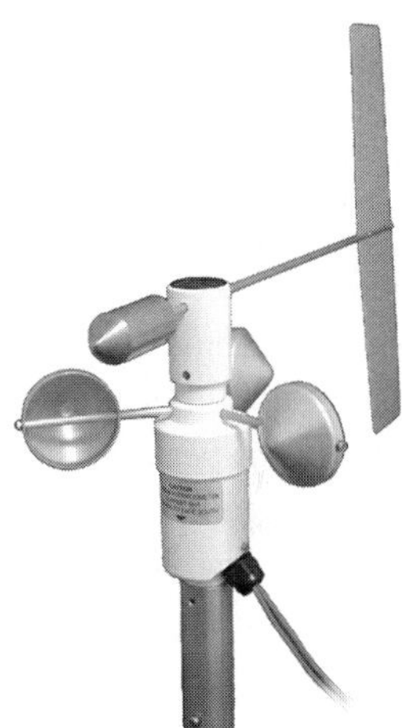

Operation:
The 034A is designed to operate within a **temperature range of** ***–30°C* to *+70°C*** and with **wind speeds up to** ***155 mph.*** Wind speed is measured using a sealed reed switch that provides a series of pulses at a rate proportional to wind speed. Aluminum cups are used for durability. Wind direction is measured using an aluminum vane directly coupled to a precision potentiometer. Variations in wind direction produce a corresponding change of voltage, which lends itself to both digital and analog measurement systems.

Construction:
High-quality aluminum and stainless steel components, and high-gloss white powder coating ensure long life with minimal maintenance. Cable length may extend hundreds of feet without affecting measurement performance.

WSD-PB Specifications

Wind Speed
- **Range:** 0–155 mph (0–69 m/s)
- **Starting Threshold:** 0.9 mph (0.4 m/s)
- **Accuracy < 22.7 mph:** 0.25 mph (0.1 m/s)
- **Accuracy > 22.7 mph:** ±1.1% of true

Wind Direction
- **Range**
  - **Mechanical:** 0–360°
  - **Electrical:** 0–356°
- **Starting Threshold:** 0.9 mph (0.4 m/s)
- **Accuracy:** ±4°
- **Damping Ratio:** 0.25 standard (0.4 to 0.6 optional)
- **Resolution:** 0.5°

Output Signal
- **Wind Speed:** Pulsed contact closure
- **Wind Direction:** Potentiometer output (0–10 kohms)

Temperature Range: –30° C to +70° C
Sensor Weight: 1 lb 12.5 oz (0.81 kg)

Reprint courtesy of Forest Technology Systems, Inc.

Figure 5–9 *WSD-PB Dual Wind Speed/Direction Sensor*

microprocessor without any de-bounce circuitry or filtering applied, erroneous data will result, because the timer receives three to ten times the number of transitions the switch actually made based on the mph of wind. Consequently, some de-bounce circuitry will be needed to provide accurate wind speed measurement, as shown in Figure 5–11, the outdoor unit schematic.

The specification for this device shows that it will measure wind speed as slow as 0.9 mph and is accurate to 1.1% of true wind speed. At the highest speed we specified (120 mph), the accuracy would be 1.1% of 120, or 1.32 mph. This exceeds our specified accuracy of +/–2 mph.

The wind speed input in this project is a pulse train. Our plan is to count the pulses on one of the timer inputs and get a total every second, in order to compute wind speed. To check what we are capable of measuring, we need to know how fast the fan or cups of the anemometer spin with respect to a certain wind speed. For now, we can assume that the anemometer turns 0.15 revolutions for every 1 inch of lineal air that passed through it. That would mean that 1 foot per second would translate into 1.8 revolutions per second, which is 108 rpm. At one mile per hour, the wind would be traveling at 1.46 feet per second. This would turn the anemometer at 158 rpm. At 100 miles per hour, the anemometer would spin at 15,768 rpm.

If we place an encoder on the system that delivers four pulses for each revolution of the anemometer, a one-mile-per-hour wind would yield 10.5 pulses per second (PPS or Hertz). At 100 miles per hour, a pulse train of 1051.2 PPS would be generated. If this signal were sampled at one-second intervals, using an 8-bit counter, the maximum wind speed that could be measured would be one that generates 255 PPS. We can find the maximum wind speed by working this backward:

$$\frac{\text{255 pulses per second}}{\text{4 pulses per revolution}} = \text{63.75 revolutions per second}$$

If the anemometer turns 0.15 revolutions per one inch of air motion, then the air speed required to turn the anemometer would be

$$\frac{\text{63.75 revolutions per second}}{\text{0.15 revolutions per inch}} \bullet \frac{\text{1 foot}}{\text{12 inches per foot}} = \text{35.41 feet per second}$$

$$\frac{\text{35.41 feet per second}}{\text{5280 feet per mile}} = \text{0.006707 miles per second}$$

$$\text{0. 006707 miles per second} \bullet \text{3600 seconds per hour} = \text{24.15 mph}$$

Our initial specification is to measure up to 120 mph, which is an extremely strong wind! What this crosscheck shows us is that we will need to sample the counter 5 times per second, or every 200 ms, to prevent the counter from rolling over during our measurement.

It also shows that the anemometer can have quite a variation in the relationship between wind speed and rpm and we will still be safe from a design aspect.

If the unit gets all put together and we find out that counts generated at 120 mph will not fit within the counter in 200 ms samples, we will just read the counter more often, maybe every 150 ms instead. On the other hand, if we find that the relationship is much less, we can read the counter less often, reducing the system overhead.

Another method for improving the measurement capacity is to simply get a bigger counter. Timer 1 is a 16-bit counter and can hold a value 256 times larger. This means that we can effectively sample at one-fiftieth the rate, or once every ten seconds! If the sampling rate is even once per second, the collected number in Timer 1 would effectively be five times larger than what could be measured within the 8-bit counter.

5.6.3.5 Wind Direction

Wind direction is usually determined by the use of a "wind vane" or "weather vane," our specialty at Wind Vanes R-Us. A weather vane is generally a finned shaft, something like an arrow, that is constructed to pivot in a horizontal fashion. The vane is free to turn. The fins create drag and are effectively pulled by the wind such that the point of the vane indicates the direction of origin of the wind. Determining the wind direction is a matter of encoding this position such that the compass direction—from 0° to 360°, where 0° is north, 90° is east, and so on—can be read electronically.

This can be accomplished using an optical encoder, a continuous turn potentiometer, or a simple mechanical switch encoder. Optical and switch encoding is generally considered a relative measurement. This means that a "home" or zero point is established, and the encoder indicates the distance from that established home position. There are absolute versions of encoders such as a Gray Scale encoder; these encoders offer a different code for each position within a single rotation of the encoder. The disadvantage to this type of encoding is limited resolution. A four-channel Gray Scale encoder offers a 4-bit word of information regarding the position within a 360° rotation. This limits the discernible number of positions to 16, or 22.5° of rotation.

Using a continuous turn *potentiometer* offers higher resolution. A potentiometer, also known as a rheostat or variable resistor, offers an absolute position effectively at an infinite resolution. The limiting factor for encoding the position then becomes the resolution of the ADC or other measuring device used to read the resistance value. Potentiometers are typically not continuous-turn devices but are more commonly single-turn or 10-turn devices. Some searching may be required to locate one if you are planning to build your own wind-direction sensor.

The Forest Technology Systems, Inc. WSD-PB Dual Wind Speed/Direction Sensor, as shown in Figure 5–9, uses a 10K ohm continuous-turn potentiometer. Note that in the specifications the measurement range is 360° but the usable range is 356°. This is because of a small "dead spot" where a single-turn pot would normally stop. This "dead spot" could be used to establish a North or South reference when the weather monitor is set up.

The output of this device can be tied directly to the ADC input of the Mega16 processor. The ADC is 10 bits (or 1024 counts), and the voltage from the sensor will go rail-to-rail, or 0–5 V DC. So that means that each 1° of position is approximately 2.876 counts on the ADC. We arrive at this number by

```
    1024 counts
----------------- = 2.876 counts per degree
      356°
```

where the positional range is 0 to 356°. This means we have a resolution of approximately 0.347° for each count of the ADC. This result is far better than our specified accuracy of +/− 7°.

5.6.3.6 Rainfall

Rainfall is measured in cubic inches. A typical rain gauge consists of a funnel and a graduated cylinder. The definition of "one inch" of rainfall according to Ref 031.02A of the *American Farm and Home Almanac* is as follows:

> An acre of ground contains 43,560 square feet, therefore one inch of rainfall over one acre of land would equal 6,272,640 cubic inches of water, which is the equivalent of 3,630 cubic feet. One cubic foot of water weighs about 62.4 pounds (this varies with the density); the weight of a uniform coating of one inch of rain over one acre of ground would be 226,512 pounds (about 113 short tons). The weight of one U.S. gallon of pure water is about 8.345 pounds, therefore one inch of rainfall over one acre of ground would be 27,143 gallons of water.

The toughest part of measuring rainfall is collecting and reading a volume of water over what could be a long period of time. There are a few common methods.

One method involves a funnel with a very small spout that produces a uniform drop of water. As the funnel fills up at a random rate with rainwater, the spout produces uniform drops that can be counted over time. This can then be converted to a volume, over time, which can be converted to inches of rainfall. This method's shortcoming is that the variation in the rate at which the rain is falling cannot be determined. It also could produce an invalid reading if the volume of the funnel were not sufficient to contain all the rain collected until it could be dribbled out and accounted for.

Another method is a paddle wheel arrangement. A funnel is used to fill known volume cups on a wheel. As the cups fill, the wheel turns. Revolutions of the wheel indicate the rate and the amount of rain collected, as well as emptying the samples taken as the wheel turns.

A reduced version of the paddle wheel is a seesaw configuration. The concept is basically the same, except there are only two cups and when a cup becomes full, it tips over to empty, bringing the other cup into position to gain a rain sample from the funnel. As the second cup fills up, it then tips to empty, bringing the first cup back into position again. This method is probably the most common one used in small electronic rain gauges, because it gives all the information needed with a minimal amount of mechanics. Each time the cups tip from one position to the other, a volume of rainwater is accounted for and an electrical signal is generated.

RG-T Rain Gauge

The RG-T Rain Gauge features a chemically milled tipper bucket for consistent balance and easy calibration. The tipper bucket pivots on shock-mounted jeweled bearings providing a sensitive yet rugged low friction pivot to track rainfall accurately.

A precision adjustment for the reed switch ensures one switch closure for each tip of the bucket. The tipper, bearings and magnetic reed switch are held in sturdy machined aluminum mounting blocks. These are seated on a rigid cast aluminum base. The calibration adjustment screws are protected against loss of calibration during handling.

Screens in the funnel and over the base plate water outlets seal out spiders and insects. The cable, which terminates in a waterproof circular bayonet connector, is protected from rodents by braided stainless steel armor. A built-in level coupled with a precision adjustment leveling arm and mounting base make installing the RG-T fast and easy.

The FTS Rain Gauge is considered by a number of outside experts, including those with a lot of field experience, to be one of the best tipping bucket rain gauges in the world, with an unbeatable combination of quality, accuracy, and long lasting calibration.

The redesign of this quality product includes investing in a special casting process so that we can precision cast the gauge's internal structure as a single piece, rather than a number of separate posts and supports. The result will be a more rigid gauge with even greater resistance to vibration and increased repeatability from one gauge to the next.

RG-T Specifications

Resolution: 0.01 in per tip (0.254 mm)
Accuracy: ±2% at 2 in per hour (50 mm)
Cylinder Dimensions: 10.5 in x 8 in diameter (25.7 cm x 20.3 cm diameter)
Base Plate Dimensions: 12 in x 12 in (30.5 cm x 30.5 cm)
Cable Length: 20 ft (6.1 m)
Weight: 10 lb (4.5 kg)

Reprint courtesy of Forest Technology Systems, Inc.

Figure 5–10 *RG-T Rain Gauge*

In all of these methods, a form of optical or mechanical encoding is used to bring the information in for analysis. The amount of information is dependent on the size of the samples taken and the rate at which a minimum sample can be met. Again, for the purposes of reduced mechanical design, we are using an off-the-shelf sensor from Forest Technology Systems, Inc., as shown in Figure 5–10. As mentioned with the wind speed sensor, if you are mechanically inclined, a less expensive solution could be built.

As in the anemometer, this unit uses a magnetic reed-switch for encoding the rainfall. Basically, a small metallic switch is mounted in the base, and a magnet is mounted on the seesaw. As a seesaw tips back and forth, the reed-switch contacts are closed and then opened again. Because a reed-switch is a mechanical device, the same rules of switch bounce apply as they do in the anemometer. Some additional circuitry is needed to prevent erroneous readings.

Whether the measuring system is purchased or you construct it yourself, even if great care is taken in creating the funnel, sampling cups, and the wheel or seesaw system, until you put water in it you never really know what kind of values the measuring system might produce. This build-and-test process in a proof of concept stage is invaluable, particularly when there are somewhat complicated mechanics involved and the "gotchas" of physics apply. Things like gravity, friction, and inertia can really mess up an engineer's day.

We can assume that in this weather monitor project we plan to use a seesaw arrangement that is fed by a funnel that is four inches in diameter, instead of the really nice, but pretty expensive, purchased sensor that we have selected. The concept is that as water enters the funnel it fills up a cup on one side of the seesaw. When the weight of the cup exceeds the leverages and frictions in the system, it will tip over. This action not only dumps the contents of the cup but also places the cup from the other side of the seesaw under the funnel to collect more water. When this cup fills, the process repeats. The speed at which the rainfall is collected is very slow compared to the microcontroller that will be observing the sampling. If the cups were smaller than sewing thimbles, the sampling would occur faster with higher resolution. But if the cups are too small, the system could be overwhelmed in a heavy rain scenario. Cups the size of coffee mugs have the opposite problem. You could go for weeks before you get one filled up! This would cause the samples to come in more slowly and the resolution of the measurement to be much less.

For the sake of the "sanity" check, we assume that the cups each hold about one cubic inch of water. With this system built up, we could actually measure the sample sizes in a simple and controlled fashion. One gallon of water weighs 8.345 pounds. A cubic foot of water weighs 62.4 pounds. Therefore, a gallon of water is

$$1 \text{ gallon} = \frac{8.345 \text{ pounds per gallon}}{62.4 \text{ pounds per cubic foot}} = 0.1337 \text{ cubic feet}$$

$$1 \text{ gallon} = \frac{0.1337 \text{ cubic feet}}{1728 \text{ cubic inches per cubic foot}} = 231.036 \text{ cubic inches}$$

Since the funnel is four inches in diameter, its collecting surface is 12.56 square inches. So

$$\frac{231.036 \text{ cubic inches}}{12.56 \text{ square inches}} = 18.39 \text{ inches of rainfall}$$

is represented by one gallon of water. If the cups hold approximately one cubic inch of water, then the seesaw should transition approximately 231 times as the gallon of water is poured into the funnel. That would mean that each transition of the seesaw would represent

```
18.39 inches of rainfall
---------------------------- = 0.0796 inches of rainfall per transition
   231 transitions
```

You can see that by simply building it up and pouring some water through it, we could uncover what is really going on. At the end of the gallon, you might come out with more or fewer counts than were expected. That is okay, as long as you know what they are.

Our preliminary specification calls for measuring up to 99.9 inches of rainfall in a day. That means approximately 4.16 inches of rainfall would occur each hour. That means, in this kind of rainfall, that

```
99.9 inches       1 transition
-----------  •  ---------------- = 52.25 transitions per hour
24 hours          0.0796 inches
```

would occur. This is less than one transition of the seesaw per minute in a downpour while still providing a resolution of 0.0796 inches of rainfall per transition! A very acceptable situation when compared with the +/– 4% error specification for rainfall monitoring.

5.6.3.7 Dew Point Computation

Dew point is a calculated number based on air temperature and humidity that tells when moisture will begin to condense on objects that are at a cooler temperature, that is, when the dew will collect on the grass in the morning or when moisture will collect inside your convertible even though the top is up.

Dew point computation is actually quite complicated. So to reduce the effort, charts have been established that provide the translations from temperature and humidity to dew point, as shown in Table 5–2.

In Table 5–2, find the air temperature in the listing across the top of the chart, and then find the relative humidity (in percent) in the listing down the left side. Follow the column and row down and across to the box where they intersect. This is the dew point. For temperature and humidity values between those shown, use interpolation to compute values between the dew points in the column or row. For example, with a temperature of 65°F and a humidity of 45% RH, the dew point temperature value will be half way between that for 60 and 70°F, or

```
37 + (47-(37 ÷ 2))  =  37 + 5  =  42°F.
```

Objects that are at a temperature of 42°F or below will collect condensed moisture given an air temperature of 65°F and a relative humidity of 45% RH.

In the weather station being designed, the table data will be included in the software and used to “look up” the dew point based on the reading for temperature and humidity.

	Temperature in °F										
	20°	30°	40°	50°	60°	70°	80°	90°	100°	110°	120°
%RH											
30	−6	4	13	20	28	36	44	52	61	69	77
35	−2	8	16	23	31	40	48	57	69	74	83
40	1	11	18	26	35	43	52	61	69	78	87
45	4	13	21	29	37	47	56	64	73	82	91
50	6	15	23	31	40	50	59	67	77	86	94
55	9	17	25	34	43	53	61	70	80	89	98
60	11	19	27	36	45	55	64	73	83	95	101
65	12	20	29	38	47	57	66	76	85	93	103
70	13	22	31	40	50	60	68	78	88	96	105
75	15	24	33	42	52	62	71	80	91	100	108
80	16	25	34	44	54	63	73	82	93	102	110
85	17	26	36	45	55	65	75	84	95	104	113
90	18	28	37	47	57	67	77	87	97	107	117

Table 5–2 *Dew Point Calculation Guide*

5.6.3.8 Wind Chill Computation

The wind chill temperature is a measure of "relative discomfort" due to combined cold and wind. It is the perceived temperature, due to the moisture evaporation from the skin that is exposed to the open air—it "feels" colder in a high wind than it does in calm air. It was developed in 1941 by Paul A. Siple and Charles F. Passel and is based on physiological studies of the rate of heat loss for various combinations of ambient temperature and wind speed. The wind chill temperature equals the actual air temperature when the wind speed is 4 mph or less. At higher wind speeds, the wind chill temperature is lower than the air temperature and measures the increased cold stress and discomfort associated with wind.

On November 1, 2001, the National Weather Service (NWS) implemented a replacement Wind Chill Temperature Index. The equation for this new index is as follows:

$$\text{Wind Chill (°F)} = 35.74 + 0.6215T - 35.75(V^{0.16}) + 0.4275T(V^{0.16})$$

where T is the air temperature in °F and V is the wind speed in miles per hour. The changes made by the NWS improve upon the Siple and Passel data, which had been used since 1945. Table 5–3 will be used in the software to determine wind chill.

The nature of the equation, with its fractional exponents, does not make it very conducive to a quick, inline calculation. Just as in the previously described dew point feature, the table data will be included in the software and used to "look up" the wind chill based on the reading for temperature and wind speed.

	Wind Speed in Miles Per Hour (MPH)							
	5	10	15	20	25	30	35	40
Temp °F								
40	36	34	32	30	29	28	28	27
30	25	21	19	17	16	15	14	13
20	13	9	6	4	3	1	0	–1
10	1	–4	–7	–9	–11	–12	–14	–15
0	–11	–16	–19	–22	–24	–26	–27	–29
–10	–22	–28	–32	–35	–37	–39	–41	–43
–20	–34	–41	–45	–48	–51	–53	–55	–57
–30	–46	–53	–58	–61	–64	–67	–69	–71
–40	–57	–66	–71	–74	–78	–80	–82	–84

Table 5–3 *NWS-2001 Siple and Passel Wind Chill Calculation Guide*

5.6.3.9 Battery Health

Battery health can be as exotic as one chooses it to be. It can be as simple as measuring the voltage or as complicated as monitoring the charging currents and determining the battery's internal impedance changes. In most cases, an embedded system is concerned with a simple requirement: "Is there enough voltage to operate correctly?" This can be accomplished in most cases with a resistive divider, to scale the voltage being measured, and an ADC. Alternately, we could use the analog comparator peripheral in the AVR device, if we are interested in simply a "Go/No-Go" reading.

Resolution is not a really important issue for this measurement in that we are only establishing a threshold to denote a problem. As long as we set the threshold carefully, resolution is a nonissue.

5.6.3.10 Real Time

Real time is what we all live by. In an embedded system, real time can be as simple as a periodic event or as complicated as a full-blown clock and calendar. When we make measurements, and particularly when we log the measurements, a clock and calendar help enhance the usefulness of the data. If an event occurs where the data looks out of the ordinary, in many cases it helps to know when it happened. This allows a human to link the data to a particular event or outside circumstance.

There are several companies that manufacture real-time clock integrated circuits (ICs). Some are completely self-contained, like the Maxim DS1302 or DS1307, which maintain the time without the microprocessor, can be powered for years on a small lithium battery, and usually interface to the microprocessor using a 3-wire SPI or 2-wire I^2C communications link. Most integrated circuits of this type require a 32.767KHz watch crystal for proper operation.

Some microcontrollers, like the Atmel AVR, are designed to support a second crystal input source, such as a 32.767KHz watch crystal, for the purposes of maintaining real time within the microcontroller without an external clock IC. As long as the processor has power, this clock can be maintained. There are even features that allow the processor to "sleep," a mode of operation of very low power consumption, nearly stopped, and to be awakened by the timer in order to update the real time and then go back to sleep. In this weather monitor, the indoor unit will be powered full time by an external plug-in supply and backed up through power outages by batteries. The display will always be active, so there is not much point in "sleeping" the microprocessor to save battery life. The outdoor unit, however, might profit from the use of the power-down mode.

5.6.4 HARDWARE DESIGN, OUTDOOR UNIT

Now that we have proven to ourselves that the project really is possible, we can start designing the final hardware. By using the block diagrams, specifications, and feasibility research from the definition phase, we can quickly pull together the schematics for the hardware. First, each block of the outdoor unit is developed, and then the blocks are combined together in one schematic.

The outdoor unit is designed to be completely wireless. A 900 MHz transmitter sends the collected information to the inside unit every second or so. The unit is battery and solar powered, with the idea that the sun will power the unit and charge the battery by day and the unit will run completely from battery at night and on cloudy days.

Low power consumption devices are utilized for measuring the parameters of temperature, humidity, wind speed, wind direction, and rainfall. The 900 MHz transmitter is powered up only to send information, thus conserving power. The battery/solar-panel voltage is monitored by the microprocessor such that the "health" of the power system can be reported to the indoor unit. The schematic developed for the outdoor unit is shown in Figure 5–11. Specific portions are discussed in the following paragraphs.

Wind Speed Input

The wind speed detector, as mentioned earlier, is a magnetically activated reed-switch arrangement. Each time the anemometer turns, a magnet is brought into proximity of the switch, and the switch is closed for the duration of the presence of the magnet. The problem with mechanical switches is switch bounce. Since the input from the anemometer is going into a counter, it would be very possible to get many more counts in the counter than there were revolutions of the anemometer.

The passive circuitry incorporated into the outdoor unit helps to reduce the possibility of "extra" counts. The reed-switch is pulled up with a resistor to Vcc (5 V DC) so that a TTL-level input is available for the microprocessor. A series capacitor is used to couple the signal into the microprocessor. The pull-down resistor located at the microprocessor's input, in combination with the internal protection diodes within the AVR, provides a pulse-shaping network. As the reed-switch closes, a narrow pulse is generated at the Timer 1 input,

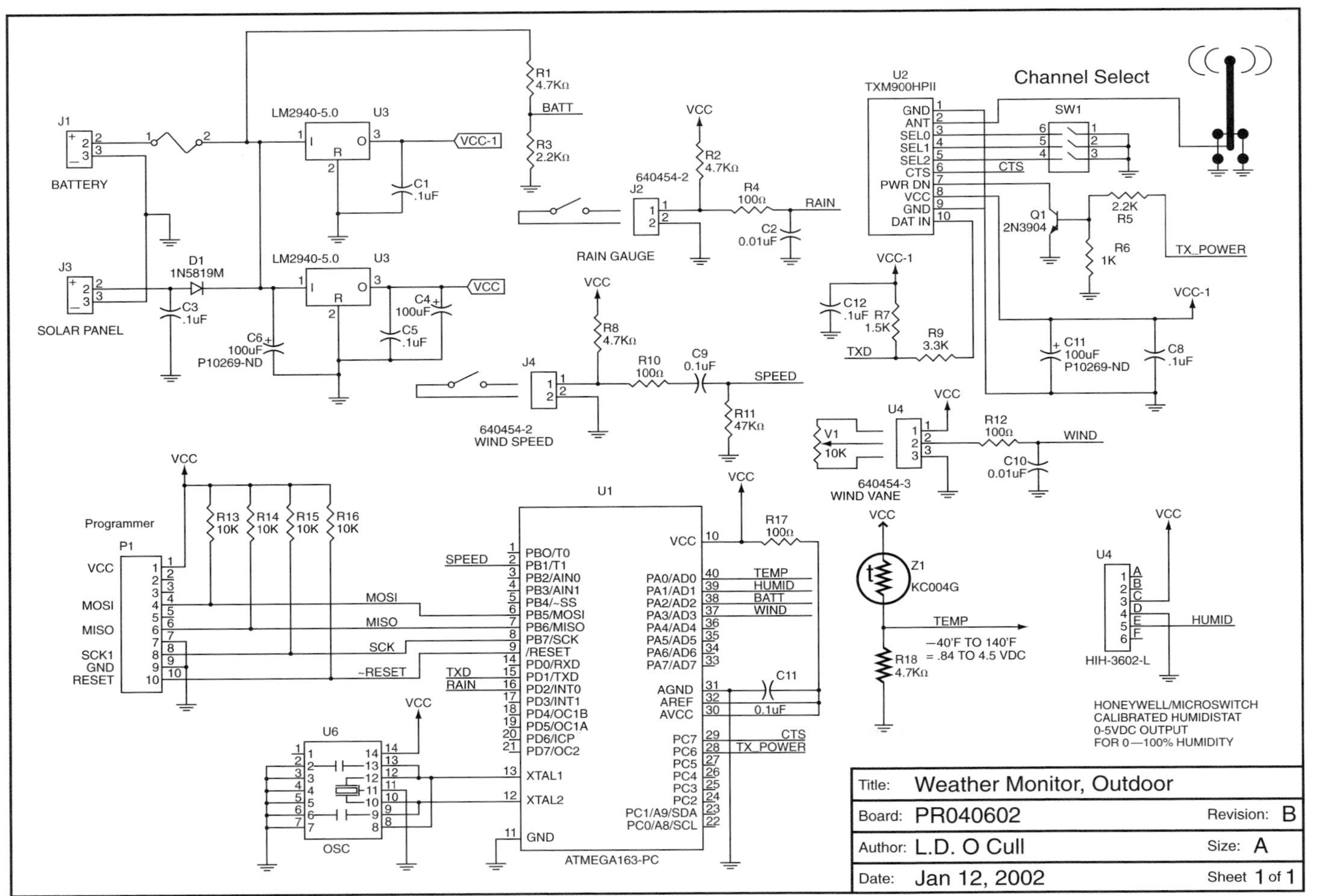

Figure 5–11 *Schematic, Weather Monitor, Outdoor Unit*

enough to cause it to count one pulse. The RC time constant of the pull-down resistor and the series capacitor allow the signal to recover slowly enough from the first edge to not pass the switch bounce but fast enough that the pulses are not missed in high wind conditions.

Rain Gauge Input

The rain gauge input is mildly filtered as well. Since we are primarily concerned with the level (on or off) at the microprocessor input, the main purpose of the filter is to eliminate noise and to "clean up" the signal coming from the reed-switch within the rain gauge.

900 MHz Transmitter

The 900 MHz transmitter is an off-the-shelf FM transmitter module by Linx Technologies. The TXM900HPII module supports eight different channels in the "portable telephone" band. A DIP switch is used to select which channel we wish to use, allowing us to move the radio frequency away from that of a local portable phone or interfering signals. The transmitter supports a power-down feature that will be used to conserve power, because the outdoor unit runs from battery and solar power.

The input to the module is the USART TXD signal directly from the Mega16. A 3.9K resistor is used to prevent overdriving the modulation input of the transmitter, which would cause it to "over modulate." The maximum modulation voltage (per the device specification) is not TTL; it is only about 1 V AC. The input of the module has a fairly low resistance. Some experimentation proved that 3.9K gave the best performance with our system. Over modulation forces the transmitter output to swing outside the proper FM modulation range, which would cause both the receiver to not "hear" the transmitted signal and the transmitter to generate unwanted RF interference.

Power Supply

The power supply is pretty basic. You will note that there are actually two 5 V supplies, one for the radio and one for everything else. This is done to reduce the amount of digital noise from the processor to the transmitter, as well as to lower the amount of conducted RF from the transmitter back into everything else. The battery is a 12 V, 2200 mAh, gel-cell type unit. These are extremely rugged and rechargeable. They are considered a zero-maintenance type of battery.

The solar panel is diode coupled into the system such that on a sunny day it will not only completely power the outside unit but charge the battery as well. You may also note that there is no current limiting between the solar panel and the system. The solar panel has a limited current capability due to its high internal resistance. Adding more resistance would simply reduce the charging rate of the battery as well as the solar panel's contribution in running the system.

5.6.5 SOFTWARE DESIGN, OUTDOOR UNIT

Designing the software is really just planning out how the software will function. When the design of the software is complete, we have a list of tasks and their priorities and a set of flowcharts describing how the tasks are tied together.

The software in the outdoor unit, by description, could be considered somewhat straightforward. The basic concept is to collect the data and send it to the indoor unit. It is not quite that simple, but it is more basic in many respects than what is going on with the indoor unit and is therefore a good place to get started.

The task list in order of top priority to least priority is as follows:

1. Read the wind speed counts from Timer 1.
2. Update the rain gauge state.
3. Handle the ADC, starting conversions and storing results.
4. Format and handle the data being sent to the transmitter.

These four tasks are all handled on an interrupt basis. The only tasks left for the main operating loop of the software are to calculate when to send information to the transmitter and to put the information into a buffer for the transmitter.

Figure 5–12 shows a basic flow diagram for the outdoor unit software. Note that there is a main process and four interrupt or exception processes: one for the rain gauge interrupt (INT0), another for sampling the temperature, humidity, wind direction, and voltage readings (ADC), another for USART transmissions (USART_TXE), and another for sampling the wind speed timer (Timer 1) and initiating transmissions (TMR0).

The basics of the actual measurements have been covered in the description of the hardware. The ADCs that are built into the AVR measure the temperature, humidity, and battery health. Timer 1 samples wind speed. INT0 is used to keep track of rainfall. The telemetry is even kept simple by using the USART of the AVR such that standard I/O routines could be used “right out of the box.”

5.6.6 HARDWARE DESIGN, INDOOR UNIT

Now we repeat the design process for the indoor unit. When the hardware design is done, we will have full schematics for the indoor unit, as shown in Figure 5–13.

The indoor unit has a 900 MHz receiver module for gathering information from the outdoor unit. The indoor unit comprises the same basic power supply, microprocessor, temperature sensing, and humidity sensing as the outdoor unit. The additions of a barometric pressure sensor, buttons, LEDs, a beeper, a 32.767KHz watch crystal for a real-time clock, and an LCD display really set it functionally apart from the outdoor unit.

There is no battery-charging system on the indoor unit, and the 900 MHz receiver is left active all the time so that the outdoor unit reports are heard. The input-power/battery voltage is monitored so that the indoor unit can report a low-battery condition.

900 MHz Receiver

The 900 MHz receiver is an off-the-shelf FM receiver module by Linx Technologies. The RXM900HPII module supports eight different channels in the “portable telephone” band. A DIP-switch is used to select which channel we wish to use, allowing us to move the radio frequency away from that of a local portable phone or interfering signals. The output from

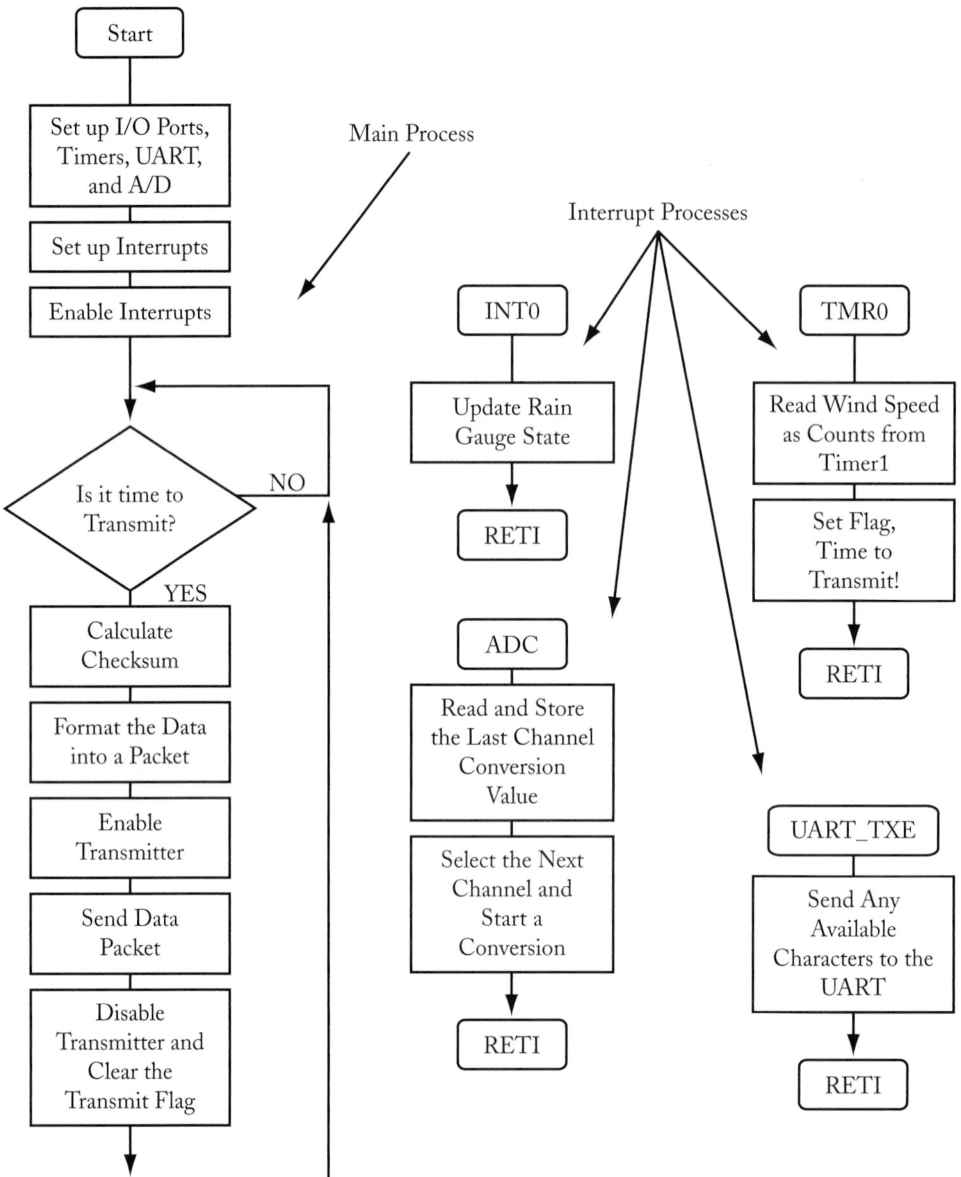

Figure 5–12 *Basic Software Flowchart, Outdoor Unit*

the module is connected to the USART RXD signal of the Mega16. A 1K resistor is used to provide a small amount of RF isolation from the microprocessor by reducing RF currents that may be passing from the microprocessor back to the receiver. These currents are a result of the capacitance that appears on all the inputs of the microprocessor. The resistor works with this capacitance to form a simple filter network.

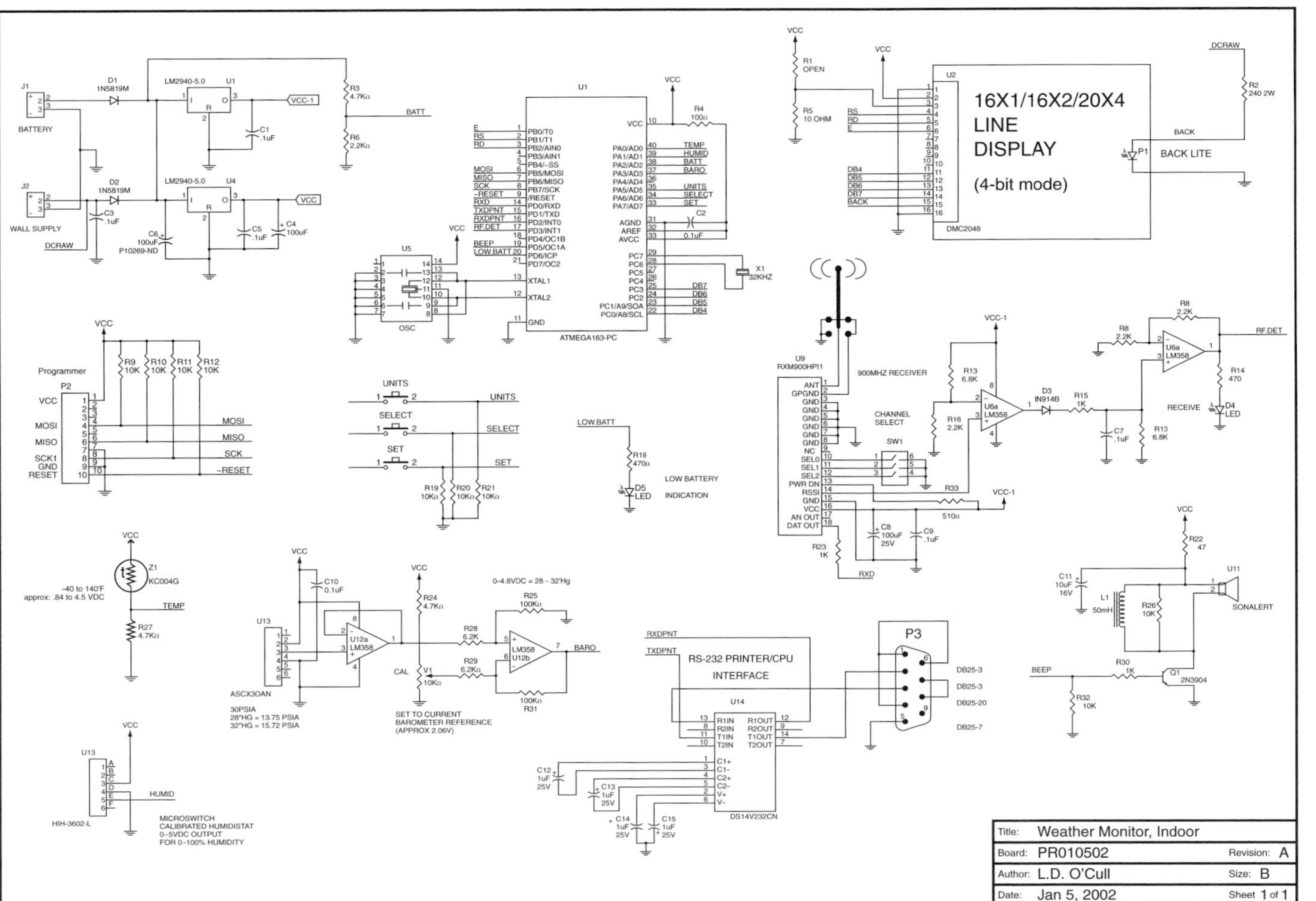

Figure 5–13 *Schematic, Weather Monitor, Indoor Unit*

The RSSI output of the receiver provides a signal-strength indication. This output is processed and lights an LED to indicate when data is being received from the outside unit. (The LED may also indicate interference and can be helpful when selecting a channel.) This processed RSSI signal is also used to "qualify" the data coming in on the USART—indicate that the data is valid, not just noise. Spurious or atmospheric noise could appear to be data. It would be left totally up to the software to discern what is signal and what is noise. By creating an "RF detected" signal, *RF.DET,* we can use an interrupt as a gating signal to enable the USART and lower the amount of software overhead required to decode the messages from the outside unit.

Power Supply

Just as in the outdoor unit, the power supply creates two 5 V supplies, one for the radio and one for everything else. This is done to reduce the amount of digital noise from the processor to the receiver. The battery is a 6 V pack, consisting of four AA alkaline batteries in series. The off-the-shelf power supply is diode-coupled into the system, allowing instant switching from power supply to battery backup in the event of a power outage.

The LCD backlight is driven directly from the wall supply. The backlight requires too much current to run from the battery, and it consumes enough current to produce a large amount of heat in the regulators. Therefore, we opted to drive the backlight directly and bypass all of these issues. In the event of a power failure, the unit will continue to operate (on battery power), but the display backlight will be out, making the display harder to see—we can shine a flashlight on it if we need to know what it says.

5.6.7 SOFTWARE DESIGN, INDOOR UNIT

The indoor unit has two basic functions: collect and convert data, and interface with a human. The collection process is fairly straightforward in that the indoor unit measures the indoor temperature, humidity, barometric pressure, and battery health on its own ADCs in exactly the same manner as the outdoor unit. The outdoor parameters of temperature, humidity, rainfall, wind speed, wind direction, and battery health are collected in the same units as the indoor unit (ADC counts, timer counts, etc.).

The toughest part to developing the indoor unit software is defining the human interface. Referring to "Operational Specification" in Section 5.6.2, "Definition Phase," this weather monitor uses a four row by twenty column LCD and an arrangement of three buttons. A 4X20 LCD may seem like a lot of room for information, but it can quickly become limiting. In this system, all four lines are used to display parameters. The parameters are displayed in groups of three, as shown in Table 5–4.

Temperature Indoor	Temperature Outdoor	Wind Chill (Outdoor)
Barometric Press. (▲, ▼, –)	Wind Speed	Wind Direction, ° & Head.
Humidity Indoor	Humidity Outdoor	Dew Point (Outdoor)
Rainfall this (H, D, M, Y)	Time (* flashes seconds)	Date

Table 5–4 *Parameter Triads to Be Displayed*

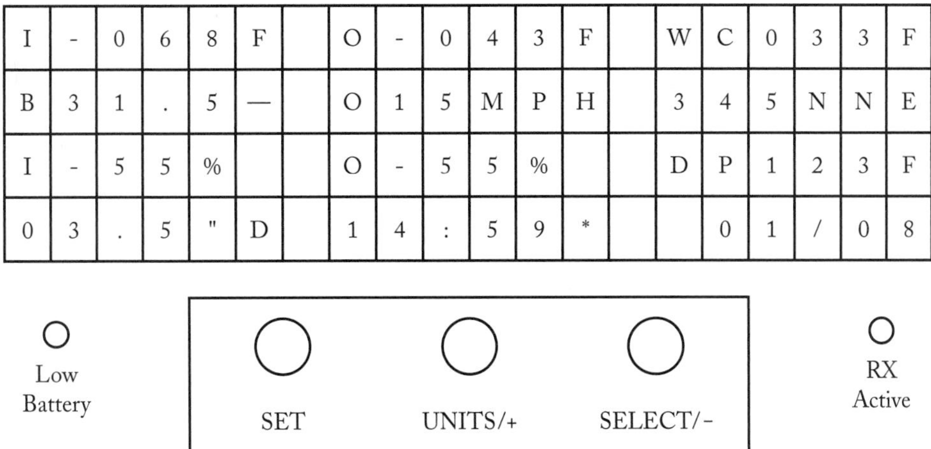

Figure 5–14 *Example Indoor Display Unit LCD and Button Layout*

Figure 5–14 shows an example layout for the LCD, LED, and buttons for the indoor unit.

The button functions are SET, UNITS/+, and SELECT/–. The SET button allows the user to set the time. The UNITS/+ and SELECT/– buttons increase or decrease the time or date component, respectively, and the SET button takes the user to the next component until all have been set. Each time the user presses the UNITS button, the current units will switch from imperial to metric, or vice versa. This allows temperature to be viewed in °C or °F, and wind speed in mph or kph. The SELECT/– button allows the user to view the rainfall in inches-this-hour, inches-this-day, inches-this-month, or inches-this-year.

There are many possible alarm or alert conditions that a weather monitor like this could detect. Temperatures that are too high or too low, extreme wind and rain, high indoor humidity that could damage furniture or equipment, and sudden large swings in barometric pressure that could indicate a storm is on its way. Table 5–5 shows some possible warning/alert conditions.

For the purposes of this text, these warnings go a bit beyond the scope of what we are trying to cover, but these messages would make a great additional feature if you choose to implement them in your weather monitor. For our weather monitor, we will key in on a couple of basic system errors that can prevent the monitor from performing correctly:

- Low battery conditions, indoors or outdoors
- No data communications from the outdoor unit

An LED indicator is located near the display to signal a detected low-battery or low-voltage condition on the indoor unit. Whenever a low-voltage condition appears, software will flash the LED until the SET button is pressed. This will inform the user that during a power outage the batteries were found to be low—it is time for a change. The user could put in new

High Indoor Temp
Low Indoor Temp
High Outdoor Temp
Low Outdoor Temp
High-Pressure Alert
Low-Pressure Alert
High Indoor Humidity
High-Wind Alert
High-Rainfall Alert
Low Batt Outdoor
Low Batt Indoor
No Outdoor Comm

Table 5–5 *Possible Warning/Alert Messages*

batteries and press the SET button to clear the error. The batteries can be checked at any time, simply by unplugging the power supply from the unit.

This low-voltage detection can also be used for an intelligent shutdown of the system, in either the outdoor or the indoor unit. If the voltage is detected to be below normal, data such as the accumulated rainfall can be saved to EEPROM, and then the unit could simply hold in a loop waiting for the power to disappear completely or to return to an acceptable level. Handling EEPROM storage in this as-needed manner also preserves its life, because continuously writing to EEPROM will eventually lead to its failure. As discussed in Chapter 1, the potential for failure is due to the way that EEPROM memory itself is constructed, a function of electro-chemistry. In many cases, this memory area will have a rating of 10,000 or 100,000 write operations, *maximum*.

A low-battery condition on the outside unit will eventually lead to no data reports from the outside unit. In this case, the low-battery LED indicator on the inside unit will be turned on solid until the outside unit voltage goes back up to normal. When the outside unit voltage goes back to normal, the LED will go back off.

When valid data has not been received from the outdoor unit within 15 seconds, a "No Communications" error exists. This could be due to the outdoor unit's power going away, or a lengthy period of RF interference. While this condition exists, all outdoor-related parameters will be replaced with question marks. For example, wind speed would read:

```
??? MPH
```

5.6.8 TEST DEFINITION PHASE

The test definition phase of the project outlines specifically how we intend to test both the individual pieces of the project and the complete project. Some measurements are very easy to test, while others require much more work and creativity. Some may also require writing software for the microcontroller.

The test definitions for various parts of the project are shown below:

Wind Direction

Input: 0 to 5 V DC through a 10K ohm potentiometer.

Expected Results: 0 to 5 V input to microcontroller.

Method: Apply different voltages to the input through a potentiometer to simulate a weather vane and measure the voltage input to the microcontroller with a voltmeter.

Wind Speed

Input: 30 mph wind—might require a friend, a reliable method of transportation, and a calm day!

Expected Results: 75 Hz signal to the microprocessor.

Method: Cause the anemometer to be blown at a known rate and measure the pulse train to the microcontroller using a frequency counter, an oscilloscope, or the microcontroller itself. You might need to write software to count the pulses and transmit them to the indoor unit for display if a portable method of counting the pulses is required. The selected anemometer for this project is factory calibrated and produces a single pulse per revolution. The output is approximately 2.515 Hz/mph. Only minor adjustments may be required.

Rain Gauge

Input: One cup (8 oz.) of water.

Expected Results: A representative number of transitions presented to the microcontroller input.

Method: *Slowly* pour the water into the rain gauge and, using a counter or the microcontroller itself, count the transitions. The RG-T transitions every 0.01 inch of rainfall, so a counter should indicate 722 low-to-high transitions, or 1444 changes of state. You may need to write software to count the transitions and transmit them to the indoor unit for display. The selected rain gauge for this project is factory calibrated and should need no adjustments.

Air Temperature

Input: Ice water, room temperature, warm hands.

Expected Results: A representative voltage presented to the microcontroller input.

Method: Place the NTC device on some long wire leads, 24 inches or so. Seal the device in a small plastic bag (remove as much air a possible). Place the device in ice water 32°F (0°C); after a minute or so, 2.39 V DC should be present at the microcontroller's input. At room temperature, 70°F (21°C) a voltage of approximately 3.3 V should be present.

Barometric Pressure

Input: The atmosphere.

Expected Results: A representative voltage presented to the microcontroller input.

Method: Using another barometer as a reference (or the local weather station if it transmits continuously through the TV), measure the voltage input to the microcontroller with a voltmeter—0 V DC represents 28 inHg, and 5 V DC represents 32 inHg.

Relative Humidity

Input: The atmosphere.

Expected Results: A representative voltage presented to the microcontroller input.

Method: Using another humidistat as a reference, measure the voltage input to the microcontroller with a voltmeter and compare the values—0 V DC represents 0% RH and 5 V DC represents 100% RH. The selected humidistat for this project is factory calibrated and should need no adjustments.

System Test for Complete Project

The extent of the system-test definition will vary according to the end user of the project. If the weather station project were to be destined for home use, the test might be no more extensive than comparing the weather station readouts to the local weather station to see the results are "about right." On the other hand, if the project is for commercial usage or for a customer, more extensive testing is required. In this case, it would be appropriate to use temperature and humidity-controlled chambers to run the outside and inside units through the entire range of weather conditions and record the results. For instance, it would be appropriate to test the temperature reading at, say, every 5°F and plot the results to ensure that the unit meets specifications over the entire range of temperature. Each of the weather parameters would be tested in the same thorough manner and the results graphed for presentation to the customer.

Whatever the extent of the tests imposed for system test, the intent is the same: to give the end user confidence that the unit is performing up to specifications as set down during the definition phase.

5.6.9 BUILD AND TEST PROTOTYPE HARDWARE PHASE

The project hardware is completely designed, and the expected results of each section of the hardware are specified. It is now time to build the hardware. After the hardware is assembled, the tests from the previous phase are executed. This section briefly discusses some real-life methods of executing the tests and measuring the results. The software required to execute the tests is also discussed.

Fundamental tests of the hardware are best performed with voltmeters and oscilloscopes. When an AVR microcontroller is completely erased, all of its pins are inputs, and it has no impact on the surrounding circuitry. This allows for some basic checking of the hardware functionality. No matter how good the software is, it can never make up for hardware that just does not work. Listed below are some step-by-step checks for both the outdoor-unit and the indoor-unit hardware. This example method of validation applies to many types of designs. Knowing for certain that the signals are as expected greatly reduces the software development and integration time, as well as the amount of time you will spend "chasing your tail" when things do not make sense.

Outdoor Unit Checkout

1. Set the lab supply for 12 V DC with a current limit of approximately one ampere. At this point, no battery or solar panel should be connected.
2. Connect a lab supply to the battery or solar panel connection.
3. Make sure that Vcc and Vcc-1 are at 5 V DC. If the power supply is not correct, nothing will be quite right.
4. Using a voltmeter, verify that the BATT signal (pin 38 of the Mega16) is approximately 3.8 V DC.
5. Using a voltmeter, verify that the output of the humidistat is a level that corresponds with the humidity of the room that you are in. An office or classroom is typically 20 to 30% RH. Whatever the percentage is, you should measure about that percentage of 5 V on the HUMID signal (pin 39 of the Mega16).
6. Using a voltmeter, verify that the output of the temperature sensor is a level that corresponds with the temperature of the room that you are in. The average office or classroom temperature is about 70°F (21°C), and that should cause a voltage of approximately 3.3 V to appear at the TEMP signal (pin 40 of the Mega16). If you pinch the thermistor in your fingers while watching this voltage, you should see the voltage rise as your body heat warms the sensor.
7. Connect the anemometer and the wind vane to the appropriate inputs.
8. Using an oscilloscope, verify that the SPEED signal (pin 2 of the Mega16) has pulses of adequate level (TTL level, which is greater than 3.3 V for a "high" and less than 1.5 V for a "low"). These pulses will exist only while you spin the anemometer, so give it a whirl!
9. Using a voltmeter, verify that the WIND input (pin 37 of the Mega16) moves uniformly from 0 V to 5 V DC as you *slowly* rotate the wind vane. As you cross north, the voltage will jump from 0 V to 5 V, or from 5 V to 0 V.
10. Connect the rain gauge to the appropriate input.
11. Using a voltmeter or an oscilloscope, verify the RAIN signal (pin 16 of the Mega16). As you either tip the seesaw assembly by hand or pour water through the gauge, you should see the voltage of the RAIN signal transition from 0 V to 5 V DC and back again.
12. Using the test program listed below, program the Mega16 with the following fuse bit settings. These settings guarantee the proper operation of the oscillator and brown-out

detector used to reset the microcontroller:

- CKSEL3:0=1100, Crystal Oscillator, BOD enabled
- SUT1:0 = 10
- BODEN = ON
- BODLVL = ON
- JTAG = OFF
- OCD = OFF
- EESAVE = OFF
- BOOTRST = OFF

```
/*********************************************
      Outdoor unit test
*********************************************/
#include <Mega16.h>
#include <stdio.h>
#include <delay.h>

void main(void)
{
    PORTC=0x00; // setup port C and turn on TX_POWER..
    DDRC=0x40;

    // USART Transmitter: On
    // USART Baud rate: 9600, 8 Data, 1 Stop, No Parity
    UCSRA=0x00;
    UCSRB=0x08;
    UBRR=0x19;
    UBRRHI=0x00;

    while (1) // send test stream every 1 second
    {
            delay_ms(100);      // allow power up time
            putsf("UUU$test*QQQ\r");
            delay_ms(500);      // transmit message
            PORTC.6 = 1;        // RF power off
            delay_ms(500);      // wait..
            PORTC.6 = 0;        // RF power on
    }
}
```

13. Using a RF signal-strength meter, a spectrum analyzer, or a Linx Technologies TXM900HPII evaluation kit with a PC, monitor the messages being transmitted. Note that the RF energy should go off between transmissions for 500 milliseconds. Transmissions should occur about one second apart.

Indoor Unit Checkout

1. Set the lab supply for 9 V DC with a current limit of approximately one ampere. At this point, no batteries should be connected.
2. Connect a lab supply to the wall supply connection.
3. Make sure that Vcc and Vcc-1 are at 5 V DC. If the power supply is not correct, nothing will be quite right. Sound familiar? It is still true.
4. Make sure the backlight is lit on the LCD display.
5. Using a voltmeter, verify that the BATT signal (pin 38 of the Mega16) is approximately 2.8 V DC.
6. Using a voltmeter, verify that the output of the humidistat is a level that corresponds with the humidity of the room that you are in. An office or classroom is typically 20 to 30% RH. Whatever the percentage is, you should measure about that percentage of 5 V on the HUMID signal (pin 39 of the Mega16).
7. Using a voltmeter, verify that the output of the temperature sensor is a level that corresponds with the temperature of the room that you are in. The average office or classroom temperature is about 70°F (21°C), and that should cause a voltage of approximately 3.3 V to appear at the TEMP signal (pin 40 of the Mega16). If you pinch the thermistor in your fingers while watching this voltage, you should see the voltage rise as your body heat warms the sensor.
8. Using a voltmeter to measure the CAL potentiometer wiper, set the CAL potentiometer such that the wiper voltage is 2.06 V DC. Once this is set, measure the voltage at the BARO signal (pin 37 of the Mega16) and verify that it is somewhat representative of the current barometric pressure. Remember that 0 V DC at the BARO signal represents 28 inHg and that 5 V DC represents 32 inHg. On an average day, the reading should be somewhere in the middle. This might be a good time to actually calibrate the barometer. Get a current local barometric pressure reading and calculate the voltage this pressure should produce at the BARO signal. Adjust the CAL potentiometer until you read that voltage at the BARO signal.
9. With the outdoor unit running its test program as an RF signal source, check the RF.DET signal by monitoring the RX Active LED. The RX Active LED should go on for 600 milliseconds and then off for 500 milliseconds, again and again. Verify that data is getting to the processor by looking at the RXD input (pin 14) of the Mega16 with an oscilloscope. The waveform should look something like the TXD output (pin 15) of the outdoor unit.
10. Using the test program listed below, program the Mega16 with the following fuse bit settings. These settings guarantee the proper operation of the oscillator and brown-out detector used to reset the microcontroller:
    - CKSEL3:0=1100, Crystal Oscillator, BOD enabled
    - SUT1:0 = 10
    - BODEN = ON

- BODLVL = ON
- JTAG = OFF
- OCD = OFF
- EESAVE = OFF
- BOOTRST = OFF

```
/*********************************************
      Indoor Unit Test Code
*********************************************/

#include <Mega16.h>
#include <delay.h>

#define LCD_E    PORTB.0        /* LCD Control Lines */
#define LCD_RS   PORTB.1
#define LCD_RW   PORTB.2
#define LCD_PORT PORTC          /* LCD DATA PORT */
unsigned char LCD_ADDR;

/***** Display Functions *****/
void wr_half(unsigned char data)
{
    LCD_RW = 0;           // set WR active

    LCD_E = 1;            // raise E
    LCD_PORT = (data & 0x0F); // put data on port
    LCD_E = 0;            // drop E

    LCD_RW = 1;           // disable WR

    delay_ms(3);          // allow display time to think
}

void wr_disp(unsigned char data)
{
    LCD_RW = 0;           // enable write..

    LCD_E = 1;
    LCD_PORT= (data >> 4);
    LCD_E = 0;            // strobe upper half of data

    LCD_E = 1;            // out to the display
    LCD_PORT = (data & 0x0F);
    LCD_E = 0;            // now, strobe the lower half
```

```
    LCD_RW = 1;      // disable write

    delay_ms(3);   // allow time for LCD to react
}

void disp_char(unsigned char c)
{
    LCD_RS = 1;
    wr_disp(c);
    LCD_ADDR++;
}

void disp_cstr(unsigned char flash *sa)     /* display string
from ROM */
{
    while(*sa != 0)
      disp_char(*sa++);
}

void init_display(void)
{
    char i;

    LCD_RW = 1;
    LCD_E = 0;        // preset interface signals..
    LCD_RS = 0;             // command mode..
    delay_ms(50);

    wr_half(0x33);     // This sequence enables the display
    wr_half(0x33);     // for 4-bit interface mode.
    wr_half(0x33);     // These commands can be found
    wr_half(0x22);     // in the manufacturer's data sheets.

    wr_disp(0x28);     // Enable the internal 5x7 font
    wr_disp(0x01);
    wr_disp(0x10);     // Set cursor to move
                       // (instead of display shift).
    wr_disp(0x06);     // Set the cursor to move right, and
                       // not shift the display.
    wr_disp(0x0c);     // Turns display on, cursor off, and
                       // cursor blinking off.

      // init character font ram to "00"
    for(i=0x40; i<0x5F; i++)
    {
```

```
            delay_ms(10);
                LCD_RS = 0;
            wr_disp(i);
            delay_ms(10);
            disp_char(0);
        }
        LCD_RS = 1;                        // data mode
    }

    void clear_display(void)
    {
        LCD_RS = 0;
        wr_disp(0x01);        // clear display and home cursor
        delay_ms(10);
          LCD_RS = 1;
    }

    void main(void)
    {
        // Port B
        PORTB=0x00;
        DDRB=0x07;      // outputs for LCD signals

        // Port C
        PORTC=0x00;
        DDRC=0x0F;      // outputs for LCD data

        // Port D
        PORTD=0x00;
        DDRD=0x60;      // beeper and LED outputs

        // Enable 32K Clock source: TOSC1 pin
        ASSR=0x08;


        delay_ms(100);

        init_display();       // initialize LCD
        delay_ms(100);

        clear_display(); // clear display
        delay_ms(100);

        disp_cstr("Indoor Test"); // print test message..
```

```
    while (1)
    {
            if(PINA & 0xE0) // if any button pressed...
            {
                    PORTD.6 = 1;    // LOW_BATT LED ON
                    PORTD.5 = 1;
                    delay_ms(1);
                    PORTD.5 = 0;    // beep the beeper
                    delay_ms(1);
            }
            else
            {
                    PORTD.6 = 0;    // LOW_BATT LED OFF
                    PORTD.5 = 0;    // beeper OFF
            }
    }
}
```

11. Verify the message "Indoor Test" appears on the LCD.
12. Press the UNITS, SELECT, and SET buttons. The beeper should sound at approximately 500 Hz, and the LOW BATTERY LED should light as long as one of the buttons is pressed.
13. Using an oscilloscope, verify that the 32 KHz oscillator is running by probing pins 28 and 29 of the Mega16.

5.6.10 SYSTEM INTEGRATION AND SOFTWARE DEVELOPMENT PHASE, OUTDOOR UNIT

The first step in putting together the software for the outdoor unit is to define the unit operation. In this case, a system decision can be made as to where the processing of the actual data is to occur, indoors or outdoors. If the data were processed outdoors, then each of the parameters would have to be calibrated, and common units, like °F or °C for temperature, would need to be defined for the data that is to be transmitted. Rainfall gets a bit messy in that it basically takes forever, and the outside unit would have to know how long forever is. This thought process leads us to leaving the data collected in the most basic form possible and letting the indoor unit do the conversions at the end. Besides, the indoor unit has a clock and interfaces with the human anyway.

So in this system the most basic units would be ADC counts, Timer 1 counts, and the state of the rain gauge and the weather vane. This data can be combined and transmitted as a packet of values on a periodic basis. The transmitted period of the packets is not really critical. There should be enough data to catch the peaks of wind and get an accurate representation of rainfall. These are not particularly fast-moving parameters, but humans like to feel like they are being updated fairly frequently. For instance, a typical digital voltmeter offers a one-second sampling rate. The only critical sampling that is taking place is the wind speed.

It is the time between readings that is important for calculating the mph of the wind. Even though it will be sampled quickly, an average of the readings is all that is necessary for an adequate wind speed representation.

One consideration is the battery consumption during the transmission of the data. It is during the transmission of the data to the indoor unit that the highest current requirements of the system are achieved. The duty cycle, or ratio of time active to time inactive, ultimately determines the average current draw on the power supply, whether it is the solar panel, the battery, or both.

The next step in getting the software started is to define the inputs and outputs of the microprocessor that are to be used. This is a great way to get started, since another round of sanity checking gets performed on the schematic, and once the definitions are made, the need for constantly referring back to the schematic is reduced. The definitions for the outdoor unit may appear as shown in Figure 5–15. A reminder: comments following the **#define** statements need "/*" and "*/" as delimiters.

Once the pins have been labeled, the actual inputs, outputs, and peripherals need to be configured. This is usually performed at or called from the top of *main()*. In the initialization routine "`init_AVR()`" (see code in Figure 5–16), note that Timer 0 is set to create an overflow interrupt. Based on a 4.0 MHz oscillator, the interrupt will be issued every 4.096 ms. This real-time tick will be used to sample the anemometer as well as determine when it is time to transmit the data to the indoor unit. Timer 1 is configured as a counter to capture the switch transitions in the anemometer to measure wind speed. INT0 is set to interrupt on "`any change`", which allows each tipping of the seesaw in the rain gauge to be captured. The USART is configured for 9600 baud, 8 data bits, and 1 stop bit for the RF transmissions. The ADC is set up to convert at the slowest rate, which is generally the best, and to generate an interrupt upon completion of each conversion.

```
#define TX_CTS              PINC.7     /* transmitter clear to send */
#define TX_POWER            PORTC.6    /* transmitter power enable */
#define WIND_SPEED_INPUT  PINB.1     /* anemometer input (Timer1) */
#define RAIN_INPUT          PIND.2     /* rain gauge input  */

#define FIRST_ADC_INPUT 0             /* A/D converter parameters */
#define LAST_ADC_INPUT 3
#define ADC_VREF_TYPE 0x40
unsigned int adc_data[LAST_ADC_INPUT-FIRST_ADC_INPUT+1];

#define TEMPERATURE        adc_data[0] /* analog chan 0, int */
#define HUMIDITY           adc_data[1] /* analog chan 1, int */
#define BATTERY            adc_data[2] /* analog chan 2, int */
#define WIND_DIRECTION     adc_data[3] /* analog chan 3, int */
```

Figure 5–15 *Example Definitions, Outdoor Unit I/O*

```
#define FIRST_ADC_INPUT 0      /* A/D converter parameters */
#define LAST_ADC_INPUT 3
#define ADC_VREF_TYPE 0x40

void init_AVR(void)
{
    /* Input/Output Ports initialization  */
    /* Port A  */
    PORTA=0x00;
    DDRA=0x00;

    /* Port B */
    PORTB=0x00;
    DDRB=0x00;

    /* Port C */
    PORTC=0x00;
    DDRC=0x40;

    /* Port D  */
    PORTD=0x00;
    DDRD=0x02;

    /* Timer/Counter 0 initialization */
    /* Clock source: System Clock     */
    /* Clock value: 62.500 kHz        */
    /* Mode: Output Compare           */
    /* OC0 output: Disconnected       */
    TCCR0=0x03;
    TCNT0=0x00;

    /* Timer/Counter 1 initialization    */
    /* Clock source: T1 pin Rising Edge  */
    /* Mode: Output Compare              */
    /* OC1A output: Discon.              */
    /* OC1B output: Discon.              */
    /* Noise Canceler: Off               */
    /* Input Capture on Falling Edge     */
    TCCR1A=0x00;
    TCCR1B=0x07;
    TCNT1H=0x00;
    TCNT1L=0x00;
    OCR1AH=0x00;
    OCR1AL=0x00;
    OCR1BH=0x00;
    OCR1BL=0x00;

    /* Timer/Counter 2 initialization  */
    /* Clock source: System Clock      */
```

Figure 5–16 *Outdoor Unit, I/O Initialization (Continues)*

```
    /* Clock value: Timer 2 Stopped    */
    /* Mode: Output Compare            */
    /* OC2 output: Disconnected        */
    TCCR2=0x00;
    ASSR=0x00;
    TCNT2=0x00;
    OCR2=0x00;

    /* External Interrupt(s) initialization */
    /* INT0: On                             */
    /* INT0 Mode: Any change                */
    /* INT1: Off                            */
    /* INT2: Off                            */
    GICR|=0x40;
    MCUCR=0x01;
    MCUCSR=0x00;
    GIFR=0x40;

    /* Timer(s)/Counter(s) Interrupt(s) initialization */
    TIMSK=0x01;

    /* UART initialization                                    */
    /* Communication Parameters: 8 Data, 1 Stop, No Parity */
    /* UART Receiver: Off                                     */
    /* UART Transmitter: On                                   */
    /* UART Baud rate: 9600                                   */
    UCSRA=0x00;
    UCSRB=0x48;
    UBRRL=0x19;
    UBRRH=0x00;

    /* Analog Comparator initialization                       */
    /* Analog Comparator: Off                                 */
    /* Analog Comparator Input Capture by Timer/Counter 1: Off */
    ACSR=0x80;
    SFIOR=0x00;

    /* ADC initialization               */
    /* ADC Clock frequency: 31.250 kHz */
    /* ADC Voltage Reference: AVCC pin */
    ADMUX=FIRST_ADC_INPUT|ADC_VREF_TYPE;
    ADCSRA=0xCF;

    /* Watchdog Timer initialization      */
    /* Watchdog Timer Prescaler: OSC/2048 */
    WDTCR=0x0F;

    /* Global enable interrupts           */
    #asm("sei")
}
```

Figure 5–16 *Outdoor Unit, I/O Initialization (Continued)*

Temperature, Humidity, Wind Direction, and Battery Health

Temperature, humidity, wind direction, and battery health can all be measured in the background using an interrupt routine servicing the ADC. The values of the array "adc_data" are automatically loaded. The definitions, "TEMPERATURE", "HUMIDITY", "WIND_DIRECTION", and "BATTERY" are aliases to help us remember which channel represents which signal. The general-purpose ADC interrupt subroutine is listed in Figure 5–17.

Rainfall

Rainfall is handled by simply keeping track of the state of the I/O pin. Using an interrupt in this manner is probably a little overdone, but there are still some questions about the "real world" operation of the rain gauge, and this is a good way to leave some options open. The INT0 function of PIND.2 has been setup to generate an interrupt in "any change". This means when the signal at PIND.2 changes from high-to-low or low-to-high, the interrupt service routine "ext_int0_isr()" (shown in Figure 5–18) gets called. In this interrupt routine, the character "rain_state" is set to reflect the rain gauge input

```
#define FIRST_ADC_INPUT 0           /* A/D converter parameters */
#define LAST_ADC_INPUT 3
#define ADC_VREF_TYPE 0x40
unsigned int adc_data[LAST_ADC_INPUT-FIRST_ADC_INPUT+1];
char  input_index=0;

#define TEMPERATURE      adc_data[0] /* analog chan 0, int */
#define HUMIDITY         adc_data[1] /* analog chan 1, int */
#define BATTERY          adc_data[2] /* analog chan 2, int */
#define WIND_DIRECTION   adc_data[3] /* analog chan 3, int */

/* ADC interrupt service routine     */
/* with auto input scanning          */
interrupt [ADC_INT] void adc_isr(void)
{
    /* Read the ADC conversion result */
    adc_data[input_index]=ADCW;

    /* Select next ADC input */
    if (++input_index>(LAST_ADC_INPUT-FIRST_ADC_INPUT))
       input_index=0;

    ADMUX=input_index+FIRST_ADC_INPUT|ADC_VREF_TYPE;
    /* Start the next ADC conversion */

    ADCSRA|=0x40;
}
```

Figure 5–17 *Example Free-Running ADC Interrupt Service Routine*

```
#define RAIN_INPUT     PIND.2     /* rain gauge input   */
char rain_state;   /* current state of rain gauge as a char */

/* External Interrupt 0 service routine */
interrupt [EXT_INT0] void ext_int0_isr(void)
{
    if(RAIN_INPUT)
         rain_state = 1; /* keep change around in a variable */
    else                 /* for later transmission to indoor */
         rain_state = 0; /* unit... */
}
```

Figure 5–18 *Rain Gauge Input Interrupt Handler*

"`RAIN_INPUT`", which is the port pin itself. Earlier, we discussed the possibility of switch bounce. In the case of the rain gauge, because the seesaw moves so infrequently, and because we report to the indoor unit the state of the actual input pin, the chance of actually having an erroneous reading is negligible.

Wind Speed

In the previous sanity checks, we decided to use Timer 1 to count the wind speed pulses. Sampling the count on Timer 1 is handled within the Timer 0 interrupt routine. As mentioned previously, the Timer 0 interrupt executes every 4.096 ms. If we reduce the sampling rate to every second, at least until we can obtain actual data from the anemometer, the collected number in Timer 1 would effectively be five times larger. To obtain a sample rate of approximately one second, we will need to take a Timer 1 reading once every 244 passes through the Timer 0 interrupt routine.

```
(1 second ÷ 4.096 ms/pass) = 244.1
```

The one-second interval also provides a method of timing when the RF telemetry is to take place. Once the samples are taken, a flag "`RF_TX_Time`", as shown in Figure 5–19, is set and is picked up in *main()* to initiate a transmission.

RF Telemetry

The RF telemetry is handled in the *main()* function. The flag "`RF_TX_Time`" is used to signal when data is to be sent to the indoor unit. The data is prearranged into packets.

Packets provide a structure to the data that makes it easier to the receiving device to decode. When the data is organized in this fashion, the receiving device is able to predict what kind of data to expect and how much. The format of the data packet to be sent is as follows:

```
UUU$ttt.hhh.vvv.ddd.ssss.r.xxxx*QQQ
```

where *ttt* is the temperature ADC reading, *hhh* is the humidity ADC reading, *vvv* is the battery voltage reading at the ADC, and *ddd* is the wind direction. The data for the ADC

```
#define WIND_SPEED_INPUT PINB.1    /* anemometer input (Timer1) */
unsigned int wind_speed;           /* average counts.. */
int wind_test_pass;       /* pass count before wind speed sample */

/* Timer 0 overflow interrupt service routine */
interrupt [TIM0_OVF] void timer0_ovf_isr(void)
{
   if(++wind_test_pass > 243)
                              /* passes before wind speed sample */
   {
       wind_speed +=TCNT1;  /* read(accumulate)the timer value */

       TCNT1 = 0;              /* reset counter for next sample.. */
       wind_test_pass = 0;    /* reset the pass counter.. */

       wind_speed /= 2;     /* simple average of two.. */

       RF_TX_Time = 1;        /* time to transmit another packet */
   }
}
```

Figure 5–19 *Timer 0 Interrupt Service Routine*

readings are each sent as 3-nibble values, because the ADC values will be between 0x000 and 0xFFF. This saves transmission time by reducing the total number of characters sent. The value *ssss* is the Timer 1 reading, *r* is the rain-gauge state, and *xxxx* is a checksum made up of the sum of the data itself. The values are all in hexadecimal.

The preamble *UUU* is used to provide a somewhat symmetrical bit pattern as the packet is starting. The ASCII code for the letter "U" is a hexadecimal 55. This helps the modulation characteristics of the transmitter and is easy to detect at the indoor unit. The dollar sign ($) is used to indicate the actual start of packet. The dots (.) are used to delimit the values such that they can be easily decoded by the *sscanf()* function in the indoor unit. The asterisk (*) is used to indicate the end of the packet, and the *QQQ* is used as a postscript to again provide a somewhat symmetrical bit pattern as the packet is ending. The ASCII code for the letter "Q" is a hexadecimal 51.

The checksum is used to check the validity of the information that is received by the indoor unit. Checksums come in a variety of forms, such as the ones complement of the sum of the data, the twos complement of the sum of the data, CRCs (*cyclic redundancy checks*), and so on. In general, a checksum is the summing or combining of the data within the message such that by utilizing the values contained within the message along with the checksum, the receiving device can test the validity of the content of the message. If the checksum does not match the sum of the data received, the data can be rejected. It is not uncommon in a wireless system to have RF interference or atmospheric conditions alter the message. What is important is that we do not calculate the weather readings with bad data.

So, using a data packet of the form

```
UUU$ttt.hhh.vvv.ddd.ssss.r.xxxx*QQQ
```

with a temperature reading of 70°F, a humidity reading of 50%, a battery level of 11.5 V, a wind out of the west at a speed of 5 mph, and no rainfall, the packet of data viewed with an ASCII terminal might look like this:

```
UUU$276.800.2EE.2FF.0032.0.1095*QQQ
```

The actual values were calculated using the conversion factors for each parameter being measured. Since the example temperature is 70°F, its value to be transmitted is calculated as follows:

$$\frac{\text{Proportion of range used}}{\text{Entire range}} \cdot 750 \text{ counts in the range} + 172 \text{ counts (to account for the 0.84 V offset)} = \text{ADC output}$$

which computes to be

$$\frac{70 - (-40)}{140 - (-40)} \cdot 750 + 172 = 630.3_{10} \text{ Counts} \rightarrow 276_{16} \text{ Counts}$$

Therefore the temperature value within the packet is 0x276 as shown below:

UUU$**276**.800.2EE.2FF.0032.0.1095*QQQ

As in other examples in this text, interrupts are used to actually send the characters out of the USART. As shown in Figure 5–20, *putchar()* has been redefined to work with this interrupt service routine, and the "`#define _ALTERNATE_PUTCHAR_`" is used to signal to the standard library that this *putchar()* is to replace the built-in function.

```
#define TX_BUFFER_SIZE          48
char   TX_Buffer [TX_BUFFER_SIZE+1]; /* UART Transmitter Buffer */
char   TX_Counter;                  /* interrupt service parameters */
char   TX_Rd_Index;
char   TX_Wr_Index;   /* UART Transmitter interrupt service routine */

/* UART Transmitter interrupt service routine   */

interrupt [USART_TXC] void uart_tx_isr(void)
{
    if(TX_Counter != 0)
    {
```

Figure 5–20 *Example, Interrupt Driven USART Transmission (Continues)*

```
        if(fPrimedIt == 1)      /* only send a char if one in buffer */
        {
              fPrimedIt = 0;  /* transmision, then don't send the */

              if(++TX_Rd_Index > TX_BUFFER_SIZE)
                                  /* test and wrap the pointer */
                  TX_Rd_Index = 0;

              TX_Counter--;       /* keep track of the counter */
          }

          if(TX_Counter != 0)
          {
               UDR = TX_Buffer[TX_Rd_Index];
                                  /* otherwise, send char out port */

             if(++TX_Rd_Index > TX_BUFFER_SIZE)
                                  /* test and wrap the pointer */
                   TX_Rd_Index = 0;

             TX_Counter--;
                                  /* keep track of the counter */
          }
      }
      UCSRA |= 0x40;              /* clear TX interrupt flag */
}

/* Write a character to the UART Transmitter buffer  */
void putchar(char c)
{
    char stuffit = 0;

    while(TX_Counter > (TX_BUFFER_SIZE-1))
        ;
             /* WAIT!! Buffer is getting full!! */


    if(TX_Counter == 0) /* if buffer empty, setup for interrupt */
       stuffit = 1;

    TX_Buffer[TX_Wr_Index++] = c;  /* jam the char in the buffer.. */

    if(TX_Wr_Index > TX_BUFFER_SIZE)       /* wrap the pointer */
        TX_Wr_Index = 0;
                                     /* keep track of buffered chars */
    TX_Counter++;

    if(stuffit == 1)
```

Figure 5–20 *Example, Interrupt Driven USART Transmission (Continues)*

```
    {                                   /* do we have to "Prime the pump"? */
        fPrimedIt = 1;
        UDR = c;                        /* this char starts the interrupt.. */
    }
}

#define         _ALTERNATE_PUTCHAR_

/* Standard Input/Output functions  */
#include <stdio.h>
```

Figure 5–20 *Example, Interrupt Driven USART Transmission (Continued)*

The function *main()* is listed in Figure 5–21. *main()* tests the "RF_TX_Time" flag and performs the telemetry operation. This operation includes computing a checksum, formatting the data into a packet, enabling the RF transmitter, sending the packet, waiting for the transmission to complete, and powering down the RF transmitter. The example shown indicates the intended operation. There may be timing considerations to be taken into account when enabling and disabling the RF transmitter. Note that the watchdog timer is reset periodically in the *main()* routine. The watchdog will help the outdoor unit to continue proper operations during periods of low battery power, or any other situation that may lead to an erratic electrical or operational condition. As mentioned previously, the definitions, "TEMPERATURE", "HUMIDITY", "WIND_DIRECTION", and "BATTERY" are aliases to help us remember which channel represents which signal. In Figure 5–21, you can see how they are treated as variable names.

```
  #define TX_CTS            PINC.7    /* transmitter clear to send */
  #define TX_POWER          PORTC.6   /* transmitter power enable */

  unsigned char packet[TX_BUFFER_SIZE];     /* RF data packet */
  bit RF_TX_Time; /* time to transmit another packet */
  unsigned int checksum;


  #define TEMPERATURE       adc_data[0] /* analog chan 0, int */
  #define HUMIDITY          adc_data[1] /* analog chan 1, int */
  #define BATTERY           adc_data[2] /* analog chan 2, int */
  #define WIND_DIRECTION    adc_data[3] /* analog chan 3, int */

void main(void)
{
    init_AVR();

    while (1)
```

Figure 5–21 *Proposed main(), for the Outdoor Unit (Continues)*

```
    {
        #asm("wdr")
        if(RF_TX_Time)
        {
        /* disable interrupts to prevent data from changing */
        /* during the formatting of the packet               */
              #asm("cli")

              checksum = TEMPERATURE + HUMIDITY + BATTERY;
              checksum += WIND_DIRECTION + wind_speed;
              checksum += rain_state;

              sprintf(packet,
                    "UUU$%03X.%03X.%03X.%03X.%04X.%u.%04X*QQQ",
                          TEMPERATURE,
                    HUMIDITY,BATTERY,WIND_DIRECTION,wind_speed,
                          (int)rain_state,checksum);
              /* Note: dots were added to make sscanf */
              /* work on indoor unit.. */

              #asm("sei")
              /* reenable interrupts to allow sending the */
              /* packet and continue to gather data.. */

              #asm("wdr")
              TX_POWER = 0;        /* turn on transmitter */
              delay_ms(20);       /* allow 20ms for power up.. */
              while(TX_CTS == 0)
                          /* wait until transmitter is ready.. */
              {
                    #asm("wdr")
              }
              puts(packet);        /* send packet out */
              while(TX_Counter)
              {           /* wait until data is all gone */
                    #asm("wdr")
              }
              delay_ms(20);
                    /* hold carrier for 20ms after end of tx.. */
              RF_TX_Time = 0; /* reset the flag until next time */
              TX_POWER = 1;        /* power down until next time.. */
        }
    }
    putchar('X');  /* dummy function to eliminate warnings.. */
}
```

Figure 5–21 *Proposed main(), for the Outdoor Unit (Continued)*

5.6.11 SYSTEM INTEGRATION AND SOFTWARE DEVELOPMENT PHASE, INDOOR UNIT

The indoor unit software shares some common features with the outdoor unit in that the ADCs are used to measure humidity, temperature, barometric pressure, and battery health. In fact, the same auto-scanning and name alias structure is used to provide references to the software for later computations. There are some key differences, particularly feature-driven differences, in the indoor unit that we will focus on in this software design and integration description.

In the indoor unit, there are additional routines for keeping time, lighting an LED, and accepting the incoming data from the outdoor unit. Most of the software in the indoor unit is dedicated to formulating and maintaining a display. This is common in systems that deal with humans. Humans are a bit "high maintenance" when it comes to developing software to relate to them. The methods of human communication that we have allowed ourselves in this weather monitor are the LCD, a couple of LEDs, some buttons, and a beeper.

Keeping Time

Time is kept through the use of the Timer 2 of the Mega16. There are special features within the microcontroller that allow a 32.767 KHz watch crystal to be used to drive an internal oscillator by connecting it to PINC.6 and PINC.7. Timer 2 is initialized to have a prescale of T1OSC/128. This allows Timer 2 to count up at 256 Hz. An interrupt is generated at rollover, or once every second.

```
// Clock source: TOSC1 pin
// Clock value: PCK2/128
TCCR2=0x05;
ASSR=0x08;
```

A structure is used to unify the time variables as a block. This is a good method of tying a set of variables together referentially.

```
struct TIME_DATE {
      int   hour;
      int   minute;
      int   second;
      int   month;
      int   day;
} time;
```

When Timer 2 rolls over, an interrupt is generated, and the time is updated. Note that there is no year and that the number of days for each month is tested against a look-up-table to establish when a month is complete. This means that the user will have to manually adjust the date on leap year on this weather monitor. Figure 5–22 lists the interrupt routine.

The real-time nature of this routine also provides a method of testing for communications from the outside unit. Whenever an outside unit telemetry message is successfully decoded, the variable "`Outdoor_Okay`" is set to 15. Each second, or pass through the interrupt

```
struct TIME_DATE {
    int      hour;
    int      minute;
    int      second;
    int      month;
    int      day;
} time;

bit editing;              /* time being edited flag */

/* seconds without valid communications */
#define OUTDOOR_TIMEOUT  15
char Outdoor_Okay;    /* outdoor unit is talking.. */

/* return max day for a given month */
const char flash MAX_DAY[13] = {
        0,31,28,31,30,31,30,31,31,30,31,30,31
};

bit lowV_in_error;        /* low battery flag.. */

int rain_this_hour;       /* collected rainfall data */
int rain_this_day;
int rain_this_month;
int rain_this_year;
void backup_rainfall(void); /* prototype EEPROM backup routine */


/* Timer 2 overflow interrupt service routine    */
/* This is used to keep "Real-Time" and happens  */
/* every 1 second    */
interrupt [TIM2_OVF] void timer2_ovf_isr(void)
{
    if(editing)
        /* time is being edited.. so don't update it!!  */
        return;

    if(Outdoor_Okay)
              /* time down for valid commumnications  */
         Outdoor_Okay--;       /* from outside unit  */

    if(++time.second > 59)      /* count up seconds  */
    {
       time.second = 0;   /* enough for a minute?  */
       if(++time.minute > 59)
       {
             if(lowV_in_error == 1)
```

Figure 5–22 *Timer 2, Real-Time Clock, Interrupt Routine (Continues)*

```
                /* if running on low battery    */
                backup_rainfall();
            /* backup rainfall data every hour.. */
            rain_this_hour = 0;
            /* reset rain for the hour..    */
            time.minute = 0;
            /* enough minutes for an hour?  */
            if(++time.hour > 23)
            {
                backup_rainfall();
                /* backup data every day to eeprom  */
                rain_this_day = 0;
                /* reset rain for the day..    */
                time.hour = 0;
                /* enough hours for a day?     */
                if(++time.day > MAX_DAY[time.month])
                {
                    rain_this_month = 0;
                    /* reset rain for the month..   */
                    time.day = 1;
                    /* enough days for this month?    */
                    if(++time.month > 12)
                    {
                        rain_this_year = 0;
                    /* reset rain for the year..   */
                        time.month = 1;
                    /* another year gone by..    */
                    }
                }
            }
        }
    }
}
```

Figure 5–22 *Timer 2, Real-Time Clock, Interrupt Routine (Continued)*

routine, this variable is decremented, and as long as the value is nonzero, we know we have reasonably current weather data.

Also note that the flag "`editing`" is used as a method of holding off the updates of the time structure during the period that the user is editing the time and date.

Low-Battery Indication

The low-battery LED and low-battery conditions are monitored and controlled within a timer interrupt. This allows the test and the handling of the LED to be a time-independent process. This code could be performed in the *main()* routine, but the flashing of the LED would then be subject to other processes and exceptions, like the display update time or a user pushing the buttons. Figure 5–23 lists the code for handling this task.

```
#define BATTERY    adc_data[2] /* analog chan 2, int */

#define LOW_INDOOR_V      306 /* A/D counts that relate to 4.7V */
#define LOW_OUTDOOR_V     306 /* A/D counts that relate to 4.7V */
bit lowV_out_error,       lowV_in_error;

#define LOW_BATT_LED     PORTD.6      /* low battery indicator */

char rtc;

                                /* function prototype */
void backup_rainfall(void);     /* save rainfall to eeprom */

/*  Timer 0 overflow interrupt service routine    */
/*  this happens about every 4.096ms             */
interrupt [TIM0_OVF] void timer0_ovf_isr(void)
{
    rtc++;          /*  keep count for blinky light..  */

    if(outdoor_battery < LOW_OUTDOOR_V)
             lowV_out_error = 1;
          /*  Detect errors and "latch" them    */

    if(BATTERY < LOW_INDOOR_V)
    {            /*  the errors are cleared elsewhere  */
                 if(lowV_in_error == 0)
                       backup_rainfall();
                       /*  backup rain gauge data.. in case */
                       /*  of power failure..      */
                 lowV_in_error = 1;
    }

    if(lowV_out_error)
          LOW_BATT_LED = 1;
          /*  display errors on LED    */
          /* low outdoor.. on solid..  */
    else if(lowV_in_error)
    {
          if(rtc & 0x40) /*  low indoor.. blink..    */
                LOW_BATT_LED = 1;
          else
                LOW_BATT_LED = 0;
    }
    else
                LOW_BATT_LED = 0;

    #asm("wdr")      /* pet the dog... */
}
```

Figure 5–23 *Low-Battery Indicator Handler, Timer 0 ISR*

Another advantage to using a timer interrupt is that the battery health reading is updated by an interrupt routine, and the fact that it is tested within an interrupt routine lends protection against using a value that could change in the middle of the test. Normally, the interrupts would need to be turned off during this type of computation, but since no interrupts are allowed while one is in process, there is no need for disabling them.

The Buttons and the Beeper

The buttons are simple switches that are brought in on input pins and monitored directly by the software. The definitions for the buttons look something like this:

```
/* definitions for buttons */
#define       UNITS_BUTTON        PINA.5
#define       SELECT_BUTTON       PINA.6
#define       SET_BUTTON          PINA.7
```

The beeper management is also kept simple by utilizing the Timer 1 peripheral in PWM (*pulse width modulation*) mode. This allows us to simply set up the timer for the output frequency we want, and then, by placing a duty cycle into the output compare register (OCR1AL), we get a tone at a volume related to that duty cycle. The PWM is initialized like this:

```
/* Timer/Counter 1 initialization           */
/* Clock source: System Clock               */
/* Clock value: 500 kHz                     */
/* Mode: 8 bit Pulse Width Modulation       */
/* OC1A output: Non-Inv.                    */
TCCR1A=0x61;
TCCR1B=0x03;
```

Then to control the beeper, a couple of macro functions are used:

```
#define BEEP_ON()    {TCCR1A=0x81;TCCR1B=0x0A;OCR1AL=0x40;}
#define BEEP_OFF()  {TCCR1A=0x00;TCCR1B=0x00;OCR1AL=0x00;}
```

This is done to make the code a little easier to read. The text "`BEEP_ON();`" is replaced with the text "`{TCCR1A=0x81;TCCR1B=0x0A;OCR1AL=0x40;}`" during compilation. Figure 5–24 lists the button/beeper handler routine.

```
void check_buttons(void)
{
    if(UNITS_BUTTON)          /* toggle units: Imperial <-> Metric */
    {
       if(last_units == 0)
       {
             units ^= 1;          /* toggle units.. */
```

Figure 5–24 *Example Button/Beeper Handler (Continues)*

```
                last_units = 1;     /* remember that button is down..   */
                BEEP_ON();
                delay_ms(25);
        }
    }
    else
        last_units = 0;             /* finger off of button? */

    if(SELECT_BUTTON)     /* toggle rainfall period: H, D, M, Y   */
    {
        if(last_select == 0)
        {
                which_rain++;       /* toggle rainfall period.. */
                which_rain &= 3;    /* 0-3 tells which period..   */
                last_select = 1;    /* remember that button is down.. */
                BEEP_ON();
                delay_ms(25);
        }
    }
    else
        last_select = 0;            /* finger off of button?  */

    if(SET_BUTTON)                  /* Set time and date??  */
    {
        if(lowV_out_error || lowV_in_error)
        {
                lowV_out_error = 0; /* clear LED errors and return  */
                lowV_in_error = 0;
                BEEP_ON();
                delay_ms(25);
        }
     else
                if(last_set == 0)  /* otherwise, edit time   */
                {
                        set_time_date();
                        last_set = 1; /* remember that button is down.. */
                        BEEP_ON();
                        delay_ms(25);
                }
    }
    else
        last_set = 0;               /* finger off of button? */

    BEEP_OFF();
}
```

Figure 5–24 *Example Button/Beeper Handler (Continued)*

While many systems rely only on a tactile, or "clicky," button to provide feedback to the user, the beeper improves the "feel" of the system. The beeper provides a confirmation from the software that the button press was acknowledged. In the future, the functionality of the beeper in this weather monitor could easily be expanded to provide audible weather alarms, or even an alarm clock feature, simply by software embellishment.

Decoding the RF Telemetry

The telemetry from the outside unit is handled very similarly to the way it was constructed for the outdoor unit. The RF.DET input is used as a gating signal to indicate that an RF signal is on frequency, that there is a good possibility it is from the outside unit, and that there may be data as well. The INT1 interrupt is configured for rising-edge. This causes an interrupt when the RF is first detected. The INT1 interrupt routine simply enables the USART receiver and interrupt:

```
// External Interrupt 1 service routine
// This occurs on RF reception..
interrupt [EXT_INT1] void ext_int1_isr(void)
{
  UCSRB=0x98;       // turn on RX enable and IRQ
  RX_Wr_Index = UDR;  // clear any chars from input of USART

  RX_Wr_Index = 0;   //reset index of next char to be put
  RX_Rd_Index = 0;   //reset index of next char to be fetched
  RX_Counter = 0;    //reset the total count of characters
  RX_Buffer_Overflow = 0; // ..and any errors
}
```

This method of gating reduces the parsing of spurious noise as data. The receive function of the USART is handled by interrupt much in the way the transmit function of the outdoor unit is handled. This automates the reception and buffering of characters and allows us to utilize the standard I/O functions *getchar()* and *scanf()* to parse the information packets from the outdoor unit. Figure 5–25 shows the code for the interrupt service routine.

```
/* UART Receiver interrupt service routine   */
interrupt [USART_RXC] void uart_rx_isr(void)
{
    char c = 0;

    c = UDR;

    Rx_Buffer[RX_Wr_Index] = c;  /* put received char in buffer */

    if(++RX_Wr_Index > RX_BUFFER_SIZE) /* wrap the pointer */
        RX_Wr_Index = 0;
```

Figure 5–25 *Interrupt-Driven USART Receive (Continues)*

```
    if(++RX_Counter > RX_BUFFER_SIZE) /* keep a character count */
    {                                 /* overflow check.. */
      RX_Counter = RX_BUFFER_SIZE; /* if too many chars came */
      RX_Buffer_Overflow = 1;  /* in before they could be used */
    }                          /* that could cause an error!! */
}

/* Get a character from the UART Receiver buffer  */
char getchar(void)
{
    char c = 0;
    int i = 0;

    i = 0;
    while(RX_Counter == 0)     /* if empty, wait for a character... */
       if(i++ > 2000)
             return -1;

    c = Rx_Buffer[RX_Rd_Index]; /* get one from the buffer..*/

    if(++RX_Rd_Index > RX_BUFFER_SIZE) /* wrap the pointer */
             RX_Rd_Index = 0;

    if(RX_Counter)
       RX_Counter--;          /* keep a count (buffer size) */

    return c;
}

/* This define tells the compiler to replace the stdio.h    */
/* version of getchar() with ours..                         */
/* That way, all the other stdio.h functions can use them!! */
#define      _ALTERNATE_GETCHAR_

/* now, we include the library and it will understand our   */
/* replacements */

#include <stdio.h>
```

Figure 5–25 *Interrupt-Driven USART Receive (Continued)*

The main loop, in function *main()*, watches the variable "`RX_Counter`", which indicates the number of characters in the receive buffer. While there are characters in the buffer, the function *getchar()* is called and the returned characters are checked for a start of message.

As defined in the outdoor unit software, the format of a telemetry packet is as follows:

```
UUU$ttt.hhh.vvv.ddd.ssss.r.xxxx*QQQ
```

where *ttt* is the temperature ADC reading, *hhh* is the humidity ADC reading, *vvv* is the battery voltage reading at the ADC, and *ddd* is the wind direction. The data for the ADC readings are each sent as 3-nibble values, because the ADC values will be between 0x000 and 0xFFF. This saves transmission time by reducing the total number of characters sent. The value *ssss* is the wind speed (Timer 1) reading, *r* is the rain gauge state, and *xxxx* is a checksum made up of the sum of the data itself. The values are all in hexadecimal.

The dollar sign ($) is used to indicate the beginning of a data packet, and the asterisk (*) is used to signal the end of packet. The *UUU* and *QQQ* are preamble and postscript data that are used to stabilize the message for RF transmission and are to be ignored by the receiver. Once the start of message ($) is detected, the routine "`get_outdoor_info()`" is called to retrieve and decode the data (refer to Figure 5–26) . A temporary buffer called "`packet`" is loaded with all the data between the $ and the * markers. Once the * is received, the USART receiver is disabled to prevent bogus input until the next transmission from the outdoor unit begins. The standard library function *sscanf()* is then used to parse out the values and place them into the appropriate variables.

```
#define OUTDOOR_TIMEOUT   15  /* seconds without valid communications */

char Outdoor_Okay; /* outdoor unit is talking.. */
int out_t,w_speed,out_batt;
int w_dir,out_h,rain,checksum; /* temporary variables */
int last_rain,rainfall,which_rain;
bit dp_valid, wc_valid;  /* values are valid flag.. */

char packet[48];   /* buffer for incoming outdoor data */

void get_outdoor_info(void)   /* UUU$276.800.2EE.2FF.0032.0.1095*QQQ  */
{
    char *p  = 0;
    char c = 0;
    int chk = 0;

    p = packet;

    chk = 0;
    c = getchar();
    while(c != '*')        /* gather data until end of message.. */
    {
       *p++=c;
       if(++chk > 46)      /* too much garbage? */
       {
              c = -1;
```

Figure 5–26 *Telemetry Parsing Example (Continues)*

```
            break;
      }
      c = getchar();
      if(c == -1)       /* not enough characters? */
            break;
   }
   *p = 0;                   /* null terminate the string..  */

   UCSRB=0x08;     /* disable receiver until next time..   */

   if(c == -1)     /* packet was junk.. toss it.. */
      return;
                   /* parse out the parameters into variables  */
   c = (char)sscanf(packet,"%x.%x.%x.%x.%x.%u.%x",&out_t,
      &out_h,&out_batt,&w_dir,&w_speed,
            &rain,&checksum);
      /* c now contains the count of assigned parameters.. */

   chk = out_t + out_h + out_batt + w_dir;
   chk += w_speed + rain;

/* test the number of parameters and the checksum for a valid message  */
   if((chk == checksum) && (c == 7))
   {
      Outdoor_Okay = OUTDOOR_TIMEOUT;   /* reset comm timeout.. */
      convert_outdoor_data(); /* update data for display  */
   }
}
```

Figure 5–26 *Telemetry Parsing Example (Continued)*

Once *sscanf()* has pulled the values from the text stream in the buffer "`packet`", the values are added together and tested against the value "`checksum`" to make sure that the data is free from corruption. (This method of summing the data mimics that found in the outdoor unit software.) If the data is complete and intact, the variable "`Outdoor_Okay`" is set to the timeout value of 15 seconds, resetting the "no outdoor data" condition. The data is pulled into temporary variables because it all must be scaled into real units at a later point.

Collecting and Protecting Rainfall Data

Rainfall is collected as transitions of a tipping bucket or seesaw over time. In our weather monitor, we are using the RG-T rain gauge. Each transition of the tipping bucket within the rain gauge is 0.01 inches of rainfall. The values "`rain`" and "`last_rain`" are flags used to determine when a transition has occurred at the rain gauge, and the data is allowed to be collected only at the transition. Variables are set up to accumulate hourly, daily, monthly, and yearly. Each time the gauge transitions, 0.01 inches is added to all of the accumulations.

The real-time clock ISR (Timer 2) resets the accumulations at the beginning of each hour, day, month, and year, respectively, as those times change.

The math for rainfall, as in all of our measurements, is handled using fixed-point numbers. This means that basic integer arithmetic is performed and the decimal point is "mentally" fixed in one place. Floating-point math is available in the CodeVisionAVR compiler as well as in others, but in general, floating-point support is big and slow and is not usually necessary for doing simple unit conversions. Using rainfall as an example, each time the gauge state changes, 0.01 inches is added, but as you can see in the example code in Figure 5–27, each value is incremented by one. This says that our decimal point is to the right of the 100s place, which means that one inch ("1.00") of rain is represented by the integer number 100. For us to display the rainfall in one-tenth inch units, we simply divide the accumulated value by 10 before displaying it.

The rainfall information comes in very slowly. A power outage with a low-battery condition would cause the loss of data that could have been collected for as much as a year! We could simply have declared the variables used for rainfall to be located in EEPROM, but the potential of wearing out the EEPROM from excessive writes also exists. In this weather-monitor

```
int last_rain, rain, which_rain;
int rain_this_hour;
int rain_this_day;
int rain_this_month;
int rain_this_year;

void get_rainfall(void)
{
    if(rain != last_rain)/* bucket (seesaw) has transitioned  */
    {
      rain_this_hour++; /* Each tip of the bucket = 0.01" of rain. */
      rain_this_day++;   /* These values are all integers and */
      rain_this_month++;  /* are treated as if the value is a fraction  */
      rain_this_year++;    /* with the decimal at the 100s place. */

      switch((int)which_rain) /* convert selected value for display */
      {
          case  0:
             rainfall = rain_this_hour/10; /* we only dislay to 0.1"  */
             break;                      /* so we scale the number  */
          case  1:                       /* down..                  */
             rainfall = rain_this_day/10;
             break;
          case  2:
             rainfall = rain_this_month/10;
```

Figure 5–27 *Rainfall Collection Example (Continues)*

```
                break;
            case  3:
                rainfall = rain_this_year/10;
                break;
        }
        /* develop fixed decimal point, 1/10ths of a inch */
        rain_mantissa = rainfall / 10;  /* now the value is "mant.frac"   */
        rain_frac = rainfall - (rain_mantissa * 10);
        last_rain = rain;
    }
}
```

Figure 5–27 *Rainfall Collection Example (Continued)*

example, we "file" the rain data to EEPROM in order to protect it. This is done in a couple of ways, daily and intelligently.

The daily method is shown in the real-time ISR (Timer 2). Basically, all the accumulations are saved to EEPROM at the end of each day. The intelligence comes in the form of a "smart save." At the moment a low-battery condition is detected (or within 4 milliseconds of it), the current accumulations are saved away. If the unit is running on battery and this low-battery condition continues to exist, the real-time ISR will save the data every hour, until the system quits from loss of power.

This filing of the data greatly reduces the activity on the EEPROM, making it last longer. There are a couple of other considerations, though. When the system is powered up from a failed battery condition or for the first time, the rainfall accumulations must be pulled from the file before we continue to add to them. The other consideration is the validity of that data. Here is an example of how it is handled in our weather monitor:

```
if(tagbyte != 0x55)     /* if eeprom uninitialized.. then */
{                       /* clear it out before using it.. */
        backup_rainfall();
        tagbyte = 0x55;
}

                                /* get values from eeprom.. */
rain_this_hour = rain_hour_save;
rain_this_day = rain_day_save;
rain_this_month = rain_month_save;
rain_this_year = rain_year_save;
```

The value "`tagbyte`" is set into EEPROM in order to determine that the EEPROM is in a known state. When the monitor is powered for the first time, the EEPROM default, or erased value, is 0xFF. Therefore, "`tagbyte`" will read 0xFF. We compare this value against a known like 0x55. If it is not equal to this known, the EEPROM is initialized and the "`tagbyte`" is set to 0x55.

CONVERTING FROM COUNTS TO REAL UNITS

As stated in "Operational Specification" in Section 5.6.2, "Definition Phase," the data for all of the weather parameters are collected in the base measurement units, such as ADC and timer counts. This keeps the actual collection process clean and simple. But here we are back at the human interface again. The problem is that most people cannot mentally convert from ADC counts to temperature in °F. In fact, most people have a tough enough time going from °F to °C. So as developers, it is our task to do this fancy thinking and present the information to the human in a nice prepackaged form.

Figure 5–28 shows the code used to convert the inside temperature, humidity, and barometric pressure. Just as in the rainfall measurements, fixed-point arithmetic is used. Most unit conversions involve two parts, a slope (or gain) and an offset. It depends on the type of conversion to define which part is applied when.

```
#define IMPERIAL   0
#define METRIC     1

char units;                 // current display units

const int flash I_K_it = 12;
                /* (* 0.12) scale indoor temperature, Imperial  */
const int flash I_Offset_it = 171;    /* (-.84V) offset for indoor
temperature, Imperial */

const int flash K_b = 26;
                /* (1/26) scale for barometric pressure   */
const int flash Offset_b = 280;
                /* (28.0"Hg)offset for barometric pressure   */

const int flash K_ih = 10;      /* (1/10) scale, indoor humidity   */
const int flash Offset_ih = 0;  /* offset, indoor humidity    */

int indoor_temp;
int barom_mantissa,barom_frac;
int indoor_humidity;

void convert_indoor_data(void)
{
    int bar = 0;

    /* disable interrupts to prevent data from changing */
    /* while we are performing the calculations    */
    #asm("cli")

    indoor_temp = ((TEMPERATURE-I_Offset_it)*I_K_it)/100;
```

Figure 5–28 *Unit Conversion Example (Continues)*

```
    if(units == METRIC)
       indoor_temp = ((indoor_temp - 32)*5)/9; /* scale to degrees C   */

    bar = BAROMETER / K_b + Offset_b;
    barom_mantissa = bar / 10;
    barom_frac = bar - (barom_mantissa * 10);

    indoor_humidity = HUMIDITY / K_oh + Offset_oh;

    /* reenable interrupts..  */
    #asm("sei")
}
```

Figure 5–28 *Unit Conversion Example (Continued)*

In the case of temperature, there is a voltage offset of 0.84 V (171 counts of the ADC) that should be removed before scaling the number. Once the offset is removed, the temperature in °F is computed by multiplying the counts by 0.24. The example shows fixed-point arithmetic performed by multiplying the ADC counts by 24, which would make the value have two places after the decimal (that is, 61.44°F), and then scaling the value back into whole °F by dividing by 100. Once the temperature is converted to °F, it is simple to re-convert it to metric if need be.

The barometric pressure and humidity are different in that the scaling occurs before the offset. This is because the values of the ADC go rail-to-rail (0 to 5 V DC) and the offset applies to the units directly instead of the counts. For example, with barometric pressure, the ADC value will be 0 V when the pressure is 28.0 inHg. The ADC value is divided by a constant (*K_b*) to get the proper number of fractional inches of mercury. The result is then offset by 280 (or 28.0 inHg) to arrive at the proper value representation, a number between 28.0 and 32.0 inHg.

The same processes described hold true for the temperature, humidity, and wind speed measurements delivered by the outdoor unit. Wind chill and dew point are treated differently in that they are products of a look-up-table. In Figure 5–29, you can see how temperature, humidity, and wind speed are converted and then those converted values are used as indices into a table to return wind chill or dew point in °F. Just as before, if the system is displaying metric values, the converted units are rescaled to metric.

Routines for Controlling the LCD

Figure 5–30 shows the routines for controlling the 4X20 LCD. The display is an Optrex-compatible device operating in 4-bit mode. This means that all of the data going to the device is transferred using four data lines. This method of interface saves precious I/O pins on the microcontroller but still allows the display to be updated quickly. There are three control signals E, RD (or R/W), and RS. E is used as a strobe to validate the data (DB4-7) to the display. RD is used to indicate that the operation is to read data from the display

```
#define IMPERIAL   0
#define METRIC     1
char units;              /* current display units */

int outdoor_temp, wind_chill;
int wind_speed, wind_degrees;
int outdoor_humidity, dew_point;
int outdoor_battery;
int out_t,w_speed,out_batt;
int w_dir,out_h,rain,checksum; /* temporary variables */
bit dp_valid, wc_valid;  /* values are valid flag.. */

/* Dew Point Look-Up-Table   - Imperial */
const char flash DEW_LUT[13][11] = {
    -6, 4,  13, 20, 28, 36, 44, 52, 61, 69, 77,
    -2, 8,  16, 23, 31, 40, 48, 57, 69, 74, 83,
    1,  11, 18, 26, 35, 43, 52, 61, 69, 78, 87,
    4,  13, 21, 29, 37, 47, 56, 64, 73, 82, 91,
    6,  15, 23, 31, 40, 50, 59, 67, 77, 86, 94,
    9,  17, 25, 34, 43, 53, 61, 70, 80, 89, 98,
    11, 19, 27, 36, 45, 55, 64, 73, 83, 95, 101,
    12, 20, 29, 38, 47, 57, 66, 76, 85, 93, 103,
    13, 22, 31, 40, 50, 60, 68, 78, 88, 96, 105,
    15, 24, 33, 42, 52, 62, 71, 80, 91, 100, 108,
    16, 25, 34, 44, 54, 63, 73, 82, 93, 102, 110,
    17, 26, 36, 45, 55, 65, 75, 84, 95, 104, 113,
    18, 28, 37, 47, 57, 67, 77, 87, 97, 107, 117
};

/* Wind Chill Look-Up-Table  - Imperial */
const char flash WC_LUT[9][8] =  {
    36, 34,   32,   30,   29,  28,  28,  27,
    25, 21,   19,   17,   16,  15,  14,  13,
    13,  9,    6,    4,    3,   1,   0,  -1,
    1,  -4,   -7,   -9,  -11, -12, -14, -15,
    -11, -16, -19, -22,  -24, -26, -27, -29,
    -22, -28, -32, -35,  -37, -39, -41, -43,
    -34, -41, -45, -48,  -51, -53, -55, -57,
    -46, -53, -58, -61,  -64, -67, -69, -71,
    -57, -66, -71, -74,  -78, -80, -82, -84
};

const int flash I_K_ot = 12;
       /* (* 0.12) scale outdoor temperature, Imperial */
const int flash I_Offset_ot = 171;
       /* (-.84V) offset for outdoor temperature, Imperial */

const int flash I_K_ws = 25;        /* scale wind speed, Imperial */
const int flash I_Offset_ws = 0;    /* offset wind speed, Imperial*/
```

Figure 5–29 *More Unit Conversions (Continues)*

```
const int flash K_wd = 35;          /* (* 0.35) scale wind direction */
const int flash Offset_wd = 0;      /* offset wind direction */

const int flash K_oh = 10;          /*(1/10) scale, outdoor humidity */
const int flash Offset_oh = 0;      /* offset, outdoor humidity */

void convert_outdoor_data(void)
{
    int dt = 0, dh = 0, dw = 0;
    long temp = 0;

    outdoor_temp = ((out_t-I_Offset_ot)*I_K_ot)/100; /* degrees F  */

    wind_speed = w_speed * I_K_ws + I_Offset_ws;   /* MPH  */

    temp = (long)w_dir;        /* use long so we don't overflow!!  */
    temp *= K_wd;
    temp /= 100;               /* after this, int is safe.. */
    wind_degrees = temp + Offset_wd; /* degrees from North  */

    outdoor_humidity = out_h / K_ih + Offset_ih;  /* % RH   */

    outdoor_battery = out_batt;        /* just counts */

    if((outdoor_temp >= 20) && (outdoor_temp <= 120) &&
          (outdoor_humidity >= 30) && (outdoor_humidity <= 90))
    {
       dt = (outdoor_temp - 20) / 10; /* scale for table index  */
       dh = (outdoor_humidity - 30) / 5;
       dew_point = DEW_LUT[dh][dt];
       dp_valid = 1;   /* interpolation could be added here!!!  */
    }
    else
    {
       dew_point = 0;
       dp_valid = 0;
    }

    if((outdoor_temp >= -40) && (outdoor_temp <= 40) &&
            (wind_speed >= 5) && (wind_speed <= 40))
    {
       dt = (outdoor_temp - 40) / 10; /* scale for table index */
       dw = (wind_speed - 5) / 5;
       wind_chill = WC_LUT[dt][dw];
       wc_valid = 1;   /* interpolation could be added here too!!! */
    }
    else
    {
       wind_chill = 0;
```

Figure 5–29 *More Unit Conversions (Continues)*

```
            wc_valid = 1;
        }

        if(units == METRIC)
        {
            wind_chill = ((wind_chill - 32)*5)/9;   /* scale to degrees C   */
            dew_point = ((dew_point - 32)*5)/9;     /* scale to degrees C   */
            outdoor_temp = ((outdoor_temp - 32)*5)/9;
                                                   /* scale to degrees C   */
            wind_speed = ((wind_speed)*88)/55;     /* scale to KPH    */
        }
}
```

Figure 5–29 *More Unit Conversions (Continued)*

```
#define LCD_E      PORTB.0         /* LCD Control Lines */
#define LCD_RS     PORTB.1
#define LCD_RW     PORTB.2
#define LCD_PORT   PORTC           /* LCD DATA PORT (lowest 4 bits)*/

/* this Look-Up-Table translates LCD line/cursor positions */
/* into the displays' internal memory addresses */
const unsigned char flash addLUT[4] = {0x80,0xC0,0x94,0xD4};

unsigned char LCD_ADDR,LCD_LINE;

/***** Display Functions *****/
void wr_half(unsigned char data)
{
    LCD_RW = 0;               /* set WR active   */

    LCD_E = 1;                /* raise E   */
    LCD_PORT = (data & 0x0F); /* put data on port   */
    LCD_E = 0;                /* drop E    */

    LCD_RW = 1;               /* disable WR   */

    delay_ms(3);              /* allow display time to think   */
}

void wr_disp(unsigned char data)
{
    LCD_RW = 0;               /* enable write..   */

    LCD_E = 1;
    LCD_PORT= (data >> 4);
    LCD_E = 0;                /* strobe upper half of data   */
```

Figure 5–30 *LCD Control Routines (Continues)*

```
    LCD_E = 1;              /* out to the display  */
    LCD_PORT = (data & 0x0F);
    LCD_E = 0;              /* now, strobe the lower half */

    LCD_RW = 1;             /* disable write */

    delay_ms(3);            /* allow time for LCD to react  */
}

void init_display(void)
{
    char i = 0;

    LCD_RW = 1;
    LCD_E = 0;        /* preset interface signals..  */
    LCD_RS = 0;       /* command mode.. */
    delay_ms(50);

    wr_half(0x33);   /* This sequence enables the display   */
    wr_half(0x33);   /* for 4-bit interface mode  */
    wr_half(0x33);   /* these commands can be found  */
    wr_half(0x22);   /* in the manufacturer's data sheets   */

    wr_disp(0x28);   /* Enable the internal 5x7 font */
    wr_disp(0x01);
    wr_disp(0x10);   /* set cursor to move (instead of display shift) */
    wr_disp(0x06);
       /* set the cursor to move right, and not shift the display  */
    wr_disp(0x0c);
       /* turns display on, cursor off, and cursor blinking off  */

    for(i=0x40; i<0x5F; i++)  /* init character font ram to "00"  */
    {
        delay_ms(10);
        LCD_RS = 0;
        wr_disp(i);
        delay_ms(10);
        disp_char(0);
    }
    LCD_RS = 1;             /* data mode  */
}

void line(char which_line)
{
    LCD_RS = 0;

    LCD_ADDR = addLUT[which_line-1];
    wr_disp(LCD_ADDR);          /* move to which_line line  */

    LCD_RS = 1;
```

Figure 5–30 *LCD Control Routines (Continues)*

```
    LCD_ADDR = 0;
    LCD_LINE = which_line;
}

void clear_display(void)
{
    LCD_RS = 0;
    wr_disp(0x01);          /* clear display and home cursor  */
    delay_ms(10);
    LCD_RS = 1;
    line(1);
}

void set_LCD_cur(char LCDrow, char LCDcol)
{
    LCD_RS = 0;

    LCD_ADDR = addLUT[LCDrow] + LCDcol;

    wr_disp(LCD_ADDR);      /* move to row and column  */

    LCD_RS = 1;
    LCD_LINE = LCDrow;
}

void disp_char(unsigned char c)
{
    LCD_RS = 1;
    wr_disp(c);
    LCD_ADDR++;             /* automatically wrap text as required  */
    if(LCD_ADDR == 20)
        line(2);
    if(LCD_ADDR == 40)
        line(3);
    if(LCD_ADDR == 60)
        line(4);
    if(LCD_ADDR == 80)
        line(1);
}

void disp_str(unsigned char *sa)     /* display string from RAM  */
{
    while(*sa != 0)
      disp_char(*sa++);
}

void disp_cstr(unsigned char flash *sa)  /* display string from ROM */
{
    while(*sa != 0)
      disp_char(*sa++);
}
```

Figure 5–30 *LCD Control Routines (Continued)*

(RD=1) or write data to the display (RD=0). The RS signal is used to indicate to the display that the data being transferred is a command (RS=0) affecting the display's mode of operation or data (RS=1) that is to be displayed.

The function "`init_display()`" configures the LCD for the appropriate operation for the weather monitor. The internal character set is used, the cursor is turned off, and the graphic characters are cleared. The manufacturer's datasheet on the LCD display describes each of the command and register settings for the display, as well as any time requirements. Some of the commands take a while for the display to execute, so delays were inserted between these commands to give the display time to complete the operations before the next command is issued. The functions "`wr_half()`" and "`wr_disp()`" are used to send only the lower half of a byte or both halves, respectively.

The remaining commands perform the text display operations that will make it easy for formatting and displaying the weather information. The "`set_LCD_cur()`" function sets the invisible cursor to a specified row and column. The "`line()`" function sets the invisible cursor to the beginning of the specified line. The "`clear_display()`" function wipes the display completely clean and sets the invisible cursor to the upper left corner (row=0, column=0).

"`disp_char()`" is the *putchar()* for the LCD. Its function is to place a character on the display. The invisible cursor is automatically moved, within the display, to the next position. Some tests have to be performed to check for the end of the display line. The cursor does not automatically "wrap" to the next line; therefore, the cursor address is checked, and, based on the address size, the appropriate "`line()`" function call is made.

The "`disp_str()` and "`disp_cstr()`" functions call "`disp_char()`" in order to place formatted "C" strings to the display. "`disp_str()`" is used to display strings located in SRAM and "`disp_cstr()`" is used to display strings located in FLASH.

Keeping the Display Up to Date

The display is maintained as a low-level task. This means that *main()* either works on the display or calls display-related functions when there is time to do so. The display items are common in form, so a simple way of updating the display is to treat each item of the display in the same way. A structure is used to indicate all the properties of a display item.

```
struct DISP_ITEM {
     char flash row;          /* row position of text */
     char flash col;          /* col position of text */
     char flash *fmtstr;      /* sprintf format   */
     int *value;              /* pointer to variable to display */
     char flash *units_I;     /* units mark imperial */
     char flash *units_M;     /* units mark metric */
     char flash source;       /* inside or outside measurement */
};
```

Each display item has a position, "`row`" and "`col`", a format, a value, a units indicator, and an indication as to whether the data came from the indoor unit or the outdoor unit. By creating an array of these items we can almost completely describe the display contents:

```
struct DISP_ITEM main_display [MAX_ITEM] =
{
        0,0,"I%3d",&indoor_temp,"F ","C ",INSIDE,
        0,7,"O%3d",&outdoor_temp,"F  ","C  ",OUTSIDE,
        0,14,"WC%2d",&wind_chill,"F  ","C  ",OUTSIDE,
        1,0,"B%02u",&barom_mantissa,".",".",INSIDE,
        1,4,"%1u",&barom_frac,"","",INSIDE,
        1,7,"%3u",&wind_speed,"MPH","KPH",OUTSIDE,
        1,14,"%03u",&wind_degrees,"","",OUTSIDE,
        2,0,"I%3u",&indoor_humidity,"%","%",INSIDE,
        2,7,"O%3u",&outdoor_humidity,"%","%",OUTSIDE,
        2,14,"DP%3d",&dew_point,"F","C",OUTSIDE,
        3,0,"%02u",&rain_mantissa,".",".",OUTSIDE,
        3,3,"%1u",&rain_frac,"\"","\"",OUTSIDE,
        3,7,"%02u",&time.hour,":",":",INSIDE,
        3,10,"%02u",&time.minute,"","",INSIDE,
        3,15,"%02u",&time.month,"/","/",INSIDE,
        3,18,"%02u",&time.day,"","",INSIDE
};
```

In Figure 5–31, the function "`update_display_item()`" formats and displays the requested item. The cursor is placed, the value is formatted into a display string using *sprintf()*, and it is put to the display using "`disp_str()`". If the values are outdoor measurements, based on "`main_display[item].source`", and the data has not been refreshed in the last 15 seconds ("`Outdoor_Okay`"), then the digits are replaced with question marks. Also note that if wind chill or dew point values are out of range, their digits are replaced with dashes. Finally, the appropriate unit text, "`main_display[item].units_I`" or "`main_display[item].units_M`", is sent to the display based on which units have been requested by the user.

```
#define IMPERIAL     0
#define METRIC       1
char units;          /* current display units */
char text_buffer[24];

char Outdoor_Okay;   /* outdoor unit is talking.. */

/*-------------------------------------------------------- */
/* display item table description for the main display    */
/*-------------------------------------------------------- */
#define INSIDE     'i'
#define OUTSIDE    'o'
#define MAX_ITEM   16
```

Figure 5–31 *Table-Driven Display Routine Example (Continues)*

```
char which_item;
struct DISP_ITEM {
    char flash row;          /* row position of text  */
    char flash col;          /* col position of text  */
    char flash *fmtstr;      /* sprintf format        */
    int *value;              /* pointer to variable to display */
    char flash *units_I;     /* units mark imperial */
    char flash *units_M;     /* units mark metric    */
    char flash source;       /* inside or outside measurement  */
};

struct DISP_ITEM main_display [MAX_ITEM] =
{
    0,0,"I%3d",&indoor_temp,"F  ","C  ",INSIDE,
    0,7,"O%3d",&outdoor_temp,"F  ","C  ",OUTSIDE,
    0,14,"WC%2d",&wind_chill,"F  ","C  ",OUTSIDE,
    1,0,"B%02u",&barom_mantissa,".",".",INSIDE,
    1,4,"%1u",&barom_frac,"","",INSIDE,
    1,7,"%3u",&wind_speed,"MPH","KPH",OUTSIDE,
    1,14,"%03u",&wind_degrees,"","",OUTSIDE,
    2,0,"I%3u",&indoor_humidity,"%","%",INSIDE,
    2,7,"O%3u",&outdoor_humidity,"%","%",OUTSIDE,
    2,14,"DP%3d",&dew_point,"F","C",OUTSIDE,
    3,0,"%02u",&rain_mantissa,".",".",OUTSIDE,
    3,3,"%1u",&rain_frac,"\"","\"",OUTSIDE,
    3,7,"%02u",&time.hour,":",":",INSIDE,
    3,10,"%02u",&time.minute,"","",INSIDE,
    3,15,"%02u",&time.month,"/","/",INSIDE,
    3,18,"%02u",&time.day,"","",INSIDE
};

void update_display_item(char item)
{
 char *s = 0;

 /* first, set the LCD cursor to the item's position  */
 set_LCD_cur(main_display[item].row,main_display[item].col);

 /* place the variable into its proper format as a string */
 sprintf(text_buffer,main_display[item].fmtstr,*main_display[item].value);

    /* if inside-data.. then always display  */
    if(main_display[item].source != INSIDE)
    {  /* outside data may not be up to date.. if it is, great! */
       if(Outdoor_Okay > 0)
       {            /* if wind chill or dew point out of range...  */
          if(((item == WIND_CHILL_ITEM) && (wc_valid == 0)) ||
                   ((item == DEW_POINT_ITEM) && (dp_valid == 0)))
```

Figure 5–31 *Table-Driven Display Routine Example (Continues)*

```
            {
            s = text_buffer;    /* point to the formatted text.. */
            while(*s)
            {           /* replace numbers with dashes...   */
                if(isdigit(*s))
                    *s = '-';
                s++;
            }
        }           /* otherwise all is well, just print it.. */
    }
    else
    {      /* the outside info is not current so, ..       */
        s = text_buffer;    /* point to the formatted text..   */
        while(*s)
        {           /* replace numbers with question marks..    */
            if(isdigit(*s))
                *s = '?';
            s++;
        }
    }
}                   /* always place units string last.. */

    disp_str(text_buffer);

    if(units == IMPERIAL)
        disp_cstr(main_display[item].units_I);
    else
        disp_cstr(main_display[item].units_M);
}
```

Figure 5–31 *Table-Driven Display Routine Example (Continued)*

The entire display is updated from *main()* by simply calling the *update_display_item()* function from a **for** loop:

```
/* update entire display */
for(which_item=0; which_item< MAX_ITEM; which_item++)
      update_display_item(which_item);

set_LCD_cur(3,5);     /* indicate which rainfall period */
disp_char(RAIN_TYPE[which_rain]);

set_LCD_cur(1,17);        /* indicate wind heading */
if(Outdoor_Okay)
      disp_cstr(WIND_HEADING[wind_degrees/22]);
else
      disp_cstr("--"); /*..if possible.. */

set_LCD_cur(3,12);    /* flash star with seconds.. */
```

```
if(time.second & 1)    /* to indicate life!! */
      disp_char('*');
else
      disp_char(' ');
```

Additional information, such as rainfall period, wind heading, and a flashing asterisk to indicate seconds, is simply added on at the end of the update. Look-up-tables are used to provide the appropriate strings for the display, eliminating the need for giant **if-else** or **switch-case** statements loaded with "`disp_cstr()`" calls:

```
const char flash RAIN_TYPE[4] = {'H','D','M','Y'};

char flash *WIND_HEADING[17] = {
      "N  ",
      "NNE",
      "NE ",
      "ENE",
      "E  ",
      "ESE",
      "SE ",
      "SSE",
      "S  ",
      "SSW",
      "SW ",
      "WSW",
      "W  ",
      "WNW",
      "NW ",
      "NNW",
      "N  "
};
```

Editing the Time and Date

The editing of time and date are also handled using a table-driven method. This allows a simple procedure to deal with all the values in the same way, even with the same code. An editing structure was designed that contains all the necessary elements, the position of the item on the LCD, the title or name of the item (telling the user what is being edited), a pointer to the value that can be changed, and some limits to prevent the user from taking the value "out-of-bounds."

```
struct EDIT_ITEM {
      char flash row;          /* row position of text */
      char flash col;          /* col position of text */
      char flash title[10];    /* item title */
      int *value;         /* pointer to variable to display */
      int flash MinValue;      /* Minimum Value */
      int flash MaxValue;      /* Maximum Value */
};
```

```
struct EDIT_ITEM edit_display[4] = {
        0,5,"Month:  ",&time.month,1,12,
        1,5,"Day:    ",&time.day,1,31,
        2,5,"Hour:   ",&time.hour,0,23,
        3,5,"Minute: ",&time.minute,0,59
};
```

The display is cleared and rebuilt to form the edit screen, as shown in Figure 5–33 (page 338). A greater than sign (>) is used to indicate the currently selected item. The buttons are processed directly and are used to change the selected item, select the next item to be edited, or exit. The flag "`editing`" is used to signal the real-time ISR (Timer 2) that editing is in process and the values are not to be updated while the user is making changes.

Figure 5–33 shows the LCD during an editing process.

```
#define UNITS_BUTTON     PINA.5   /* definitions for buttons */
#define SELECT_BUTTON    PINA.6
#define SET_BUTTON       PINA.7

struct EDIT_ITEM {
    char flash row;           /* row position of text   */
    char flash col;           /* col position of text   */
    char flash title[10];     /* item title   */
    int *value;               /* pointer to variable to display   */
    int flash MinValue;       /* Minimum Value */
    int flash MaxValue;       /* Maximum Value */
};

struct EDIT_ITEM edit_display[4] = {
    0,5,"Month:  ",&time.month,1,12,
    1,5,"Day:    ",&time.day,1,31,
    2,5,"Hour:   ",&time.hour,0,23,
    3,5,"Minute: ",&time.minute,0,59
};

bit editing;                /* time being edited flag */

void set_time_date()
{
    int cur_item = 0;

    editing = 1;        /* signal interrupt routine to skip update  */
    cur_item = 0;       /* start with first item..    */

    clear_display();   /* build the entire edit screen.. */

    for(cur_item=0; cur_item<4; cur_item++)
```

Figure 5–32 *Table-Driven Editing Example (Continues)*

```
{   /* first, set the LCD cursor to the item's position  */
  set_LCD_cur(edit_display[cur_item].row,edit_display[cur_item].col);

    /* place the variable into its proper format as a string   */
    sprintf(text_buffer,"%s %02u",edit_display[cur_item].title,
            *edit_display[cur_item].value);

    disp_str(text_buffer);
}

while(SET_BUTTON)        /* wait for finger off the button..  */
    ;

cur_item = 0;
while(1)            /* now work each item.. +/-, and SET to move.. */
{
 set_LCD_cur(edit_display[cur_item].row,edit_display[cur_item].col-1);
 disp_char('>');                /* point to the item.. */

 while(SET_BUTTON)        /* finger off? */
    ;

 delay_ms(25);

 while((!SET_BUTTON) && (!UNITS_BUTTON) && (!SELECT_BUTTON))
    ;               /* wait for a button press.. */

 if(UNITS_BUTTON)          /* increment the current item.. */
 {                         /* and do limit checks.. */
    BEEP_ON();
    time.second = 0;    /* reset the seconds to 0 when time edited */
    *edit_display[cur_item].value += 1;
    if(*edit_display[cur_item].value > edit_display[cur_item].MaxValue)
      *edit_display[cur_item].value = edit_display[cur_item].MinValue;
 }

 if(SELECT_BUTTON)            /* decrement the current item..  */
 {                            /* and do limit checks..    */
    BEEP_ON();
    time.second = 0;   /* reset the seconds to 0 when time edited */
    *edit_display[cur_item].value -= 1;
    if(*edit_display[cur_item].value < edit_display[cur_item].MinValue)
       *edit_display[cur_item].value = edit_display[cur_item].MaxValue;
 }
                          /* update the item on the screen..  */
 set_LCD_cur(edit_display[cur_item].row,edit_display[cur_item].col);
 sprintf(text_buffer,"%s %02u",edit_display[cur_item].title,
       *edit_display[cur_item].value);
```

Figure 5–32 *Table-Driven Editing Example (Continues)*

```
      disp_str(text_buffer);

      delay_ms(25);      /* a little time for switch settling..  */
      BEEP_OFF();

      while((UNITS_BUTTON) || (SELECT_BUTTON))
              ;

      if(SET_BUTTON)
      {
         BEEP_ON();
         delay_ms(25);          /* a little time for switch settling.. */
         BEEP_OFF();
         set_LCD_cur(edit_display[cur_item].row,
                     edit_display[cur_item].col-1);
         disp_char(' ');        /* erase the current cursor... */
         cur_item++;
         if(cur_item > 3)
            break;
      }
   }

   while(SET_BUTTON)  /* finger off? */
         ;

   editing = 0;
   clear_display();   /* back to business as usual..   */
   }
```

Figure 5–32 *Table-Driven Editing Example (Continued)*

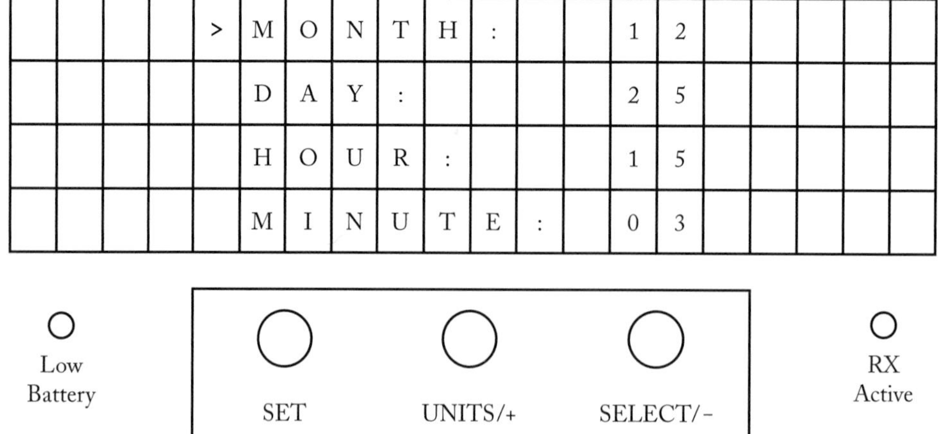

Figure 5–33 *Time/Date Edit Window*

The function "`set_time_date()`" is called from the "`check_buttons()`" procedure, which is called from *main()*. As mentioned previously, the display maintenance and user interface are all happening at the lowest level. The normal operation of device is not disturbed by the time spent converting and displaying information, but during editing, the real-time clock does not update and other data collection processes are ignored until the user exits the editing process.

5.6.12 SYSTEM TEST PHASE

At this point in the project, the tests specified for system test during the Test Definition Phase of the project are carried out to give the user confidence that the unit performs as defined by the specifications. As stated earlier, the tests may be as extensive as needed to convince a customer, or they may be as simple as comparing the results to the local weather station.

For the purposes of this text, extensive testing is above and beyond the scope of the requirements and would contribute little to expanding your knowledge of project development.

However, one important topic to be addressed here is what to do if the project does not meet specifications, that is, it fails a system test. Problems such as incorrect calibration of the temperature results, lack of linearity in the humidistat, and mismeasurement of the rain gauge will all be spotted during system test. The important issue is what to do about it, and there are really only two choices: fix it, or change the specification. Changing the specification is only a very last resort and should ***not*** be required if the Definition Phase was carried out properly. So fixing it is the only *real* choice.

As an example, consider the wind speed indicator. Its intended calibration was based on an empirically derived number from the manufacturer (2.515 Hertz/mph), and the design allows for some variation in the actual results, allowing for losses due to friction, method of mounting, and so on, so that the wind speed could be properly calibrated.

For example, suppose that when we measured the output of the anemometer in a wind tunnel at a local university, we noticed an error of 4%. We just did not get quite as many pulses per second out of the system as we expected at a wind speed of 100 mph. So when we plot the output frequency (Hz) versus the wind speed (mph), our graph looks something like "Measured A" shown in Figure 5–34.

In this case, even though there is some error, the result is linear, so a simple adjustment of the scaling would be in order. This could be accomplished by changing the scaling constant in the software from

```
const int flash I_K_ws = 25;     /* scale wind speed, Imperial */
```

to

```
const int flash I_K_ws = 24;     /* scale wind speed, Imperial */
```

This easy adjustment shows the value of putting conversion constants and calibration factors in as constant variables stored in FLASH memory. You can adjust these easily if necessary without looking through the program to find where the constants or calibration factors

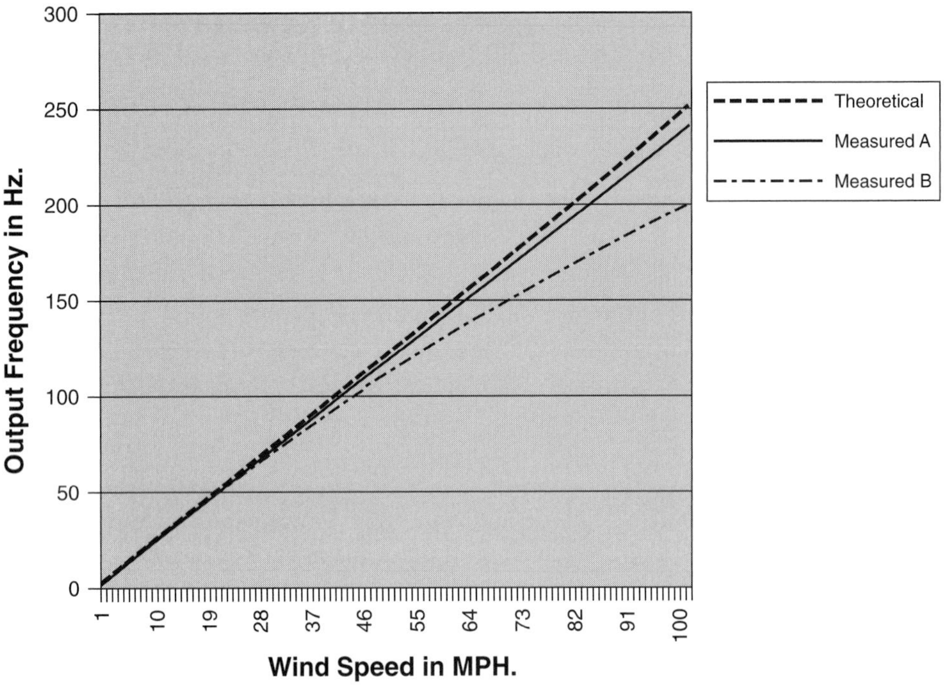

Figure 5–34 *Example, Measured Anemometer Output*

are used, and perhaps missing one or two, if they are used in multiple places. Examination of the program will show that these are inserted in the program and really do not increase or decrease code size by their use, so they are a no-cost convenience.

If the result were nonlinear, like the "Measured B" data in Figure 5–34, a more serious type of correction might be in order, requiring a complex algebraic expression or perhaps the simpler approach of using a look-up-table and some interpolation. The choice of method, in this case, greatly depends on the expected accuracy, the precision of the result, and available computing time. Overall, a look-up-table (LUT) is a tough approach to beat no matter how great someone is with "that fancy math." In real systems like the one we are describing, it is not uncommon at all to come across some really strange-looking curves and bends in the data that are collected from a device. As long as there is curve to the earth, gravity pulling us down, and the sun rising every day to cause a constant change in temperature, the data that is collected is going have a shape, and it is probably not going to be a straight line.

So we can convert the "Measured B" data to mph using a LUT and linear-interpolation approach. One of the best parts of the LUT approach is that you can use the data you actually collected. Table 5–6 shows measured frequencies.

Wind Speed (mph)	**Output Frequency (Hertz)**
1	2.5
10	24.6
20	48.3
30	68.7
40	92.6
50	113.3
60	132.9
70	151.5
80	169.2
90	185.8
100	201.5

Table 5–1 *Measured Frequencies from Wind Tunnel Test*

If the data is more straight than it is curved or S-shaped, fewer points can be used in the actual table, and the linear interpolation can fill in the values in between as they are needed. With ten frequency measurements from our wind tunnel testing, an array is formed to create the LUT.

```
const int flash Wfreqs[11] = /* freq X 10, i.e. 24.6 = 246 */
      {25,246,483,687,926,1133,1329,1515,1692,1858,2015};
```

The following function shows how the frequency detected by the counter and placed in the variable "w_speed" is checked against the values of the LUT. If the speed is lower than the LUT, it is less than one mph and nothing can be computed. If the speed is higher than the LUT, nothing can be computed either, so an error is displayed using a bogus wind speed of "888 MPH."

Once two values are selected from the table, one above "w_speed" and one below, the delta ("t - b") is calculated for the two table values to determine the slope of what is effectively a short linear segment within the long, curved data. The delta from the measured value "w_speed" to the lower selected value "b" is used to compute at what point (in percentage) the measured data would fall on the short line segment. The base mph is then computed from the index "x", and the fractional portion of the mph based on the ratio of the deltas is added back to yield the actual mph.

```
void calc_wind_speed(void) /* compute global var wind_speed */
{
    int t,b,x;
    long v1,v2,v3;

    if(w_speed <= Wfreqs[0])
```

```
        {
                wind_speed = 0;     /* too slow to measure, < 1MPH */
                return;
        }

        if(w_speed > Wfreqs[10])
        {
                wind_speed = 888;  /* too high to measure, > 100MPH */
                return;            /* "888" is an error message!! */
        }

        for(x=0; x < 9; x++)
        {
                if(w_speed > Wfreqs[x])
                                   /* find where the counts fall */
                {                          /* in the table
                        t = Wfreqs[x+1];
                        b = Wfreqs[x];     /* top and bottom values
                        break;             /* x will be our base MPH/10
                }
        }
        v1 = (long)w_speed - (long)b;  /* (note the 'casts').. */
        v1 *= 100L; /* now calculate the percentage between */
        v2 = (long)t - (long)b;        /* the two values */
        v3 = (v1/v2);  /* percentage of MPH diff * 10 */

        v1 = (long)x * 100L; /* make x into MPH * 100 (fraction!) */
        v1 += v3;    /* now add in percentage of difference */

        wind_speed = (int)(v1 / 10L); /* now scale to whole MPH */
}
```

Therefore, if "w_speed" was measured in this example at 1200 counts, the values as a result of the **for** loop for the index "x" and the top and bottom values of the linear segment, "t" and "b", would be

```
x = 5
t = Wfreqs[5+1] = 1329
b = Wfreqs[5] = 1133
```

Stepping through the program, the computations would go like this:

```
v1 = (long)w_speed -(long)b;   /* (note the 'casts').. */
```

result: v1 = 1200 – 1133 = 67

```
v1 *= 100L; /* now calculate the percentage between */
```

result: v1 = 67 * 100 = 6700

```
v2 = (long)t - (long)b;  /* the two values */
```

result: v2 = 1329 – 1133 = 196

```
v3 = (v1/v2);   /* percentage of MPH diff * 10 */
```

result: v3 = (6700/196) = 34, which is 34% of 10mph, or 3.4 mph

```
v1 = (long)x * 100L; /* make x into MPH * 100 (fraction!) */
```

result: v1 = 5 * 100 = 500, which is fixed point for 50.0 mph

```
v1 += v3;     /* now add in percentage of difference */
```

result: v1 = 500 + 34 = 534, again, which is fixed point for 53.4 mph

```
wind_speed = (int)(v1 / 10L); /* now scale to whole MPH */
finally: wind_speed  =  534 / 10  =  53 mph
```

As you may recall from the indoor unit specification, only whole mph is displayed. Otherwise, it is not difficult to see the level of accuracy that can be achieved using a method of this type. In this example, we simply throw the fractional portion away. In this instance, a simple LUT and linear interpolation can be applied effectively to linearize a nonlinear result so that the project will meet its specifications.

As shown in the example above, the purpose of the system test is to ensure, to whatever degree necessary, that the system performs to specifications. After the tests are completed and any necessary adjustments made, the project is ready for use with a high degree of confidence in the results.

5.6.13 CHANGING IT UP

The Wind Vanes R-Us Company thinks that our weather station is the best thing to come along since indoor plumbing, and now we face the task of improving the profit margin of the product by finding some places to reduce cost.

Now that the weather station product is designed and proven stable and reliable, what are the areas that costs could be reduced? In your career, you may be faced with this question time and time again. There are many areas in our weather station project where cost could be reduced. Items such as the humidity sensor, tipping rain bucket, and anemometer are all pretty expensive—but their performance is known and therefore may not be the first place to starting cutting away at the cost.

When we started the development process, we chose a processor that would give us ample room to design and debug the software—but we left ourselves the option to increase or decrease memory and features as required, without involving a large redesign effort. Let's look at changing the CPU as a cost-reduction maneuver.

Picking a Part for a Better Fit

The outdoor unit CPU has very little software on board. In fact, with the code as it a stands, only about 11% of the total flash memory is used. Looking at other members of the AVR family, we need to select a component with very similar features, but perhaps one with a smaller memory space. There are very few external connections, so even a smaller pin-count part would be acceptable.

The required features that we need to include are a USART, ADCs, and a timer to count our wind speed pulses. The ATMega48 has these features, comes in a 32-pin TQFP package, which is a surface-mount package, so no socket is required (and it can be easily assembled by a robot, saving labor costs as well), and has a 4K memory space, so it is more closely matched to our actual application size. At the time of the writing of this text, an ATMega16 has a unit cost of about $6.00. The ATMega48 costs about $2.00. That can be significant. If the Wind Vanes R-Us Company sells 100,000 units, that is a $400,000 cost savings (and you will be a hero for making the change from an ATMega16 to an ATMega48!).

Changes to the Schematic

Figure 5–35 shows the outdoor unit schematic modified to work with the ATMega48. Signals were simply moved around to the appropriate pins, picking up the required functionality of the USART, ADCs, and timer.

Changes to I/O Mapping

Because we thought ahead and used *#define* to assign functional names to specific I/O pins, it is easy to look at the software and know what few changes have to be made to support the design change. The I/O specific changes are as follows:

```
#define RAIN_INPUT          PIND.2    /* rain gauge input   */
#define WIND_SPEED_INPUT    PIND.5    /* anemometer input   */
#define TX_CTS              PINB.0    /* transmitter clear to send */
#define TX_POWER            PORTB.1   /* transmitter power enable */
```

There are other changes that go with these, including the DDRB and DDRD settings, as well as the default states of PORTB and PORTD. These changes can all be made in the initialization function.

```
void init_AVR(void)
{

    /* Crystal Oscillator division factor: 1 */
    #pragma optsize-
      CLKPR=0x80;       /* these instructions must be */
      CLKPR=0x00;       /* within a few CPU cycles, so optimization */
                        /* is temporarily disabled */
    #ifdef _OPTIMIZE_SIZE_
    #pragma optsize+
    #endif

      /* Input/Output Ports initialization   */
      /* Port B */
      PORTB=0x00;
      DDRB=0x02;       /* enable radio power control */
```

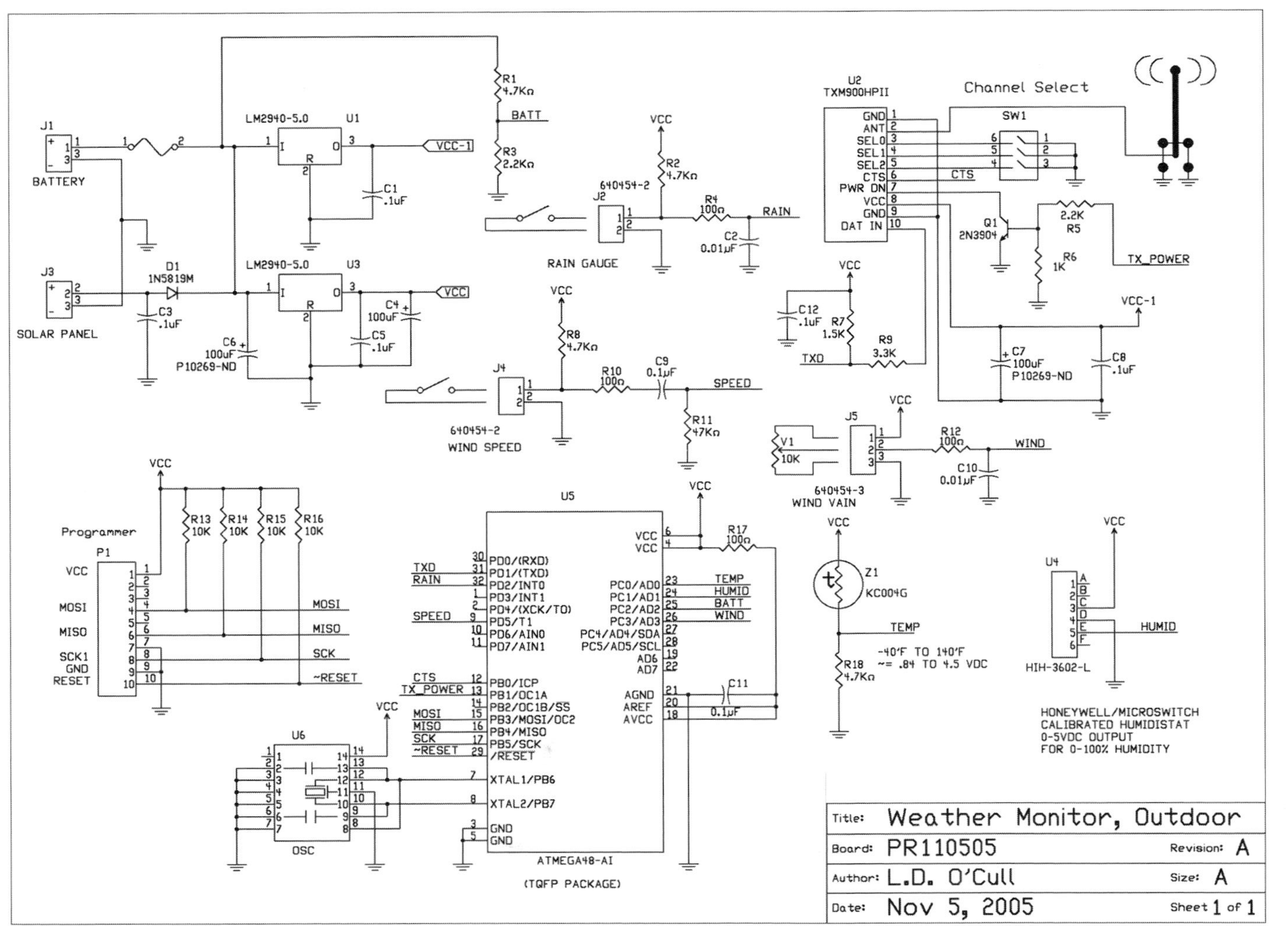

Figure 5–35 *Schematic of Outdoor Unit with ATMega48*

```
/* Port C */
PORTC=0x00;
DDRC=0x00;

/* Port D  */
PORTD=0x00;
DDRD=0x02;     /* enable USART transmitter output */

/* Timer/Counter 0 initialization */
/* Clock source: System Clock     */
/* Clock value: 62.500 kHz        */
/* Mode: Output Compare           */
/* OC0 output: Disconnected       */
TCCR0A=0x00;
TCCR0B=0x03;
TCNT0=0x00;
OCR0A=0x00;
OCR0B=0x00;

/* Timer/Counter 1 initialization    */
/* Clock source: T1 pin Rising Edge  */
/* Mode: Output Compare              */
/* OC1A output: Discon.              */
/* OC1B output: Discon.              */
/* Noise Canceler: Off               */
/* Input Capture on Falling Edge     */
TCCR1A=0x00;
TCCR1B=0x07;
TCNT1H=0x00;
TCNT1L=0x00;
ICR1H=0x00;
ICR1L=0x00;
OCR1AH=0x00;
OCR1AL=0x00;
OCR1BH=0x00;
OCR1BL=0x00;

/* External Interrupt(s) initialization               */
/* INT0: On                                           */
/* INT0 Mode: Any change                              */
/* INT1: Off                                          */
/* Interrupt on any change on pins PCINT0-7: Off      */
/* Interrupt on any change on pins PCINT8-14: Off     */
/* Interrupt on any change on pins PCINT16-23: Off    */
EICRA=0x01;
```

```
    EIMSK=0x01;
    EIFR=0x01;
    PCICR=0x00;

    /* Timer(s)/Counter(s) Interrupt(s) initialization */
    TIMSK0=0x01;
    TIMSK1=0x00;
    TIMSK2=0x00;

    /* UART initialization                               */
    /* Communication Parameters: 8 Data, 1 Stop, No Parity */
    /* UART Receiver: Off                                */
    /* UART Transmitter: On                              */
    /* UART Baud rate: 9600                              */
    UCSR0A=0x00;
    UCSR0B=0x48;
    UBRR0L=0x19;
    UBRR0H=0x00;

    /* Analog Comparator initialization                     */
    /* Analog Comparator: Off                               */
    /* Analog Comparator Input Capture by Timer/Counter 1: Off */
    ACSR=0x80;
    ADCSRB=0x00;

    /* ADC initialization              */
    /* ADC Clock frequency: 31.250 kHz */
    /* ADC Voltage Reference: AVCC pin */
    ADMUX=FIRST_ADC_INPUT|ADC_VREF_TYPE;
    ADCSRA=0xCF;

    /* Watchdog Timer initialization          */
    /* Watchdog Timer Prescaler: OSC/1024k    */
    /* Watchdog Timer interrupt: Off          */
    #pragma optsize-
    #asm("wdr")
    WDTCSR=0x39;    /* these instructions must be within */
    WDTCSR=0x29;    /* a few CPU cycles, so optimization */
                    /* is temporarily disabled */
    #ifdef _OPTIMIZE_SIZE_
    #pragma optsize+
    #endif

    /* Global enable interrupts              */
    #asm("sei")
}
```

Since the AVR processors are all very similar in capability, it is simple to move or "port" the code from one AVR to another. To move the ATMega16 code to the ATMega48, we automatically gain the functionality of the USART and ADCs with virtually no changes whatsoever. There is a small difference in how TIMER1 is handled.

On the ATMega48, TIMER1 is in an upper address within the register memory space. This means that the CodeVisionAVR compiler does not provide for a direct 16-bit access to TCNT1. Performing two 8-bit accesses and combining them for the 16-bit result can easily overcome this. To accomplish this, the interrupt routine that reads the wind speed counter is modified as follows:

```
/* Timer 0 overflow interrupt service routine */
interrupt [TIM0_OVF] void timer0_ovf_isr(void)
{
   static unsigned int TCNT1 = 0;

   TCCR1B=0x00; /* stop counter to prevent rollover during read */
   TCNT1 = ((unsigned int)TCNT1H) << 8;
   TCNT1 |= (((unsigned int)TCNT1L) & 0xFF);
   TCCR1B=0x07; /* start counter again */

   if(++wind_test_pass > 244) /* pass count before wind speed sample */
   {
      wind_speed += TCNT1;
      TCNT1 = 0;              /* reset counter for next sample.. */
      wind_test_pass = 0;     /* reset pass counter */
      wind_speed /= 2;        /* simple average of two.. */

      RF_TX_Time = 1;         /* time to transmit!! */
   }
}
```

A local variable TCNT1 is created to hold the 16-bit counter value. The TIMER1 counter is stopped during the read to prevent the counter from changing while it is being read, yielding a bogus result. Once the two 8-bit reads are performed, the TIMER1 counter is re-enabled and everything continues on as usual.

Other Considerations

There are a few other differences from the ATMega16 and the ATMega48. It is these differences that reinforce (the fact) that reading the datasheet is a critical step in the development and support of software. As Atmel finds better ways to do things with AVRs, those features will be incorporated into their parts. Atmel made a few changes to the clock-oscillator circuits on the ATMega48 that are not currently in the ATMega16. So you may have noticed a couple of unique items in the initialization process.

```
/* Crystal Oscillator division factor: 1 */
#pragma optsize-
```

```
   CLKPR=0x80;      /* these instructions must be */
   CLKPR=0x00;      /* within a few CPU cycles, so optimization */
                    /* is temporarily disabled */
#ifdef _OPTIMIZE_SIZE_
#pragma optsize+
#endif
```

The ATMega48 has many clock configurations to support a wide variety of oscillator options as well as low power modes of operation. One of the features is a clock prescaler that allows the system clock to be divided down to reduce the power consumption of the processor by running it slower. In order to guarantee a glitch-free change in the system clock while changing the prescale, it is required that the CLKPR register be enabled and modified within four system clock cycles. To guarantee this type of timing within the CodeVisionAVR compiler, the optimizer must be disabled. This causes the compiler to generate the assignment instructions in-line and to reduce the potential of the optimizer inserting a call or jump instruction between the assignments.

The same type of assignment rule applies to the watchdog timer as well:

```
/* Watchdog Timer initialization         */
/* Watchdog Timer Prescaler: OSC/1024k   */
/* Watchdog Timer interrupt: Off         */
#pragma optsize-
#asm("wdr")
WDTCSR=0x39;      /* these instructions must be within */
WDTCSR=0x29;      /* a few CPU cycles, so optimization */
                  /* is temporarily disabled */
#ifdef _OPTIMIZE_SIZE_
#pragma optsize+
#endif
```

In order to guarantee that the watchdog timer was not simply disabled by errant code running wildly through memory, the "two assignments within four cycles" rule applies here as well.

You can see that it is realistic in many cases to move from one processor to another in order to add or reduce features, improve unit cost, improve performance, or reduce power consumption. In many cases, a move like this can be performed in a few hours with very little impact to the entire design.

5.7 CHALLENGES

Listed below are various features that could be added to or changed in the software to enhance the weather monitor operation and performance:

- Add year and leap year tracking.
- Add weather alarms, as discussed in Section 5.6.7, "Software Design, Indoor Unit."

- Add averaging to the wind speed and rainfall measurements to make them more stable and more readable.
- Add linear interpolation to improve the wind chill and dew point calculations.
- Rainfall tracking: instead of today, this month, this year, make the system track rainfall in terms of the last 24 hours, the last 30 days, and the last year to date.
- Track and indicate barometric pressure change: rising, falling, and steady.
- Add software to send the display information to a PC using the serial port and connector P3.
- Enhance the software to accept commands from a PC using a software USART receive on the INT0 pin.
- Add logging capability to track peak wind, rain, temperature, and pressure conditions and when they occurred. Allow the user to view them on the display or download them to a PC.
- Make the measuring functions of the monitor a state machine, such that time/date editing and alarm set point adjustments do not impede the normal operation of the monitor.

5.8 CHAPTER SUMMARY

In this chapter, project development has been approached as a *process,* an orderly set of steps that, when followed, will virtually always lead to successful project. The process has been demonstrated by the development of a weather station based on the Atmel Mega16 microcontroller.

5.9 EXERCISES

1. List each of the steps of the process of project development and give an example of the activities that would take place during that step (Section 5.4).
2. In which step of the project development process would each of the following occur (Section 5.4)?

   A. Prototyping a sensor and its conditioning circuitry

   B. Simulating a circuit's operation

   C. Creating detailed specifications for a project

   D. Drawing a project schematic

   E. Testing individual software functions

   F. Writing software flowcharts

   G. Testing the final project to its specifications

   H. Doing proof-of-concept testing on a questionable circuit

3. Using the table data below, write a program to perform a look-up-table and linear-interpolation operation to return a compensated temperature from the ADC reading (Section 5.6.12):

ADC Values, 10 Bits (0–1023)	Temperature °C
123	–20
344	0
578	20
765	40
892	60
1002	80

5.10 LABORATORY ACTIVITY

The only laboratory activity that is really appropriate to this chapter is to demonstrate the process of project development by developing a project. Some suggested projects are shown below. Any of these ideas could be modified or expanded through the use of displays and/or additional sensors or input devices.

1. A robotic "mouse" that can follow black electrical tape on a whitish vinyl floor. This project will require controlling the speed and steering as well as sensing the black line.
2. A device composed of the microcontroller and a video camera that can detect simple shapes such as a square, a triangle, or a circle of black paper on a white surface.
3. A device using motors, sensors, and a cardboard arrow to point to and follow a heat source as it moves about the room. This device would make excellent use of stepper motors.
4. A replacement for a furnace thermostat.
5. A security system for an automobile with sensors to detect doors being opened (including the gas tank cover, the trunk, and the hood) and that would detect when the vehicle is being moved.

EXERCISE

Library Functions Reference

INTRODUCTION

Much of the programming power and convenience of the C language lies in its built-in, or library, functions. These are routines that accomplish many of the more common tasks that face a C programmer.

As with all C functions, you call the library functions from your code by using the function name and either passing values to the function or receiving the values returned by the function. In order to make use of the functions, the programmer needs to know what parameters the function is expecting to be passed to it and the nature of any values returned from it. This information is made clear by the prototype for the function. The function prototypes are contained in a set of header files that the programmer "includes" in the code.

There are many library functions available in most C compilers, gathered into groups by function. This grouping avoids having to include a whole host of unneeded function prototypes in every program. For example, the function prototypes for the group of routines concerned with the standard input and output routines are contained in a header file called stdio.h. If programmers wish to use some of these functions, they would put the following statement into the beginning of their code:

```
#include <stdio.h>
```

With the header file included, the functions concerned with standard I/O such as the *printf* function (referred to in earlier chapters) are made available to the programmer.

This reference section describes each of the library functions available in CodeVisionAVR. A comprehensive list, grouped by their header file names, is followed by a detailed description and example of each function, in alphabetical order.

FUNCTIONS LISTED BY LIBRARY FILE

bcd.h

```
unsigned char bcd2bin(unsigned char n);
unsigned char bin2bcd(unsigned char n);
```

ctype.h

```
unsigned char isalnum(char c);
unsigned char isalpha(char c);
unsigned char isascii(char c);
unsigned char iscntrl(char c);
unsigned char isdigit(char c);
unsigned char islower(char c);
unsigned char isprint(char c);
unsigned char ispunct(char c);
unsigned char isspace(char c);
unsigned char isupper(char c);
unsigned char isxdigit(char c);
unsigned char toint(char c);
char tolower(char c);
char toupper(char c);
char toascii(char c);
```

delay.h

```
void delay_us(unsigned int n);
void delay_ms(unsigned int n);
```

gray.h

```
unsigned char gray2binc(unsigned char n);
unsigned int gray2bin(unsigned int n);
unsigned long gray2binl(unsigned long n);
unsigned char bin2grayc(unsigned char n);
unsigned int bin2gray(unsigned int n);
unsigned long bin2grayl(unsigned long n);
```

math.h

```
unsigned char cabs(signed char x);
unsigned int abs(int x);
unsigned long labs(long x);
float fabs(float x);
signed char cmax(signed char a,signed char b);
int max(int a,int b);
long lmax(long a,long b);
float fmax(float a,float b);
signed char cmin(signed char a,signed char b);
int min(int a,int b);
long lmin(long a,long b);
float fmin(float a,float b);
```

```
signed char csign(signed char x);
signed char sign(int x);
signed char lsign(long x);
signed char fsign(float x);
unsigned char isqrt(unsigned int x);
unsigned int lsqrt(unsigned long x);
float sqrt(float x);
float floor(float x);
float ceil(float x);
float fmod(float x,float y);
float modf(float x,float *ipart);
float ldexp(float x,int expon);
float frexp(float x,int *expon);
float exp(float x);
float log(float x);
float log10(float x);
float pow(float x,float y);
float sin(float x);
float cos(float x);
float tan(float x);
float sinh(float x);
float cosh(float x);
float tanh(float x);
float asin(float x);
float acos(float x);
float atan(float x);
float atan2(float y,float x);
```

mem.h

```
void pokeb(unsigned int addr,unsigned char data);
void pokew(unsigned int addr,unsigned int data);
unsigned char peekb(unsigned int addr);
unsigned int peekw(unsigned int addr);
```

sleep.h

```
void sleep_enable(void);
void sleep_disable(void);
void idle(void);
void powerdown(void);
void powersave(void);
void standby(void);
void extended_standby(void);
```

spi.h

```
unsigned char spi(unsigned char data);
```

stdio.h

```
char getchar(void);
void putchar(char c);
void puts(char *str);
void putsf(char flash *str);
char *gets(char *str,unsigned int len);
void printf(char flash *fmtstr);
void sprintf(char *str, char flash *fmtstr);
signed char scanf(char flash *fmtstr);
signed char sscanf(char *str, char flash *fmtstr);
void vprintf(char flash *fmtstr, va_list argptr);
void vsprintf(char *str, char flash *fmtstr, val_list argptr);
```

stdlib.h

```
int atoi(char *str);
long atol(char *str);
float atof(char *str);
void itoa(int n,char *str);
void ltoa(long int n,char *str);
void ftoa(float n,unsigned char decimals,char *str);
void ftoe(float n,unsigned char decimals,char *str);
void srand(int seed);
int rand(void);
void *malloc(unsigned int size)
void *calloc(unsigned int num, unsigned int size)
void *realloc(void *ptr, unsigned int size)
void free(void *ptr)
```

string.h

```
char *strcat(char *str1,char *str2);
char *strcatf(char *str1,char flash *str2);
char *strchr(char *str,char c);
signed char strcmp(char *str1,char *str2);
signed char strcmpf(char *str1,char flash *str2);
char *strcpy(char *dest,char *src);
char *strcpyf(char *dest,char flash *src);
unsigned char strcspn(char *str,char *set);
unsigned char strcspnf(char *str,char flash *set);
unsigned int strlen(char *str);
unsigned int strlenf(char flash *str);
char *strncat(char *str1,char *str2,unsigned char n);
char *strncatf(char *str1,char flash *str2,unsigned char n);
signed char strncmp(char *str1,char *str2,unsigned char n);
signed char strncmpf(char *str1,char flash *str2,unsigned char n);
char *strncpy(char *dest,char *src,unsigned char n);
char *strncpyf(char *dest,char flash *src,unsigned char n);
```

```
char *strpbrk(char *str,char *set);
char *strpbrkf(char *str,char flash *set);
char strpos(char *str,char c);
char *strrchr(char *str,char c);
char *strrpbrk(char *str,char *set);
char *strrpbrkf(char *str,char flash *set);
signed char strrpos(char *str,char c);
char *strstr(char *str1,char *str2);
char *strstrf(char *str1,char flash *str2);
unsigned char strspn(char *str,char *set);
unsigned char strspnf(char *str,char flash *set);
char *strtok(char *str1,char flash *str2);
void *memccpy(void *dest,void *src,char c,unsigned n);
void *memchr(void *buf,unsigned char c,unsigned n);
signed char memcmp(void *buf1,void *buf2,unsigned n);
signed char memcmpf(void *buf1,void flash *buf2,unsigned n);
void *memcpy(void *dest,void *src,unsigned n);
void *memcpyf(void *dest,void flash *src,unsigned n);
void *memmove(void *dest,void *src,unsigned n);
void *memset(void *buf,unsigned char c,unsigned n);
```

abs

#include <math.h>

unsigned int abs(int x);

unsigned char cabs(signed char x);

unsigned long labs(long x);

float fabs(float x);

The *abs* function returns the absolute value of the integer *x*. The *cabs, labs,* and *fabs* functions return the absolute value of the signed char, long, and float variable *x* as an unsigned char, unsigned long, and float value, respectively.

Returns: Absolute value of *x* sized according to the function called

```
#include <math.h>
void main()
{
    unsigned int int_pos_val;
    unsigned long long_pos_val;

        // get absolute value of an integer
    int_pos_val = abs(-19574);
```

```
        // get absolute value of a long
    long_pos_val = labs(-125000);

    while(1)
    {
    }
}
```

Results:

```
int_pos_val = 19574
long_pos_val = 125000
```

acos

#include <math.h>

float acos(float x);

The *acos* function calculates the arc cosine of the floating point number *x*. The result is in the range of $-\pi/2$ to $\pi/2$. The variable *x* must be in the range of -1 to 1.

Returns: acos(x) in the range of $-\pi/2$ to $\pi/2$ where *x* is in the range of -1 to 1

```
#include <math.h>

void main()
{
      float new_val;

      new_val = acos(0.875);

      while(1)
      {
      }
}
```

Results: new_val = 0.505

asin

#include <math.h>

float asin(float x);

The *asin* function calculates the arc sine of the floating point number *x*. The result is in the range of $-\pi/2$ to $\pi/2$. The variable *x* must be in the range of -1 to 1.

Returns: asin(x) in the range of $-\pi/2$ to $\pi/2$ where *x* is in the range of -1 to 1

```
#include <math.h>

void main()
```

```
{
        float new_val;

        new_val = asin(0.875);

        while(1)
        {
        }
}
```

Results: new_val = 1.065

atan

#include <math.h>

float atan(float x);

The *atan* function calculates the arc tangent of the floating point number *x*. The result is in the range of $-\pi/2$ to $\pi/2$.

Returns: atan(x) in the range of $-\pi/2$ to $\pi/2$

```
#include <math.h>

void main()
{
        float new_val;

        new_val = atan(1.145);

        while(1)
        {
        }
}
```

Results: new_val = 0.852

atan2

#include <math.h>

float atan2(float y, float x);

The *atan2* function calculates the arc tangent of the floating point numbers *y/x*. The result is in the range of $-\pi$ to π.

Returns: atan(y/x) in the range of $-\pi$ to π

```
#include <math.h>

void main()
```

```
{
        float new_val;

        new_val = atan2(2.34, 5.12);

        while(1)
        {
        }
}
```

Results: new_val = 0.428

atof, atoi, atol

#include <stdlib.h>

float atof(char *str);

int atoi(char *str);

long atol(char *str);

The *atof* function converts the ASCII character string pointed to by *str* to its floating point equivalent. The *atoi* and *atol* functions convert the string to an integer or long integer, respectively. All three functions skip leading white space characters. Numbers may also be preceded by + and – signs. Valid numeric characters are the digits *0* through *9*. The function *atof* also accepts the decimal point as a valid numeric character.

Once a non–white space character is encountered, the conversion to a numeric equivalent starts. The conversion stops when the first non-numeric character is encountered. If no numeric characters are found, the functions return zero. All three functions return signed values.

Returns:

- atof, atoi, and atol return zero if no numeric data is found
- atof – floating point equivalent of the ASCII string pointed to by *str*
- atoi – signed integer equivalent of the ASCII string pointed to by *str*
- atol – signed long integer equivalent of the ASCII string pointed to by *str*

```
#include <mega16.h>
#include <stdio.h>      // this to include putchar and printf!
#include <stdlib.h>

/* quartz crystal frequency [Hz] */
#define xtal 7372000L

/* Baud rate */
#define baud 9600
```

```
#define MAX_ENTRY_LENGTH        10
void main(void)
{
   char mystr[MAX_ENTRY_LENGTH+1];
   int myint;
   char c;

   /* initialize the USART's baud rate */
   UBRRH=0x00;
   UBRRL=xtal/16/baud-1;
   /* initialize the USART control register
      RX & TX enabled, no interrupts, 8 data bits */
   UCSRA=0x00;
   UCSRB=0xD8;
   UCSRC=0x86;

   while (1)
   {
      c = 0;
      putsf("Enter a signed integer number followed by !\n\r");
      while (c < MAX_ENTRY_LENGTH)
      {
         mystr[c++] = getchar();      // wait for a character
         // if it is our terminating character, then quit!
         if (mystr[c-1] == '!')
               break;
      }
      mystr[c] = '\0';   // null terminate the string!

      myint = atoi(mystr);     // convert

      printf("Your integer value is: %d\n\r",myint);
   }
}
```

Results:

The following is transmitted by the USART.

```
Enter a signed integer number followed by !
```

Once '!' is received, the string is converted and the result is transmitted to the USART.

bcd2bin

#include <bcd.h>

unsigned char bcd2bin(unsigned char n);

The *bcd2bin* function converts a binary coded decimal (bcd) value to a binary value. In binary coded decimal representation, the upper nibble of the value represents the 10s digit of a decimal value. The lower nibble of the value represents the 1s digit of a decimal value. *n* must be a value between 0d and 99d.

Returns: Binary value of *n*

```
#include <bcd.h>

void main()
{
      unsigned char bcd_value, bin_value;

      bcd_value = 0x15;    /* bcd representation of decimal 15 */
      bin_value = bcd2bin(bcd_value);

      while(1)
      {
      }
}
```

Results: bin_value = 15(= 0x0F)

bin2bcd

#include <bcd.h>

unsigned char bin2bcd(unsigned char n);

The *bin2bcd* function converts a binary value to a binary coded decimal (bcd) value. In binary coded decimal representation, the upper nibble of the value represents the 10s digit of a decimal value. The lower nibble of the value represents the 1s digit of a decimal value. *n* must be a value between 0d and 99d.

Returns: Binary coded decimal value of *n*

```
#include <bcd.h>

void main()
{
      unsigned char bin_value, bcd_value;

      bin_value = 0x0F;  /* 15 decimal */
      bcd_value = bin2bcd(bin_value);

      while(1)
      {
      }
}
```

Results: bcd_value = 0x15

bin2grayc, bin2gray, bin2grayl

<include <gray.h>

unsigned char bin2grayc(unsigned char n);

unsigned int bin2gray(unsigned int n);

unsigned long bin2grayl(unsigned long n);

The *bin2gray* functions convert the binary value *n* to a Gray-coded value. Gray codes were developed to prevent noise in systems where analog-to-digital conversions were being performed. Gray codes have the advantage over binary numbers in that only one bit in the code changes between successive numbers. The *bin2gray* functions *bin2grayc, bin2gray,* and *bin2grayl* are tailored for unsigned char, unsigned int, and unsigned long variables, respectively.

Table A–1 lists the Gray codes and their binary and decimal equivalents for values 0 through 15.

Decimal Number	Binary Equivalent B_3 B_2 B_1 B_0	Gray Code Equivalent B_3 B_2 B_1 B_0
0	0 0 0 0	0 0 0 0
1	0 0 0 1	0 0 0 1
2	0 0 1 0	0 0 1 1
3	0 0 1 1	0 0 1 0
4	0 1 0 0	0 1 1 0
5	0 1 0 1	0 1 1 1
6	0 1 1 0	0 1 0 1
7	0 1 1 1	0 1 0 0
8	1 0 0 0	1 1 0 0
9	1 0 0 1	1 1 0 1
10	1 0 1 0	1 1 1 1
11	1 0 1 1	1 1 1 0
12	1 1 0 0	1 0 1 0
13	1 1 0 1	1 0 1 1
14	1 1 1 0	1 0 0 1
15	1 1 1 1	1 0 0 0

Table A–1 *Gray Code Equivalents of Decimal and Binary Values*

Returns: Binary value of *n*

```
#include <gray.h>

void main()
{
        unsigned char gray_char, bin_char;

        bin_char = 0x07;
        gray_char = bin2gray(bin_char);

        while(1)
        {
        }
}
```

Results: gray_char = 4

calloc

#include <stdlib.h>

void *calloc(unsigned int num, unsigned int size);

The *calloc* function allocates a memory block in the heap for an array of *num* elements, each element having the length of *size* bytes. On return, the function returns a pointer to the start of the memory block which is filled with zeros. The allocated memory block occupies *size*+4 bytes in the heap. This must be taken into account when specifying the heap size in the **Project|Configure|C Compiler|Code Generation** menu.

Returns: If successful in finding contiguous free memory of the appropriate size, returns a pointer to the memory block. If unsuccessful, returns a null pointer.

```
#include <mega16.h>
/* include the standard input/output library */
#include <stdio.h>
/* include the variable length argument lists macros */
#include <stdlib.h>

#define xtal 6000000L
#define baud 9600

// declaring a pointer, but not reserving memory for the array!
struct two_d *two_d_p;
struct two_d *start_two_d_p;
int memory_size;
```

```
struct two_d
{
    int index;
    int value;
};

void main(void)
{
   int i;

   // USART initialization
   // Communication Parameters: 8 Data, 1 Stop, No Parity
   // USART Receiver: On
   // USART Transmitter: On
   // USART Mode: Asynchronous
   // USART Baud rate: 9600
   UCSRA=0x00;
   UCSRB=0xD8;
   UCSRC=0x86;
   UBRRH=0x00;
   UBRRL=xtal/16/baud-1;

   while (1)
   {
      putsf("\n\rHow much memory should I fill?(ENTER)\n\r");
      if (scanf("%d",&memory_size) != -1)
      {
         if ((memory_size*sizeof(struct two_d)) < (_HEAP_SIZE_ - 4))
         {
            printf("\n\rThanks, I'll try!\n\r");

            // try to get enough memory
            start_two_d_p =
                    calloc(memory_size,sizeof(struct two_d));
            two_d_p = start_two_d_p;
            if (two_d_p != NULL)
            {
               printf("\n\rInitial Values!\n\r");
               for (i=0;i<memory_size;i++)
               {
                  printf("%04X %04X\n\r",two_d_p->index,
                        two_d_p->value);
                  two_d_p->index = i;
                  two_d_p->value = i + 1000;  // change the value
                  two_d_p++;      // move the pointer forward!
               }
```

```
                    printf("\n\rModified Values!\n\r");
                    two_d_p = start_two_d_p;
                    // print the new values!
                    for (i=0;i<memory_size;i++)
                    {
                        printf("%04X %04X\n\r",two_d_p->index,
                             two_d_p->value);
                        two_d_p++;        // move the pointer forward!
                    }

                    // free the calloc memory for next time!
                    free(start_two_d_p);
                }
                else
                     printf("Failed to calloc correctly.\n\r");
            }
            else
            {
               printf("\n\rHeap size limit is %d\n\r",_HEAP_SIZE_-4);
            }
        }
    }
}
```

Results: The USART transmits, at 9600 baud,

```
How much memory should I fill?(ENTER)
```

Then it waits until a number is entered followed by a newline character. Once this is received (5 for example), it transmits

```
Thanks, I'll try!
Initial Values!
0000 0000
0000 0000
0000 0000
0000 0000
0000 0000
Modified Values!
0000 03E8
0001 03E9
0002 03EA
0003 03EB
0004 03EC
```

cabs

#include <math.h>

unsigned char cabs(signed char x);

Returns: Absolute value of *x*

See *abs*.

ceil

#include <math.h>

float ceil(float x);

The *ceil* function returns the smallest integer value that is not less than the floating point number *x*. In other words, the *ceil* function rounds *x* up to the next integer value and returns that value.

Returns: Smallest integer value that is not less than the floating point number *x*

```
#include <math.h>

void main()
{
        float new_val;

        new_val = ceil(2.531);

        while(1)
        {
        }
}
```

Results: new_val = 3

cmax

#include <math.h>

signed char cmax(signed char a, signed char b);

Returns: Maximum value of *a* or *b*

See *max*.

cmin

#include <math.h>

signed char cmin(signed char a, signed char b);

Returns: Minimum value of *a* or *b*

See *min.*

cos

#include <math.h>

float cos(float x);

The *cos* function calculates the cosine of the floating point number *x*. The angle *x* is expressed in radians.

Returns: cos(x)

```
#include <math.h>

void main()
{
        float new_val;

        new_val = cos(5.121);

        while(1)
        {
        }
}
```

Results: new_val = 0.397

cosh

#include <math.h>

float cosh(float x);

The *cosh* function calculates the hyperbolic cosine of the floating point number *x*. The angle *x* is expressed in radians.

Returns: cosh(x)

```
#include <math.h>

void main()
{
        float new_val;

        new_val = cosh(5.121);

        while(1)
        {
        }
}
```

Results: new_val = 83.754

csign

#include <math.h>

signed char csign(signed char x);

Returns: −1, 0, or 1 if *x* is negative, zero, or positive, respectively

See *sign*.

delay_ms

#include <delay.h>

void delay_ms(unsigned int n);

The *delay_ms* function generates a delay of *n* milliseconds before returning. The interrupts must be turned off around the call to the *delay_ms* function or the delay will be much longer than intended. The actual delay depends on the clock crystal frequency. Therefore, it is important to specify the correct clock frequency. This can be done either in the **Project|Configure|C Compiler** menu or with the statement *#define xtal xL*, where *xL* is the clock frequency in Hertz.

Returns: None

```
#include <delay.h>
#include <mega16.h>
#include <stdio.h>

/* quartz crystal frequency [Hz] */
#define xtal 7372000L

/* Baud rate */
#define baud 9600

void main(void)
{
   unsigned int pause_time;

   /* initialize the USART's baud rate */
   UBRRH=0x00;
   UBRRL=xtal/16/baud-1;
   /* initialize the USART control register
      RX & TX enabled, no interrupts, 8 data bits */
   UCSRA=0x00;
   UCSRB=0xD8;
   UCSRC=0x86;
```

```
        while (1)
        {
           putsf("How many milliseconds should I pause?(ENTER)\n\r");
           if (scanf("%u\n",&pause_time) != -1)
           {
              printf("Thanks! I'll pause for %u milliseconds.\n\r",
                    pause_time);

              // disable interrupts
              #asm("cli");

              delay_ms(pause_time);

              // enable interrupts
              #asm("sei");
           }
        }
     }
```

Results:

The microprocessor transmits

```
How many milliseconds should I pause? (ENTER)
```

Then it waits until a number is entered followed by a newline character. Once this is received (2000, for example), it transmits

```
Thanks! I'll pause for 2000 milliseconds.
```

Then after an appropriate pause (2 seconds in our example), the first prompt is again transmitted.

delay_us

#include <delay.h>

void delay_us(unsigned int n);

The *delay_us* function generates a delay of *n* microseconds before returning. *n* must be a constant expression. The interrupts must be turned off around the call to the *delay_us* function or the delay will be much longer than intended. The actual delay depends on the clock crystal frequency. Therefore, it is important to specify the correct clock frequency. This can be done either in the **Project|Configure|C Compiler** menu or with the statement *#define xtal xL,* where *xL* is the clock frequency in Hertz.

Returns: None

```
#include <delay.h>
#include <mega16.h>
#include <stdio.h>
```

```
/* quartz crystal frequency [Hz] */
#define xtal 7372000L
/* Baud rate */
#define baud 9600

void main(void)
{
   unsigned int pause_time;

   /* initialize the USART's baud rate */
   UBRRH=0x00;
   UBRRL=xtal/16/baud-1;
   /* initialize the USART control register
      RX & TX enabled, no interrupts, 8 data bits */
   UCSRA=0x00;
   UCSRB=0xD8;
   UCSRC=0x86;

   while (1)
   {
      putsf("How many thousands of uSeconds should I pause?\r");
      putsf("Enter a number between 1 and 5. Press ENTER.\n\r");
      if (scanf("%u\n",&pause_time) != -1)
      {
         if ((pause_time > 0) && (pause_time < 6))
         {
             printf("I'll pause %1u000 microseconds.\n\r",
             pause_time);

            // disable interrupts
            #asm("cli");
            switch (pause_time)
            {
                case 1:
                      delay_us(1000);
                      break;
                case 2:
                      delay_us(2000);
                      break;
                case 3:
                      delay_us(3000);
                      break;
                case 4:
                      delay_us(4000);
                      break;
```

```
                    case 5:
                          delay_us(5000);
                          break;
                    default:
                          break;
                }
                // enable interrupts
                #asm("sei");
             }
             else
                putsf("Huh?\n\r");
          }
       }
    }
```

Results:

The microprocessor transmits

```
How many thousands of uSeconds should I pause?
Enter a number between 1 and 5. Press ENTER.
```

Then it waits until a number is entered followed by a newline character. Once this is received (2, for example), it transmits

```
I'll pause 2000 microseconds.
```

Then after an appropriate pause (2 milliseconds in our example), the first prompt is again transmitted.

exp

#include <math.h>

float exp(float x);

The *exp* function calculates the natural (base *e*), exponential of the variable *x*.

Returns: e^x

```
#include <math.h>

void main()
{
      float new_val;

      new_val = exp(5);

      while(1)
      {
      }
}
```

Results: new_val = e^5 = 148.413

extended_standby

#include <sleep.h>

void extended_standby(void);

The *extended_standby* function puts the AVR microcontroller into extended standby mode. This is a sleep mode and similar to the *powerdown* function.

See *powerdown* in this appendix for the function use.

See the Atmel datasheets for the complete description of this sleep mode as it applies to a particular AVR device. Extended standby sleep mode is not available on all AVR devices.

>fabs

#include <math.h>

float fabs(float x);

Returns: Absolute value of *x*

See *abs*.

floor

#include <math.h>

float floor(float x);

The *floor* function returns integer value of the floating point number *x*. This is the largest whole number that is less than or equal to *x*.

Returns: Integer portion of *x*

```
#include <math.h>

void main()
{
        float new_val;

        new_val = floor(2.531);

        while(1)
        {
        }
}
```

Results: new_val = 2

fmax

#include <math.h>

float fmax(float a, float b);

Returns: Maximum value of *a* or *b*

See *max.*

fmin

#include <math.h>

float fmin(float a, float b);

Returns: Minimum value of *a* or *b*

See *min.*

fmod

#include <math.h>

float fmod(float x, float y);

The *fmod* function returns the remainder of *x* divided by *y*. This is a modulo function specifically designed for float type variables. The modulo operator (%) is used to perform the modulo operation on integer variable types.

Returns: Remainder of *x* divided by *y*

```
#include <math.h>
void main()
{
        float remains;

        remains = fmod(25.6, 8);

        while(1)
        {
        }
}
```

Results: remains = 1.6

free

#include <stdlib.h>

void free(void *ptr);

The *free* function frees memory allocated in the heap by *malloc, calloc,* or *realloc* functions and pointed to by *ptr.* After being freed, the memory block is available for new allocation. If *ptr* is null, then it is ignored.

Returns: None

See *malloc.*

frexp

#include <math.h>

float frexp(float x, int *expon);

The *frexp* function returns the mantissa of the floating point number *x,* or 0 if *x* is 0. The power of 2 exponent of *x* is stored at the location pointed to by *expon.* If *x* is 0, the value stored at *expon* is also 0.

In other words, if the following call is made,

```
y = frexp(x, expon);
```

the relationshp between *expon, x,* and the return value *y* can be expressed as

$$x = y * 2^{expon}$$

Returns: Mantissa of *x* in the range 0.5 to 1.0 or 0 if *x* is 0

```
#include <mega16.h>
#include <math.h>

void main(void)
{
        float x,z;
        int y;
        float a,c;
        int b;
        float d,f;
        int e;

        x = 3.14159;
        z=frexp(x,&y);

        a = 0.14159;
        c=frexp(a,&b);

        d = .707000;
        f=frexp(d,&e);
```

```
        while (1)
        {
        }

}
```

Results:

```
x = 3.141590
y = 2
z = 0.785398
a = 0.141590
b = -2
c = 0.566360
d = 0.707000
e = 0
f = 0.707000
```

fsign

#include <math.h>

signed char fsign(float x);

Returns: –1, 0, or 1 if *x* is negative, zero, or positive, respectively

See *sign.*

ftoa

#include <stdlib.h>

void ftoa(float n, unsigned char decimals, char *str);

The *ftoa* function converts the floating point value *n* to its ASCII character string equivalent. The value is represented with the specified number of decimal places. The string is stored at the location specified by **str.* Take care to ensure that the memory allocated for *str* is large enough to hold the entire value plus a null-terminating character.

Returns: None

```
#include <mega16.h>
#include <stdio.h>      // this to include putchar and printf!
#include <stdlib.h>

/* quartz crystal frequency [Hz] */
#define xtal 7372000L

/* Baud rate */
#define baud 9600
```

```
void main(void)
{
        char mystr[11];

        /* initialize the USART's baud rate */
        UBRRH=0x00;
        UBRRL=xtal/16/baud-1;
        /* initialize the USART control register
           RX & TX enabled, no interrupts, 8 data bits */
        UCSRA=0x00;
        UCSRB=0xD8;
        UCSRC=0x86;

        ftoa(12470.547031,2,mystr);
        putsf("The floating point value is: ");
        puts(mystr);

        while (1)
        {
        }

}
```

Results: The USART transmits, at 9600 baud,

```
The floating point value is:
                                        12470.55
```

ftoe

#include <stdlib.h>

void ftoe(float n, unsigned char decimals, char *str);

The *ftoe* function converts the floating point value *n* to its ASCII character string equivalent. The value is represented as a mantissa with the specified number of decimal places and an integer power of 10 exponent. The string is stored at the location specified by **str*. Take care to ensure that the memory allocated for *str* is large enough to hold the entire value plus a null terminating character.

Returns: None

```
#include <mega16.h>
#include <stdio.h>      // this to include putchar and printf!
#include <stdlib.h>

/* quartz crystal frequency [Hz] */
#define xtal 7372000L
```

```
/* Baud rate */
#define baud 9600

void main(void)
{
        char mystr[11];

        /* initialize the USART's baud rate */
        UBRRH=0x00;
        UBRRL=xtal/16/baud-1;
        /* initialize the USART control register
           RX & TX enabled, no interrupts, 8 data bits */
        UCSRA=0x00;
        UCSRB=0xD8;
        UCSRC=0x86;

        ftoe(0.001247,2,mystr);
        putsf("The floating point value is: ");
        puts(mystr);

        while (1)
        {
        }

}
```

Results: The USART transmits, at 9600 baud,

```
The floating point value is:
                                1.25e-3
```

getchar

#include <stdio.h>

char getchar(void);

The *getchar* function is a standard C language I/O function, but it has been adapted to work on embedded microcontrollers with limited resources. The *getchar* function returns a character received by the USART. This function uses polling to get the character from the USART. As a result, the function waits indefinitely for a character to be received before returning.

Prior to this function being used, the USART must be initialized and the USART receiver must be enabled.

If another peripheral is to be used for receiving, the *getchar* function can be modified accordingly. The source for this function is in the stdio.h file.

If you want to replace the standard library *getchar* with your own version, you must do three things. First, you must place the new *getchar* function in the *.c* file before including the standard library. Second, you must follow the new *getchar* function with a definition statement telling the compiler to ignore any other *getchar* functions that it finds. Finally, include the standard library. Put together, the code might appear as

```
char getchar(void)
{
      // your getchar routine statements here
}

#define _ALTERNATE_GETCHAR_
#include <stdio.h>

// the rest of the source file!
```

The standard *getchar()* returns: char received by the USART

```
#include <mega16.h>
#include <stdio.h>

/* quartz crystal frequency [Hz] */
#define xtal 7372000L

/* Baud rate */
#define baud 9600

void main(void)
{
      char k;

      /* initialize the USART's baud rate */
      UBRRH=0x00;
      UBRRL=xtal/16/baud-1;
      /* initialize the USART control register
         RX & TX enabled, no interrupts, 8 data bits */
      UCSRA=0x00;
      UCSRB=0xD8;
      UCSRC=0x86;

      while (1)
      {
            /* receive the character */
            k=getchar();
            /* and echo it back */
            putchar(k);
      }
}
```

Results: Characters received by the USART are echoed back out of the USART until power is removed from the processor.

*gets

#include <stdio.h>

char *gets(char *str, unsigned int len);

The **gets* function reads characters from the USART using the *getchar* function until a newline character is encountered and places them in the string pointed to by *str.* The newline character is replaced by '\0' and a pointer to *str* is returned. If *len* characters are read before the newline character is read, then the function terminates the string *str* with '\0' and returns.

Prior to this function being used, the USART must be initialized and the USART receiver must be enabled.

Returns: Pointer to the string *str*

```
#include <mega16.h>
#include <stdio.h>

/* quartz crystal frequency [Hz] */
#define xtal 7372000L

/* Baud rate */
#define baud 9600

void main(void)
{
   char your_name[11];    // room for 10 chars plus termination

   /* initialize the USART's baud rate */
   UBRRH=0x00;
   UBRRL=xtal/16/baud-1;
   /* initialize the USART control register
      RX & TX enabled, no interrupts, 8 data bits */
   UCSRA=0x00;
   UCSRB=0xD8;
   UCSRC=0x86;

   putsf("Please enter your name and press return.\r");
   putsf("(Only 10 characters are allowed)\r");
   gets(your_name,10);    // up to 10 chars!
   printf("Hi %s!\n\r",your_name);

   while (1)
   {
   }
}
```

Results: The USART transmits the prompt for a name:

```
Please enter your name and press return.
(Only 10 characters are allowed.)
```

Then the microprocessor waits for either a newline character or 10 characters to be received by the USART. Assume the string "Jane Doe" is received followed by the newline character. The following is transmitted:

```
Hi Jane Doe!
```

gray2binc, gray2bin, gray2binl

#include <gray.h>

unsigned char gray2binc(unsigned char n);

unsigned int gray2bin(unsigned int n);

unsigned long gray2binl(unsigned long n);

The *gray2bin* functions convert the Gray-coded decimal value *n* to a binary coded value. Gray codes were developed to prevent noise in systems where analog-to-digital conversions were being performed. Gray codes have the advantage over binary numbers in that only one bit in the code changes between successive numbers. The *gray2bin* functions *gray2binc*, *gray2bin*, and *gray2binl* are tailored for unsigned char, unsigned int, and unsigned long variables, respectively.

Table A–1 in the *bin2gray* function description lists the Gray codes and their binary equivalents for values 0 through 15.

Returns: Binary value of *n*

```
#include <gray.h>

void main()
{
   unsigned char gray_char, bin_char;

   gray_char = 0x04;     /* gray representation of decimal 7 */
   bin_char = gray2bin(gray_char);

   while(1)
   {
   }
}
```

Results: bin_char = 7

idle

#include <sleep.h>

void idle(void);

The *idle* function puts the AVR microcontroller into idle mode. The function *sleep_enable* must be called prior to this function being used. In idle mode, the CPU is stopped, but the timers/counters, watchdog timer, and the interrupt system continue to run. As such, the microcontroller can wake up from either internal or external interrupts. Upon waking up from an interrupt, the MCU executes the interrupt and then continues executing at the instruction following the sleep command (called by the *idle* function).

Returns: None

See *sleep_enable* for a code example.

isalnum

#include <ctype.h>

unsigned char isalnum(char c);

The *isalnum* function tests *c* to see if it is an alphanumeric character.

Returns: 1 if *c* is alphanumeric

```
#include <ctype.h>
void main()
{
    unsigned char c_alnum_flag, d_alnum_flag;
    c_alnum_flag = isalnum('1');   // test the ASCII value of 1 (0x31)
    d_alnum_flag = isalnum(1);     // test the value 1
    while(1)
    {
    }
}
```

Results:

```
c_alnum_flag = 1
d_alnum_flag = 0
```

isalpha

#include <ctype.h>

unsigned char isalpha(char c);

The *isalpha* function tests *c* to see if it is an alphabetic character.

Returns: 1 if *c* is alphabetic

```
#include <ctype.h>
```

```
void main()
{
    unsigned char c_alpha_flag, d_alpha_flag;
    c_alpha_flag = isalpha('a');   // test the ASCII character 'a'
    d_alpha_flag = isalpha('1');   // test the ASCII character '1'
    while(1)
    {
    }
}
```

Results:

```
    c_alpha_flag  = 1
    d_alpha_flag = 0
```

isascii

#include <ctype.h>

unsigned char isascii(char c);

The *isascii* function tests *c* to see if it is an ASCII character. ASCII characters range from 0d to 127d.

Returns: 1 if *c* is ASCII

```
#include <ctype.h>
void main()
{
   unsigned char c_ascii_flag, d_ascii_flag;
   c_ascii_flag = isascii('a');     // test the ASCII character a
   d_ascii_flag = isascii(153);     // test the value 153
   while(1)
   {
   }
}
```

Results:

```
    c_ascii _flag  = 1
    d_ascii _flag = 0
```

iscntrl

#include <ctype.h>

unsigned char iscntrl (char c);

The *iscntrl* function tests *c* to see if it is a control character. Control characters range from 0d to 31d and 127d.

Returns: 1 if *c* is a control character

```
#include <ctype.h>
void main()
{
   unsigned char c_iscntrl_flag, d_iscntrl_flag;
   c_iscntrl_flag = iscntrl('\t'); // test the control
                                   // character, horizontal tab
   d_iscntrl_flag = iscntrl('a');  // test the ASCII character a
   while(1)
   {
   }
}
```

Results:

```
    c_iscntrl_flag  = 1
    d_iscntrl_flag = 0
```

isdigit

#include <ctype.h>

unsigned char isdigit(char c);

The *isdigit* function tests *c* to see if it is an ASCII representation of a decimal digit.

Returns: 1 if *c* is a decimal digit

```
#include <ctype.h>
void main()
{
   unsigned char c_isdigit_flag, d_isdigit_flag;
   c_isdigit_flag = isdigit('1');  // test the ASCII character 1
   d_isdigit_flag = isdigit('a');  // test the ASCII character a
   while(1)
   {
   }
}
```

Results:

```
    c_digit_flag  = 1
    d_digit_flag = 0
```

islower

#include <ctype.h>

unsigned char islower(char c);

The *islower* function tests *c* to see if it is a lowercase alphabetic character.

Returns: 1 if *c* is a lowercase alphabetic character

```
#include <ctype.h>
void main()
{
   unsigned char c_islower_flag, d_islower_flag;
   c_islower_flag = islower('A');  // test the ASCII character A
   d_islower_flag = islower('a');  // test the ASCII character a
   while(1)
   {
   }
}
```

Results:

```
c_islower_flag  = 0
d_islower_flag = 1
```

isprint

#include <ctype.h>

unsigned char isprint(char c);

The *isprint* function tests *c* to see if it is a printable character. Printable characters are between 32d and 127d.

Returns: 1 if *c* is a printable character

```
#include <ctype.h>
void main()
{
   unsigned char c_isprint_flag, d_isprint_flag;
   c_isprint_flag = isprint('A');    // test the ASCII character A
   d_isprint_flag = isprint(0x03);   // test the control
                                     // character, backspace

   while(1)
   {
   }
}
```

Results:

```
c_isprint_flag = 1
d_isprint_flag = 0
```

ispunct

#include <ctype.h>

unsigned char ispunct(char c);

The *ispunct* function tests *c* to see if it is a punctuation character. All characters that are not control characters and not alphanumeric characters are considered to be punctuation characters.

Returns: 1 if *c* is a punctuation character

```
#include <ctype.h>
void main()
{
   unsigned char c_ispunct_flag, d_ispunct_flag;
   c_ispunct_flag = ispunct(',');    // test the ASCII character, comma
   d_ispunct_flag = ispunct('\t');  // test the horizontal tab
                                     //     character
   while(1)
   {
   }
}
```

Results:

```
c_ispunct_flag = 1
d_ispunct_flag = 0
```

isqrt

#include <math.h>

unsigned char isqrt(unsigned int x);

Returns: The square root of the unsigned integer variable *x*

See *sqrt*.

isspace

#include <ctype.h>

unsigned char isspace(char c);

The *isspace* function tests *c* to see if it is a white space character. Characters such as space, line feed, and horizontal tab are considered white space characters.

Returns: 1 if *c* is a white space character

```
#include <ctype.h>
void main()
{
   unsigned char c_isspace_flag, d_isspace_flag;
   c_isspace_flag = isspace('A');    // test the ASCII character A
   d_isspace_flag = isspace('\t');  // test the horizontal tab
                                     // character
```

```
        while(1)
        {
        }
    }
```

Results:

```
c_isspace_flag = 0
d_isspace_flag = 1
```

isupper

#include <ctype.h>

unsigned char isupper(char c);

The *isupper* function tests *c* to see if it is an uppercase alphabetic character.

Returns: 1 if *c* is an uppercase alphabetic character

```
#include <ctype.h>
void main()
{
    unsigned char c_isupper_flag, d_isupper_flag;
    c_isupper_flag = isupper('A');  // test the ASCII character A
    d_isupper_flag = isupper('a');  // test the ASCII character a
    while(1)
    {
    }
}
```

Results:

```
c_isupper_flag = 1
d_isupper_flag = 0
```

isxdigit

#include <ctype.h>

unsigned char isxdigit(char c);

The *isxdigit* function tests *c* to see if it is an ASCII representation of a hexadecimal digit.

Returns: 1 if *c* is a hexadecimal digit

```
#include <ctype.h>
void main()
{
    unsigned char c_isxdigit_flag, d_isxdigit_flag;
    c_isxdigit_flag = isxdigit('a');   // test the ASCII character a
    d_isxdigit_flag = isxdigit('z');   // test the ASCII character z
```

```
    while(1)
    {
    }
}
```

Results:

```
c_isxdigit_flag  = 1
d_isxdigit_flag = 0
```

itoa

#include <stdlib.h>

void itoa(int n, char *str);

The *itoa* function converts the signed integer value *n* to its ASCII character string equivalent. The string is stored at the location specified by **str*. Take care to ensure that the memory allocated for *str* is large enough to hold the entire value plus a termination character.

Returns: None

```
#include <mega16.h>
#include <stdio.h> // this to include putchar and printf!
#include <stdlib.h>

/* quartz crystal frequency [Hz] */
#define xtal 7372000L

/* Baud rate */
#define baud 9600

void main(void)
{
      char mystr[10];

      /* initialize the USART's baud rate */
      UBRRH=0x00;
      UBRRL=xtal/16/baud-1;
      /* initialize the USART control register
         RX & TX enabled, no interrupts, 8 data bits */
      UCSRA=0x00;
      UCSRB=0xD8;
      UCSRC=0x86;

      itoa(-1231,mystr);
      putsf("The value is: \r");
      puts(mystr);
```

```
        while (1)
        {
        }
}
```

Results: The USART transmits, at 9600 baud,

```
The value is:
-1231
```

labs

#include <math.h>

unsigned long labs(long x);

Returns: Absolute value of *x*

See *abs*.

ldexp

#include <math.h>

float ldexp(float x, int expon);

The *ldexp* function calculates the value of *x* multiplied by the result of 2 raised to the power of *expon*.

Returns: $x * 2^{expon}$

```
#include <math.h>

void main()
{
        float new_val;

        new_val = ldexp(5,3);

        while(1)
        {
        }
}
```

Results: new_val = $5 * 2^3 = 40$

lmax

#include <math.h>

long lmax(long a, long b);

Returns: Maximum value of *a* or *b*

See *max*.

lmin

#include <math.h>

long lmin(long a, long b);

Returns: Minimum value of *a* or *b*

See *min*.

log

#include <math.h>

float log(float x);

The *log* function calculates the base *e* or natural logarithm of the floating point value *x*. *x* must be a positive, nonzero value.

Returns: log(x)

```
#include <math.h>

void main()
{
        float new_val;

        new_val = log(5);

        while(1)
        {
        }
}
```

Results: new_val = 1.609

log10

#include <math.h>

float log10(float x);

The *log10* function calculates the base 10 logarithm of the floating point value *x*. *x* must be a positive, nonzero value.

Returns: log10(x)

```
#include <math.h>
```

```
void main()
{
        float new_val;
        new_val = log10(5);

        while(1)
        {
        }
}
```

Results: new_val = 0.699

lsign

#include <math.h>

signed char lsign(long x);

Returns: –1, 0, or 1 if *x* is negative, zero, or positive, respectively

See *sign*.

lsqrt

#include <math.h>

unsigned int lsqrt(unsigned long x);

Returns: The square root of the unsigned long variable *x*

See *sqrt*.

ltoa

#include <stdlib.h>

void ltoa(long n, char *str);

The *ltoa* function converts the long signed integer value *n* to its ASCII character string equivalent. The string is stored at the location specified by **str*. Take care to ensure that the memory allocated for *str* is large enough to hold the entire value plus a null-terminating character.

Returns: None

```
#include <mega16.h>
#include <stdio.h>     // this to include putchar and printf!
#include <stdlib.h>

/* quartz crystal frequency [Hz] */
#define xtal 7372000L
```

```
/* Baud rate */
#define baud 9600

void main(void)
{
        char mystr[11];

        /* initialize the USART's baud rate */
        UBRRH=0x00;
        UBRRL=xtal/16/baud-1;
        /* initialize the USART control register
           RX & TX enabled, no interrupts, 8 data bits */
        UCSRA=0x00;
        UCSRB=0xD8;
        UCSRC=0x86;

        ltoa(120031,mystr);
        putsf("The long signed integer value is: \r");
        puts(mystr);

        while (1)
        {
        }
}
```

Results: The USART transmits, at 9600 baud,

```
The long signed integer value is:
120031
```

malloc

#include <stdlib.h>

void *malloc(unsigned int size);

The *malloc* function allocates a memory block in the heap with the length of *size* bytes. On return, the function returns a pointer to the start of the memory block, which is filled with zeros. The allocated memory block occupies *size*+4 bytes in the heap. This must be taken into account when specifying the heap size in the **Project|Configure|C Compiler|Code Generation** menu.

Returns: If successful in finding contiguous free memory with *size* bytes, returns a pointer to the memory block. If unsuccessful, returns a null pointer.

```
#include <mega16.h>
/* include the standard input/output library */
#include <stdio.h>
```

```
/* include the variable length argument lists macros */
#include <stdlib.h>

#define xtal 6000000L
#define baud 9600

// declaring a pointer, but not reserving memory for the array!
char *cptr;
char *start_cptr;
int memory_size;

void main(void)
{
    int i;

    // USART initialization
    // Communication Parameters: 8 Data, 1 Stop, No Parity
    // USART Receiver: On
    // USART Transmitter: On
    // USART Mode: Asynchronous
    // USART Baud rate: 9600
    UCSRA=0x00;
    UCSRB=0xD8;
    UCSRC=0x86;
    UBRRH=0x00;
    UBRRL=xtal/16/baud-1;

    while (1)
    {
        putsf("\n\rHow much memory should I fill?(ENTER)\n\r");
        if (scanf("%d",&memory_size) != -1)
        {
            if (memory_size < (_HEAP_SIZE_ - 4))
            {
                printf("\n\rThanks, I'll try!\n\r");

                // try to get enough memory
                start_cptr = malloc(memory_size);
                cptr = start_cptr;
                if (cptr != NULL)
                {
                    printf("\n\rInitial Values!\n\r");
                    for (i=0;i<memory_size;i++)
                    {
                        //print the initial value
                        printf("%02X ",*cptr);
```

```
                    *cptr = (char)i + 1;  // change the value
                    cptr++;              // move the pointer forward!
                    if (((i+1)%10) == 0)
                        printf("\n\r");
                }
                printf("\n\rModified Values!\n\r");
                cptr = start_cptr;
                // print the new values!
                for (i=0;i<memory_size;i++)
                {
                    // print the modified value!
                    printf("%02X ",*cptr++);
                    if (((i+1)%10) == 0)
                        printf("\n\r");
                }
                // free the malloc'd memory for next time!
                free(start_cptr);
            }
            else
                printf("Failed to malloc correctly.\n\r");
        }
        else
        {
            printf("\n\rHeap size limit is %d\n\r",_HEAP_SIZE_-4);
        }
    }
  }
}
```

Results: The USART transmits, at 9600 baud,

```
How much memory should I fill?(ENTER)
```

Then it waits until a number is entered followed by a newline character. Once this is received (15 for example), it transmits

```
Thanks, I'll try!
Initial Values!
00 00 00 00 00 00 00 00 00 00
00 00 00 00 00
Modified Values!
01 02 03 04 05 06 07 08 09 0A
0B 0C 0D 0E 0F
```

max

#include <math.h>

int max(int a, int b);

signed char cmax(signed char a, signed char b);

long lmax(long a, long b);

float fmax(float a, float b);

The *max* function returns the maximum of the two integer values *a* and *b* as an integer. The *cmax, lmax,* and *fmax* functions return the maximum of the two signed char, long, and float values as a signed char, long, and float value, respectively.

Returns: Maximum value of *a* and *b* sized according to the function called

```
#include <math.h>
void main()
{
        int big_int;
        signed char a_char, b_char, big_char;

        big_int = max(1200, 3210);   //get the maximum of the values
        a_char = 23;
        b_char = 0x7A;
        big_char = cmax(a_char,b_char);

        while(1)
        {
        }
}
```

Results:

```
big_int = 3210
big_char = 0x7A
```

*memccpy

#include <string.h>

For the TINY memory model:

```
void *memccpy(void *dest, void *src, char c, unsigned char n);
```

For the SMALL memory model:

```
void *memccpy(void *dest, void *src, char c, unsigned int n);
```

The function *memccpy* copies at most *n* bytes from the memory location pointed to by *src* to the memory location *dst* until the character *c* is copied. The *dst* and *src* memory blocks must not overlap. *memccpy* returns a null pointer if the last character copied was *c*. If *c* is not copied within the *n* bytes, then a pointer to the next location in *dst* (which is calculated as *dst* + *n* + 1) is returned.

Returns: Null pointer if *c* is copied to *dst*; otherwise returns *dst*+*n*+1

```
#include <string.h>

char inputstr[] = "$11.2#";
char outputstr1[6];
char outputstr2[6];
void *a, *b;

void main(void)
{
        a = memccpy(outputstr1,inputstr,'.',3);
        b = memccpy(outputstr2,inputstr,'.',4);

        while (1)
        {
        }
}
```

Results:

```
outputstr1 = "$11"
a = outputstr1 + 3 + 1 = &(outputstr1[4])
outputstr2 = "$11."
b = NULL
```

*memchr

#include <string.h>

For the TINY memory model:

```
void *memchr(void *buf, unsigned char c, unsigned char n);
```

For the SMALL memory model:

```
void *memchr(void *buf, unsigned char c, unsigned int n);
```

The function *memchr* searches memory starting at the location pointed to by *buf* for the first occurrence of the character *c* or until *n* bytes have been searched. If *c* is found in *buf* within *n* bytes, a pointer to *c* is returned. If *c* is not found, a null pointer is returned.

Returns: Pointer to the character *c* in *buf* if *c* is found; otherwise NULL is returned

```
#include <string.h>

char inputstr[] = "$11.2#";
char *a;
char *b;

void main(void)
{
```

```
        a = memchr(inputstr,'.',3);
        b = memchr(inputstr,'.',5);

        while (1)
        {
        }
}
```

Results:

```
a = NULL
b = &(inputstr[3]) = ".2#"
```

memcmp, memcmpf

#include <string.h>

For the TINY memory model:

```
signed char memcmp(void *buf1, void *buf2, unsigned char n);
signed char memcmpf(void *buf1, void flash *buf2, unsigned char n);
```

For the SMALL memory model:

```
signed char memcmp(void *buf1, void *buf2, unsigned int n);
signed char memcmpf(void *buf1, void flash *buf2, unsigned int n);
```

The functions *memcmp* and *memcmpf* compare the contents of the memory pointed to by *buf1* to the memory pointed to by *buf2* one byte at a time. The functions return after comparing two unequal bytes or after *n* bytes are compared, whichever comes first.

Returns:

- signed char < 0 if **buf1* < **buf2*
- signed char = 0 if **buf1* = **buf2*
- signed char > 0 if **buf1* > **buf2*

```
#include <string.h>

char inputstr1[] = "name a";
char inputstr2[] = "name b";

void main(void)
{
        signed char a;
        signed char b;
        signed char c;
```

```
        a = memcmp(inputstr1,inputstr2,5);
        b = memcmp(inputstr1,inputstr2,6);
        c = memcmpf(inputstr1,"name 1",6);

        while (1)
        {
        }
    }
```

Results:

```
    a = 0
    b = 0xFF
    c = 0x30
```

*memcpy, *memcpyf

#include <string.h>

For the TINY memory model:

```
void *memcpy(void *dest, void *src, unsigned char n);
```

For the SMALL memory model:

```
void *memcpy(void *dest, void *src, unsigned int n);
```

For either model when *src* is in FLASH:

```
void *memcpyf(void *dest, void *src, unsigned int n);
```

The functions *memcpy* and *memcpyf* copy *n* bytes from the memory location pointed to by *src* to the memory location pointed to by *dst*. For the function *memcpy*, the *dst* and *src* memory blocks must not overlap. This is not a concern with *memcpyf*, because the *src* in *memcpyf* must be located in FLASH. If the memory blocks do overlap, use *memmove* instead of *memcpy*.

Returns: Pointer to *dest*

```
#include <string.h>

char inputstr[] = "$11.2#";
char outputstr[6];
void *a;
char outputstrf[6];
void *b;

void main(void)
{
   a = memcpy(outputstr,inputstr+1,4);
   outputstr[4] = '\0';   // null terminate our new string
```

```
    b = memcpyf(outputstrf,"Hello World!",5);
    outputstrf[5] = '\0';  // null terminate our new string!

    while (1)
    {
    }
}
```

Results:

```
outputstr = "11.2"
a = "11.2"
outputstrf = "Hello"
b = "Hello"
```

*memmove

#include <string.h>

For the TINY memory model:

```
void *memmove(void *dest, void *src, unsigned char n);
```

For the SMALL memory model:

```
void *memmove(void *dest, void *src, unsigned int n);
```

The *memmove* function copies *n* bytes from the memory pointed to by *src* to the memory pointed to by *dest*. Unlike *memcpy*, the *src* and *dest* may overlap when calling *memmove*.

Returns: *dest*

```
#include <string.h>

char inputstr1[] = "abc1";
char *a;

void main(void)
{
      // move the string one place to the right
      a = memmove(&(inputstr1[1]),inputstr1,3);

      while (1)
      {
      }
}
```

Results:

```
a = "abc"
inputstr1 = "aabc"
```

*memset

#include <string.h>

For the TINY memory model:

```
void *memset(void *buf, unsigned char c, unsigned char n);
```

For the SMALL memory model:

```
void *memset(void *buf, unsigned char c, unsigned int n);
```

The *memset* function fills *n* bytes of the memory pointed to by *buf* with the character *c*.

Returns: *buf*

```
#include <string.h>

char inputstr1[] = "abc1";
char *a;

void main(void)
{
        // starting after a, fill in with some 2's
        a = memset(&(inputstr1[1]),'2',3);
        while (1)
        {
        }
}
```

Results:

```
a = "222"
inputstr1 = "a222"
```

min

#include <math.h>

int min(int a, int b);

signed char cmin(signed char a, signed char b);

long lmin(long a, long b);

float fmin(float a, float b);

The *min* function returns the minimum of the two integer values *a* and *b* as an integer. The *cmin, lmin,* and *fmin* functions return the minimum of the two signed char, long, and float values as a signed char, long, and float value, respectively.

Returns: Minimum value of *a* and *b* sized according to the function called

```
#include <math.h>
void main()
{
   int little_int;
   signed char a_char, b_char, little_char;

   little_int = min(1200, 3210);    //get the minimum of the values
   a_char = 23;
   b_char = 0x7A;
   little_char = cmin(a_char,b_char);

   while(1)
   {
   }
}
```

Results:

```
little_int = 1200
little_char = 23
```

modf

#include <math.h>

float modf(float x, float *ipart);

The *modf* function splits the floating point number *x* into its integer and fractional components. The fractional part of *x* is returned as a signed floating point number. The integer part is stored as a floating point number at *ipart.* Notice that the **address** of the variable to hold the integer portion, not the variable itself, is passed to *modf.* Both the integer and the floating point results have the same sign as *x.*

Returns:

- Fractional portion of the floating point number *x* as a signed floating point number
- Sets the value at the address pointed to by **ipart* to the integer part of *x*

```
#include <math.h>

void main()
{
      float integer_portion, fract_portion;

      fract_portion = modf(-45.7, &integer_portion);
```

```
        while(1)
        {
        }
}
```

Results:

```
fract_portion = -.7
integer_portion = -45
```

peekb, peekw

#include <mem.h>

unsigned char peekb(unsigned int addr);

unsigned int peekw(unsigned int addr);

The *peekb* function reads the value of SRAM at the address *addr*. The *peekw* function reads an unsigned integer value from the SRAM at the address *addr*. The *peekw* function reads the LSB first from address *addr* and then reads the MSB of *data* from *addr+1*. Values can be directly written to the SRAM by using the *pokeb* and *pokew* functions.

Returns: None

See *pokeb, pokew* for the code example.

pokeb, pokew

#include <mem.h>

void pokeb(unsigned int addr, unsigned char data);

void pokew(unsigned int addr, unsigned int data);

The *pokeb* function writes the value of *data* to the SRAM at the address *addr*. The *pokew* function writes the integer value of *data* to the SRAM at the address *addr*. The *pokew* function writes the LSB of *data* first, to address *addr*, and then writes the MSB of *data* to *addr+1*. Values can be read from the SRAM by using the *peekb* and *peekw* functions.

Returns: None

```
#include <mega16.h>
#include <mem.h>


/* the structure "alfa" is stored in SRAM at address 260h */
struct x {
      unsigned char b;
      unsigned int w;
} alfa @0x260;
```

```
void main(void)
{
        unsigned int read_b;
        unsigned int read_w;
        unsigned int read_b2;
        unsigned int read_w2;

        MCUCR = 0xC0; // enable external SRAM with 1 wait state

        alfa.b = 0x11;     // initialize the value at 0x260
        alfa.w = 0x2233;   // initialize the value at 0x261

        read_b = (unsigned int) peekb(0x260);
        read_w = peekw(0x261);

        pokeb(0x260,0xAA);     // place 0xAA at address 0x260
        pokew(0x261,0xBBCC);   // place 0xCC at address 0x261 and
                               // 0xBB at address 0x2662

        read_b2 = (unsigned int) alfa.b;
        read_w2 = alfa.w;

        while (1)
        {
        }

}
```

Results:

```
read_b = 0x0011
read_w = 0x2233
read_b2 = 0x00AA
read_w2 = 0xBBCC
```

pow

#include <math.h>

float pow(float x, float y);

The *pow* function calculates *x* raised to the power of *y*.

Returns: x^y

```
#include <math.h>

void main()
{
```

```
          float new_val;

          new_val = pow(2,5);

          while(1)
          {
          }
    }
```

Results: new_val = 31.9

powerdown

#include <sleep.h>

void powerdown(void);

The *powerdown* function puts the AVR microcontroller into power-down mode. The function, *sleep_enable*, must be called prior to this function being used. In power-down mode, the external oscillator is stopped, while the external interrupts and the watchdog (if enabled) continue operating. Only an external reset, a watchdog reset (if enabled), or an external level interrupt can wake up the MCU. Upon waking up from an interrupt, the MCU executes the interrupt and then continues executing at the instruction following the sleep command (called by *powerdown*).

Returns: None

```
#include <mega16.h>
#include <stdio.h>
#include <sleep.h>
#include <delay.h>

/* quartz crystal frequency [Hz] */
#define xtal 6000000L

/* Baud rate */
#define baud 9600

interrupt [EXT_INT1] void int1_isr(void)
{
      putsf("I was interrupted!\r");
}

void main(void)
{
   PORTD = 0xFF;   // turn on internal pull-up on pin3
   DDRD = 0x00;    // make INT1 (PORTD pin 3) an input
```

```
    // low level interrupt on INT1
    // turn on INT1 interrupts
    GICR|=0x80;
    MCUCR=0x00;
    MCUCSR=0x00;
    GIFR=0x80;

    /* initialize the USART's baud rate */
    UBRRH=0x00;
    UBRRL=xtal/16/baud-1;
    /* initialize the USART control register
       RX & TX enabled, no interrupts, 8 data bits */
    UCSRA=0x00;
    UCSRB=0xD8;
    UCSRC=0x86;

    #asm("sei");     // enable interrupts
    sleep_enable(); // enable us to go to sleep when we are ready!

    putsf("Reset Occurred\r");
    while (1)
    {
       putsf("Good Night!\r");
       putsf("I am going to sleep until you bug me!\r");
       delay_ms(100);     // wait for string to be transmitted!
       powerdown();   // enter powerdown until INT1 wakes us up
    }
}
```

Results:

The microprocessor transmits

```
Reset Occurred
Good Night!
I am going to sleep until you bug me!
```

Then it enters powerdown sleep mode and wakes up only when an INT1 interrupt occurs by PORTD pin 3 being held low or an external reset occurs. Upon waking up, it continues executing at the instruction following the powerdown. Upon waking up, it will transmit the following continuously while PORTD pin 3 is low:

```
I was interrupted!
```

The microprocessor continues to run the **while** loop until power is removed.

powersave

#include <sleep.h>

void powersave(void);

The *powersave* function puts the AVR microcontroller into powersave mode. This is a sleep mode and very similar to the *powerdown* function.

See *powerdown* in this appendix for the function use.

See the Atmel datasheets for the complete description of this sleep mode as it applies to a particular AVR device. Powersave sleep mode is not available on all AVR devices.

printf

#include <stdio.h>

void printf(char flash *fmtstr [, arg1, arg2, ...]);

The *printf* function transmits formatted text according to the format specifiers in the *fmtstr* string. The transmittal is performed using the *putchar* function. The standard *putchar* function defaults to transmitting using the USART. However, it is possible to use an alternate *putchar* function to redirect the data. See *putchar* for details.

The implementation of *printf* is a reduced version of the standard C function. This was necessary due to the specific needs of an embedded system and because the full implementation would require a large amount of memory space. In order to reduce code size, the user can specify what options the *printf* is required to support for their specific application. These options can be accessed under the **Project|Configure|C Compiler|Code Generation (s)printf Features** option.

The format specifier string *fmtstr* is a constant and must be located in FLASH memory and has the following format:

```
%[flags][width][.precision][l]type_char
```

The optional *flags* characters are:

–	Left justifies the result, padding on the right with spaces. If this flag is not present, the result is right-justified, padded on the left with zeroes or spaces.
+	Forces a plus or minus sign to preceded the numerical value.
(space character)	A space character forces a space to precede a positive number. If the value to be printed is negative, a minus sign precedes the value.

The optional *width* specifier sets the minimal width of an output value. If the result of the conversion is wider than the field width, the field is expanded to accommodate the result. The following *width* specifiers are supported:

n	Forces at least *n* characters to be output. If the result has less than *n* characters, then its field is padded with spaces. If the – flag is used, the result field is padded on the right, otherwise it is padded on the left.

0n	Forces at least *n* characters to be output. If the result has fewer than *n* characters, it is padded on the left with zeroes.

The optional *precision* specifier sets the maximal number of characters or minimal number of integer digits that may be outputted. The *precision* specifier always begins with a '.' in order to separate it from the *width* specifier. The following *precision* specifiers are supported:

.0	Sets the precision to 1 for the 'i', 'd', 'u', 'x', and 'X' type characters.
.n	Forces *n* characters or *n* decimal places to be output. Specifically for the 'i', 'd', 'u', 'x', and 'X' conversion type characters, if the value has fewer than *n* digits, then it is padded on the left with zeros. If the value has more than *n* digits, then it will not be truncated. For the 's' and 'p' conversion type characters, no more than *n* characters from the string are output. The 'e', 'E', and 'f' conversion type characters are output with *n* digits to the right of the decimal point. The precision specifier has no effect on the 'c' conversion type character.

If no *precision* specifier is entered, the precision is set to 1 for the 'i', 'd', 'u', 'x', and 'X' conversion type characters. For the 's' and 'p' converstion type characters, the char string is output up to the first null character.

The optional 'l' (lower case 'L') input size modifier specifies that the function argument must be treated as a long integer for the 'i', 'd', 'u', 'x', and 'X' conversion type characters.

The following conversion type characters, *type char,* are supported:

c	Outputs the next argument as an ASCII character
d	Outputs the next argument as a decimal integer
i	Outputs the next argument as a decimal integer
u	Outputs the next argument as an unsigned decimal integer
x	Outputs the next argument as an unsigned hexadecimal integer using lowercase letters
X	Outputs the next argument as an unsigned hexadecimal integer using uppercase letters
e	Outputs the next argument as a float formatted in scientific notation, [-]d.ddddd e[-]dd
E	Outputs the next argument as a float formatted in scientific notation, [-]d.ddddd E[-]dd
f	Outputs the next argument as a float formatted as [-]ddd.ddddd
s	Outputs the next argument as a null terminated character string, located in SRAM
p	Outputs the next argument as a null terminated character string, located in FLASH
%%	Outputs the % character

Returns: None

```
#include <mega16.h>
#include <stdio.h>

/* quartz crystal frequency [Hz] */
#define xtal 7372000L

/* Baud rate */
#define baud 9600

void main(void)
{
   unsigned int j;
   char c;

   /* initialize the USART's baud rate */
   UBRRH=0x00;
   UBRRL=xtal/16/baud-1;
   /* initialize the USART control register
      RX & TX enabled, no interrupts, 8 data bits */
   UCSRA=0x00;
   UCSRB=0xD8;
   UCSRC=0x86;

   for (j=0;j<=500;j+=250)
   {
      // print the current value of j
      printf("Decimal: %u\tHexadecimal: %X\n\r",j,j);
      printf("Zero Padded Decimal: %0u\n\r",j);
      printf("Four Digit Lower Case Hexadecimal: %04x\r\n\n",j);
   }

   while (1)
   {
      /* receive the character */
      c=getchar();
      /* and echo it back */
      printf("The received character was %c\n\r",c);
   }
}
```

Results:

The following is output by the microprocessor to the USART at startup.

```
Decimal: 0   Hexidecimal: 0
Zero Padded Decimal: 00000
Four Digit Lower Case Hexadecimal: 0000
```

```
Decimal: 250     Hexidecimal: FA
Zero Padded Decimal: 00250
Four Digit Lower Case Hexadecimal: 00fa

Decimal: 500      Hexidecimal: 1F4
Zero Padded Decimal: 00500
Four Digit Lower Case Hexadecimal: 01f4
```

Then any character received by the USART is transmitted back preceded by the specified string. For example, if the character 'z' is received, this text is sent to the transmitter.

```
The received character was z.
```

putchar

#include <stdio.h>

void putchar(char);

The *putchar* function is a standard C language I/O function, but it has been adapted to work on embedded microcontrollers with limited resources. The *putchar* function transmits a character using the USART. This function uses polling to transmit the character using the USART. As a result, the function waits indefinitely for a character to be transmitted before returning.

Prior to this function being used, the USART must be initialized, and the USART transmitter must be enabled.

If another peripheral is to be used for transmitting, the *putchar* function can be modified accordingly. The source for this function is in the stdio.h file.

If you want to replace the standard library *putchar* with your own version, you must do three things. First, you must place the new *putchar* function in the *.c* file before including the standard library. Second, you must follow the new *putchar* function with a definition statement telling the compiler to ignore any other *putchar* functions that it finds. Finally, include the standard library. Put together, the code might appear as

```
void  putchar (void)
{
      // your putchar routine statements here
}

#define _ALTERNATE_PUTCHAR_
#include <stdio.h>

// the rest of the source file!
```

The standard *putchar()* returns: None

```
#include <mega16.h>
#include <stdio.h>
```

```
/* quartz crystal frequency [Hz] */
#define xtal 7372000L

/* Baud rate */
#define baud 9600

void main(void)
{
      char k;

      /* initialize the USART's baud rate */
      UBRRH=0x00;
      UBRRL=xtal/16/baud-1;
      /* initialize the USART control register
         RX & TX enabled, no interrupts, 8 data bits */
      UCSRA=0x00;
      UCSRB=0xD8;
      UCSRC=0x86;

      while (1)
      {
            /* receive the character */
            k=getchar();
            /* and echo it back */
            putchar(k);
      }
}
```

Results: Characters received by the USART are echoed back from the USART until power is removed from the processor.

puts

#include <stdio.h>

void puts(char *str);

The *puts* function transmits the string *str* using the standard *putchar* function, or an alternate *putchar* function if one has been defined. The string must be null-terminated. A newline character is appended to the end of the string when it is transmitted. Finally, the string must be located in SRAM. (See *putsf* for strings located in FLASH.) The *putchar* function defaults to transmitting characters using the USART.

Returns: None

```
#include <mega16.h>
#include <stdio.h>
```

```
/* quartz crystal frequency [Hz] */
#define xtal 7372000L

/* Baud rate */
#define baud 9600

char hello_str[] = "Hello World";

void main(void)
{
        /* initialize the USART's baud rate */
        UBRRH=0x00;
        UBRRL=xtal/16/baud-1;
        /* initialize the USART control register
           RX & TX enabled, no interrupts, 8 data bits */
        UCSRA=0x00;
        UCSRB=0xD8;
        UCSRC=0x86;

        puts(hello_str);
        while (1)
        {
        }
}
```

Results: "Hello World", with line feed but not carriage return, is transmitted using the USART.

putsf

#include <stdio.h>

void putsf(char flash *str);

The *putsf* function transmits the contant string *str* using the standard *putchar* function, or an alternate *putchar* function if one has been defined. The string must be null-terminated. A newline character is appended to the end of the string when it is transmitted. Finally, the string must be located in FLASH. (See *puts* for strings located in RAM.) The *putchar* function defaults to transmitting characters using the USART.

Returns: None

```
#include <mega16.h>
#include <stdio.h>

/* quartz crystal frequency [Hz] */
#define xtal 7372000L
```

```
/* Baud rate */
#define baud 9600

char flash hello_str[] = "Hello World";

void main(void)
{
        /* initialize the USART's baud rate */
        UBRRH=0x00;
        UBRRL=xtal/16/baud-1;
        /* initialize the USART control register
           RX & TX enabled, no interrupts, 8 data bits */
        UCSRA=0x00;
        UCSRB=0xD8;
        UCSRC=0x86;

        putsf(hello_str);
        while (1)
        {
        }
}
```

Results: "Hello World", with line feed but not carriage return, is transmitted using the USART.

rand

#include <stdlib.h>

int rand(void);

The *rand* function returns a pseudo-random number between 0 and 32767.

Returns: None

```
#include <mega16.h>
#include <stdio.h>     // this to include putchar and printf!
#include <stdlib.h>

/* quartz crystal frequency [Hz] */
#define xtal 7372000L

/* Baud rate */
#define baud 9600

void main(void)
{
```

```
    int seed;
    int rand_num;

    /* initialize the USART's baud rate */
    UBRRH=0x00;
    UBRRL=xtal/16/baud-1;
    /* initialize the USART control register
       RX & TX enabled, no interrupts, 8 data bits */
    UCSRA=0x00;
    UCSRB=0xD8;
    UCSRC=0x86;

    putsf("Enter a seed value followed by ENTER. ");
    scanf("%d\n",&seed);
    srand(seed);  // seed the generator

    putsf("\n\rSend anything to get a pseudo-random number!\n\r");

    while (1)
    {
       getchar();    // this will return when SOMETHING is received
       rand_num = rand();
       printf("%d\n\r",rand_num);
    }
}
```

Results: The USART transmits, at 9600 baud,

```
Enter a seed value followed by ENTER.
```

Once a value is entered, the following is printed:

```
Send anything to get a pseudo-random number!
```

Upon the receipt of any character by the USART, the next random number is transmitted. Notice that if you start with the same seed, you get the same number sequence each time.

realloc

#include <stdlib.h>

void *realloc(void *ptr, unsigned int size);

The *realloc* function changes the size of a memory block previously allocated in the heap by *malloc* or *calloc*. The *size* argument specifies the new size of the memory block. On success, the function returns a pointer to the start of the newly allocated memory block, the contents of which are a copy of the previously allocated block of memory. If the newly allocated memory block is larger in size than the old one, the size difference is not filled with zeros. The allocated memory block occupies *size*+4 bytes in the heap. This must be taken into account when specifying the heap size in the **Project|Configure|C Compiler|Code Generation** menu.

Returns: If successful in finding contiguous free memory with *size* bytes, returns a pointer to the memory block. If unsuccessful, returns a null pointer.

```
void main(void)
{
   int i;

   cptr = NULL;     // initialize the pointers to null!
   start_cptr = NULL;

   // USART initialization
   // Communication Parameters: 8 Data, 1 Stop, No Parity
   // USART Receiver: On
   // USART Transmitter: On
   // USART Mode: Asynchronous
   // USART Baud rate: 9600
   UCSRA=0x00;
   UCSRB=0xD8;
   UCSRC=0x86;
   UBRRH=0x00;
   UBRRL=xtal/16/baud-1;

   while (1)
   {
      putsf("\n\rHow much memory should I fill?(ENTER)\n\r");
      if (scanf("%d",&memory_size) != -1)
      {
         if (memory_size < (_HEAP_SIZE_ - 4))
         {
            printf("\n\rThanks, I'll try!\n\r");

            // try to get enough memory
            if (start_cptr == NULL)
                start_cptr = malloc(memory_size);
            else
                start_cptr = realloc(start_cptr,memory_size);
            cptr = start_cptr;
            if (cptr != NULL)
            {
               printf("\n\rInitial Values!\n\r");
               for (i=0;i<memory_size;i++)
               {
                  printf("%02X ",*cptr);//print the initial value
                  *cptr = (char)i + 1;  // change the value
                  cptr++;               // move the pointer forward!
                  if (((i+1)%10) == 0)
```

```
                    printf("\n\r");
                }
                printf("\n\rModified Values!\n\r");
                cptr = start_cptr;
                // print the new values!
                for (i=0;i<memory_size;i++)
                {
                    // print the modified value!
                    printf("%02X ",*cptr++);
                    if (((i+1)%10) == 0)
                        printf("\n\r");
                }
            }
            else
                printf("Failed to realloc correctly.\n\r");
        }
        else
        {
            printf("\n\rHeap size limit is %d\n\r",_HEAP_SIZE_-4);
        }
    }
  }
  // if we were going on to do other things we would 'free' the
  //    memory here!
  free(start_cptr);
}
```

Results: The USART transmits, at 9600 baud,

```
How much memory should I fill?(ENTER)
```

Then it waits until a number is entered followed by a newline character. Once this is received (15, for example), it transmits

```
Thanks, I'll try!
Initial Values!
xx xx xx xx xx xx xx xx xx xx
xx xx xx xx xx
Modified Values!
01 02 03 04 05 06 07 08 09 0A
0B 0C 0D 0E 0F
```

Note that the *xx*'s above represent the fact that these values are dependent upon what was previously in the memory locations allocated using *realloc*.

scanf

#include <stdio.h>

signed char scanf(char flash *fmtstr);

The *scanf* function inputs values from a text string according to the format specifiers in the *fmtstr* string and places the values in memory locations also specified by *fmtstr*. The formatting section of the *fmtstr* string is enclosed in quotes and uses format specifiers, listed below, to tell *scanf* what kinds of values are to be read. The second part of *fmtstr* lists the addresses of where the read values are to be stored.

ATTENTION! It is important to always specify a ***pointer*** to the variable to receive the value in the *fmtstr* of the *scanf* function, ***not*** the variable itself. Failing to pass pointers to the *scanf* function causes erratic results, because values are being stored in the memory locations related to the value of the variable passed instead of being stored at the address of the variable.

The receiving of characters is performed using the *getchar* function. The *getchar* function defaults to receiving using the USART. An alternate *getchar* function can be defined to receive the data from an alternate source. See *getchar* for details.

The format specifier string *fmtstr* is a constant and must be located in FLASH memory.

The implementation of *scanf* is a reduced version of the standard C function. This was necessary due to the specific needs of an embedded system and because the full implementation would require a large amount of memory.

The following format specifiers are available:

%c	Inputs the next argument as an ASCII character
%d	Inputs the next argument as a decimal integer
%i	Inputs the next argument as a decimal integer
%u	Inputs the next argument as an unsigned decimal integer
%x	Inputs the next argument as an unsigned hexadecimal integer
%s	Inputs the next argument as a null terminated character string

Returns: Number of successful entries or a –1 if an error occurred

```
#include <mega16.h>
#include <stdio.h>

/* quartz crystal frequency [Hz] */
#define xtal 7372000L

/* Baud rate */
#define baud 9600

void main(void)
{
    char your_initial;
    int your_number;
```

```
    /* initialize the USART's baud rate */
    UBRRH=0x00;
    UBRRL=xtal/16/baud-1;
    /* initialize the USART control register
       RX & TX enabled, no interrupts, 8 data bits */
    UCSRA=0x00;
    UCSRB=0xD8;
    UCSRC=0x86;

    while (1)
    {
       putsf("\n\r");
       putsf("Please enter your first initial followed by \r");
       putsf("a comma then your favorite number.\r");
       putsf("Press Enter to finish.\r");
       // tell scanf what to look for
       // NOTICE the ADDRESSES of the variables are sent to scanf!
       if (scanf("%c,%d\n",&your_initial,&your_number) == 2)
       {
          printf("%c, congrats! You got a %d%% on your exam!\n\r",
                your_initial, your_number);
       }
       else
           putsf("I didn't understand. Please try again.\n\r");

    }
}
```

Results: The USART transmits the prompt for an initial and a number.

```
Please enter your first initial followed by
a comma then your favorite number.
Press Enter to finish.
```

The *scanf* function waits for the initial, comma, number, and newline character before continuing. If the user enters

```
S, 32
```

the following is printed:

```
S, congrats! You got a 32% on your exam!
```

If the user enters something like

```
S, AC
```

Then an error occurs and this is printed:

```
I didn't understand. Please try again.
```

sign

#include <math.h>

signed char sign(int x);

signed char csign(signed char x);

signed char lsign (long x);

signed char fsign (float x);

The *sign* function returns the sign of the integer *x*. It returns −1 if the sign of *x* is negative, 0 if *x* is zero, or 1 if *x* is a positive value. The *csign, lsign,* and *fsign* functions return the sign of the signed char, long, and float variable *x*, respectively.

Returns:

- −1 if *x* is a negative value
- 0 if *x* is zero
- 1 if *x* is a positive value

```
#include <math.h>
void main()
{
        signed char pos_sign;
        signed char neg_sign;
        signed char zero_sign;

        neg_sign = sign(-19574);   // get the sign of an integer
        pos_sign = lsign(125000);  // get the sign of a long
        zero_sign = csign(0);      // get the sign of a char

        while(1)
        {
        }
}
```

Results:

```
pos_sign = 1

neg_sign = -1

zero_sign = 0
```

sin

#include <math.h>

float sin(float x);

The *sin* function calculates the sine of the floating point number *x*. The angle *x* is expressed in radians.

Returns: sin(x)

```
#include <math.h>

void main()
{
        float new_val;

        new_val = sin(5.121);

        while(1)
        {
        }
}
```

Results: new_val = –0.918

sinh

#include <math.h>

float sinh(float x);

The *sinh* function calculates the hyperbolic sine of the floating point number *x*. The angle *x* is expressed in radians.

Returns: sinh(x)

```
#include <math.h>

void main()
{
        float new_val;

        new_val = sinh(5.121);

        while(1)
        {
        }
}
```

Results: new_val = 83.748

sleep_disable

#include <sleep.h>

void sleep_disable(void);

The *sleep_disable* function clears the SE bit in the MCUCR register. This prevents the microcontroller from accidentally entering sleep mode.

Returns: None

```
#include <mega16.h>
#include <stdio.h>
#include <sleep.h>
#include <delay.h>

/* quartz crystal frequency [Hz] */
#define xtal 6000000L

/* Baud rate */
#define baud 9600

interrupt [EXT_INT1] void int1_isr(void)
{
   putsf("I was interrupted!\r");
}

void main(void)
{
   PORTD = 0xFF;   // turn on internal pull-up on pin3
   DDRD = 0x00;    // make INT1 (PORTD pin 3) an input

   MCUCR = 0x00;   // low level interrupt on INT1
   GIMSK = 0x80;   // turn on INT1 interrupts

   /* initialize the USART's baud rate */
   UBRRH=0x00;
   UBRRL=xtal/16/baud-1;
   /* initialize the USART control register
      RX & TX enabled, no interrupts, 8 data bits */
   UCSRA=0x00;
   UCSRB=0xD8;
   UCSRC=0x86;

   #asm("sei");     // enable interrupts
   sleep_enable();  // enable us to go to sleep when we are ready!

   putsf("Reset Occurred\r");
   while (1)
   {
      putsf("Good Night!\r");
      putsf("I am going to sleep until you bug me!\r");
```

```
            delay_ms(100);     // wait for string to be transmitted!
            idle();      // enter idle until INT1 wakes us up
            sleep_disable();   // stop future sleeps
        }

    }
```

Results:

The microprocessor transmits

```
Reset Occurred
Good Night!
I am going to sleep until you bug me!
```

Then it enters idle sleep mode and wakes up only when an INT1 interrupt occurs by PORTD pin 3 being held low or an external reset occurs. Upon waking up, it continues executing at the instruction following the *idle* function. It transmits the following continuously while PORTD pin 3 is held low:

```
I was interrupted!
```

The next line disables entering sleep mode. So, although the microprocessor tries to call *idle* to go to sleep, it does not. The microprocessor continuously prints the above lines without a pause until power is removed.

sleep_enable

#include <sleep.h>

void sleep_enable(void);

The *sleep_enable* function sets the SE bit in the MCUCR register. This bit must be set to allow the microcontroller to enter sleep mode when the SLEEP instruction is executed. When in sleep mode, some features of the microcontroller are stopped, allowing it to consume less power. Depending on the type of sleep, the microcontroller can be awakened by internal or external interrupts. The particular sleep modes available depend upon the microcontroller being used. Refer to Atmel AVR datasheets for the types of sleep modes and their availability.

Returns: None

```
#include <mega16.h>
#include <stdio.h>
#include <sleep.h>
#include <delay.h>

/* quartz crystal frequency [Hz] */
#define xtal 7372000L

/* Baud rate */
#define baud 9600
```

```
void main(void)
{
   PORTD = 0x04;   // turn on internal pull-up on pin2
   DDRD = 0xFB;    // make INT0 (PORTD pin 2) an input

   MCUCR = 0x00;   // low level interrupt on INT0
   GIFR = 0x40;    // turn on INT0 interrupts

   /* initialize the USART's baud rate */
   UBRRH=0x00;
   UBRRL=xtal/16/baud-1;
   /* initialize the USART control register
      RX & TX enabled, no interrupts, 8 data bits */
   UCSRA=0x00;
   UCSRB=0xD8;
   UCSRC=0x86;

   #asm("sei");    // enable interrupts

   sleep_enable();  // enable us to go to sleep when we are ready!


   putsf("Reset Occurred");
   while (1)
   {
      putsf("Good Night!");
      putsf("I am going to sleep until you bug me!");

      idle();      // enter idle until INT0 wakes us up
      putsf("I was interrupted!");
      sleep_disable();   // stop future sleeps
   }

}
```

Results: The microprocessor transmits

```
Reset Occurred
Good Night!
I am going to sleep until you bug me!
```

Then it enters idle sleep mode and wakes up only when an INT0 interrupt occurs by PORTD pin 2 being held low or an external reset occurs. Upon waking up, it continues executing at the instruction following the *idle* function. Upon waking up, it will transmit

```
I was interrupted!
```

The microprocessor continues to run the while loop until power is removed.

spi

#include <spi.h>

unsigned char spi(unsigned char data);

Serial Peripheral Interface (SPI) allows high-speed synchronous data transfer between devices. Many of the Atmel AVR devices have hardware SPI ports. SPI ports allow data to be simultaneously transmitted and received over three connections: clock, master-in-slave-out (MISO), and master-out-slave-in (MOSI). The master generates the clock signal and places data on the MOSI pin synchronized to the clock signal. The slave places data on the MISO line, also synchronized with the clock. In this way, data is transmitted and received by both the master and the slave at the same time.

Once the SPI port is initialized, data transfer begins by placing data on the SPI data register if the microcontroller is the master. Otherwise, the microcontroller must wait for the clock from the master to transfer the data. After the data transfer is complete, the SPI data register contains the value read from the other device. See the Atmel AVR datasheets for the requirements for initializing the SPI port.

The SPI functions are intended for easy interfacing between C programs and various peripherals using the SPI bus. The function *spi* loads the SPI data register with the byte *data*, and then waits for the hardware to complete clocking the data out and reading the data from the peripheral device. Once the transfer is complete, the *spi* function returns the data from the SPI data register. Polling is used by *spi* to determine when the transfer is complete, so the SPI interrupts do not need to be enabled before this function is called.

Returns: Character read from the SPI data register after the completion of the transfer

```
#include <mega16.h>
#include <delay.h>
#include <spi.h>
#include <stdio.h>

// This example reads the status from an Atmel AT45D081 Flash
// memory chip
// over SPI.
#define xtal 7372000L

int stats;

void main(void)
{
   char junk;
   PORTA = 0xFF;
   PORTB = 0x00;
   PORTC = 0x00;        // clear port registers
   PORTD = 0x10;
```

```
    DDRA = 0x03;        // all inputs, except the lights!
    DDRB = 0xBF;        // all outputs, except MISO pin!
    DDRC = 0x03;        // all inputs except for bit 0 and 1
    DDRD = 0xFF;        // all outputs

    /* Serial Peripheral interface setup */
    SPCR = 0x5E;

    SREG = 0x00;        //Disable interrupts
    GIMSK = 0x00;       //no external interrupts

    /* USART */
    /* initialize the USART's baud rate */
    UBRRH=0x00;
    UBRRL=xtal/16/baud-1;
    /* initialize the USART control register
       RX & TX enabled, no interrupts, 8 data bits */
    UCSRA=0x00;
    UCSRB=0xD8;
    UCSRC=0x86;
    delay_ms(2000);    // wait for things to get stable

    PORTD.4 = 0;    // pull the chip select line low
    delay_ms(5);
    // to get status send command 0x57 0x00
    // status is returned while the second byte of the command is
    // being sent
    spi(0x57);                            //byte 1
    stats = (int) spi(0x00);                //byte 2

    PORTD.4 = 1;    // release the chip select line
    printf("Status: %X\n\r",stats);

    while(1)
    {
    }
}
```

Results: The microprocessor transmits at 57600 baud on power-up

```
Status: A0
```

sprintf

#include <stdio.h>

void sprintf(char *str, char flash *fmtstr [, arg1, arg2, . . .]);

The *sprintf* function copies the formatted text according to the format specifiers in the *fmtstr* string to the string *str*. A null termination is appended to the end of *str* after the formatted text is copied to it. The memory for *str* should be large enough to accommodate the copied text and the null termination.

The format specifier string *fmtstr* is a constant and must be located in FLASH memory. *sprintf* interprets *fmtstr* in the exact same manner as *printf* interprets *fmtstr*. See *printf* for more information on format specifiers.

sqrt

#include <math.h>

float sqrt(float x);

unsigned char isqrt(unsigned int x);

unsigned int lsqrt (unsigned long x);

The *sqrt* function returns, as a floating point variable, the square root of the positive floating point variable *x*. The *isqrt* and *lsqrt* functions return the square root of unsigned integer and unsigned long variables *x*, respectively. Notice the reduction in size from unsigned integer to unsigned char and from unsigned long to unsigned integer in the *isqrt* and *lsqrt* functions.

Returns:

sqrt – the square root of the positive floating point variable *x*

isqrt – the square root of the unsigned integer variable *x* as an unsigned char

lsqrt – the square root of the unsigned long variable *x* as an unsigned integer

```
#include <math.h>
void main()
{
    unsigned char my_i_sqrt;
    unsigned int my_l_sqrt;
    float my_f_sqrt;

    my_f_sqrt = sqrt(6.4);     // get the square root of a float
    my_l_sqrt = lsqrt(250000); // get the square root of a long value
    my_i_sqrt = isqrt(81);     // get the square root of an int value

    while(1)
    {
    }
}
```

Results:

```
my_f_sqrt = 2.530

my_l_sqrt = 500

my_i_sqrt = 9
```

srand

#include <stdlib.h>

void srand(int seed);

The *srand* function sets the seed value used by the pseudo-random number generator *rand.*

Returns: None

See *rand* for a code example.

sscanf

#include <stdio.h>

signed char sscanf(char *str, char flash *fmtstr);

The *sscanf* function inputs values from the text string *str*, located in SRAM, according to the format specifiers in the *fmtstr* string and places the values into memory locations also specified by *fmtstr.* The formatting section of the *fmtstr* string is enclosed in quotes and uses format specifiers, listed below, to tell *sscanf* what kinds of values are to be read. The second part of *fmtstr* lists the addresses of where the read values are to be stored.

ATTENTION! It is important to always specify a ***pointer*** to the variable to receive the value in the *fmtstr* of the *sscanf* function, and ***not*** the variable itself. Failing to pass pointers to the *sscanf* function causes erratic results, because values are being stored in the memory locations related to the value of the variable passed instead of being stored at the address of the variable.

The format specifier string *fmtstr* is a constant and must be located in FLASH memory.

The implementation of *sscanf* is a reduced version of the standard C function. This was necessary due to the specific needs of an embedded system and because the full implementation would require a large amount of memory space.

The following format specifiers are available:

%c	Inputs the next argument as an ASCII character
%d	Inputs the next argument as a decimal integer
%i	Inputs the next argument as a decimal integer
%u	Inputs the next argument as an unsigned decimal integer
%x	Inputs the next argument as an unsigned hexadecimal integer
%s	Inputs the next argument as a null-terminated character string

Returns: Number of successful entries or a –1 if an error occurred

```
#include <mega16.h>
#include <stdio.h>
```

```
/* quartz crystal frequency [Hz] */
#define xtal 7372000L

/* Baud rate */
#define baud 9600

char SOFTWARE_VERSION[] = "3.5b";

void main(void)
{
   char version_letter;
   int major_version;
   int minor_version;
   char results;

   /* initialize the USART's baud rate */
   UBRRH=0x00;
   UBRRL=xtal/16/baud-1;
   /* initialize the USART control register
      RX & TX enabled, no interrupts, 8 data bits */
   UCSRA=0x00;
   UCSRB=0xD8;
   UCSRC=0x86;

   // tell sscanf what to look for
   // NOTICE the ADDRESSES of the variables are sent to sscanf!

   results = sscanf(SOFTWARE_VERSION,"%d.%d%c",&major_version,
         &minor_version,&version_letter);

   if (results != -1)
   {
      printf("Major Version: %d, Minor Version %d, Letter %c.\n\r",
            major_version, minor_version, version_letter);
   }
   else
       putsf("An error occurred. Something is not right!\r");


   while (1)
   {
   }
}
```

Results: The USART transmits the following:

```
Major Version: 3, Minor Version 5, Letter b.
```

standby

#include <sleep.h>

void standby(void);

The *standby* function puts the AVR microcontroller into standby mode. This is a sleep mode and similar to the *powerdown* function.

See *powerdown* in this appendix for the function use.

See the Atmel datasheets for the complete description of this sleep mode as it applies to a particular AVR device. Standby sleep mode is not available on all AVR devices.

*strcat, *strcatf

#include <string.h>

char *strcat(char *str1, char *str2);

char *strcatf(char *str1, char flash *str2);

The *strcat* and *strcatf* functions concatenate string *str2* onto the end of string *str1*. The memory allocated for *str1* must be long enough to accommodate the new, longer string plus the null-terminating character or else unexpected results occur. For the function *strcatf*, *str2* must point to a string located in FLASH. A pointer to *str1* is returned.

Returns: **str1* (a pointer to the null-terminated concatenation of strings *str1* and *str2*)

```
#include <mega16.h>
#include <stdio.h>
#include <string.h>

/* quartz crystal frequency [Hz] */
#define xtal 7372000L

/* Baud rate */
#define baud 9600

void main(void)
{
   char stra[] = "abc";
   char strb[10];

   /* initialize the USART's baud rate */
   UBRRH=0x00;
   UBRRL=xtal/16/baud-1;
   /* initialize the USART control register
      RX & TX enabled, no interrupts, 8 data bits */
   UCSRA=0x00;
```

```
    UCSRB=0xD8;
    UCSRC=0x86;

    strcpyf(strb,"xyz");    // initialize strb!

    strcat(strb,stra);      // add stra to strb
    strcatf(strb,"123");    // add a flash string too!

    puts(strb);             // put it out to the USART!
    while (1)
    {
    }
}
```

Results: The USART transmits, at 9600 baud,

```
xyzabc123
```

*strchr

#include <string.h>

char *strchr(char *str, char c);

The *strchr* function locates the first occurrence of the character *c* in the string *str*. If the character *c* is not found within the string, then a null pointer is returned.

Returns: A pointer to the first occurrence character *c* in string *str*; if *c* is not found in *str*, a null pointer is returned

```
#include <mega16.h>
#include <stdio.h>
#include <string.h>

/* quartz crystal frequency [Hz] */
#define xtal 7372000L

/* Baud rate */
#define baud 9600

void main(void)
{
    char stra[] = "123.45";
    char *strb;     // no need to allocate space, pointing
                    // into stra - already allocated!

    /* initialize the USART's baud rate */
    UBRRH=0x00;
    UBRRL=xtal/16/baud-1;
```

```
    /* initialize the USART control register
       RX & TX enabled, no interrupts, 8 data bits */
    UCSRA=0x00;
    UCSRB=0xD8;
    UCSRC=0x86;

    strb = strchr(stra,'.');
    printf("Full String: %s\n\rNew String: %s\n\r",stra,strb);
    *strb = '?';    // replace the decimal point with a ?
    printf("Modified string: %s\n\r",stra);

    while (1)
    {
    }
}
```

Results: The USART transmits, at 9600 baud,

```
Full String: 123.45
New String: .45
Modified String: 123?45
```

strcmp, strcmpf

#include <string.h>

signed char strcmp(char *str1, char *str2);

signed char strcmpf(char *str1, char flash *str2);

The *strcmp* and *strcmpf* functions compare string *str1* with string *str2*. For the function *strcmpf,* the *str2* must point to a string located in FLASH. The functions start comparing with the first character in each string. When the character in *str1* fails to match the character in *str2,* the difference in the character values is used to determine the return value of the function. The return value of the function is the result of the subtraction of the ASCII code of the character from *str2* from the ASCII code of the character from *str1.*

Returns:

- negative value if *str1* < *str2*
- zero if *str1* = *str2*
- positive value if *str1* > *str2*

```
#include <string.h>

char stra[] = "george";
char strb[] = "georgie";
signed char result;
signed char resultf;
```

```
void main(void)
{

        result = strcmp(stra, strb);
        resultf = strcmpf(stra, "george");

        while (1)
        {
        }
}
```

Results:

```
result = 0xFC    ('e' - 'i' < 0)
resultf = 0
```

*strcpy, *strcpyf

#include <string.h>

char *strcpy(char *dest, char *src);

char *strcpyf(char *dest, char flash *src);

The *strcpy* and *strcpyf* functions copy the string pointed to by *src* to the location pointed to by *dest*. The null-terminating character of the *src* string is the last character copied to the *dest* string. The memory allocated for the *dest* string must be large enough to hold the entire *src* string plus the null-terminating character.

Returns: Pointer to *dest*

```
#include <mega16.h>
#include <stdio.h>
#include <string.h>

/* quartz crystal frequency [Hz] */
#define xtal 7372000L

/* Baud rate */
#define baud 9600

void main(void)
{
        char stra[] = "Hello";
        char strb[6];
        char strc[6];
```

```
        /* initialize the USART's baud rate */
        UBRRH=0x00;
        UBRRL=xtal/16/baud-1;
        /* initialize the USART control register
           RX & TX enabled, no interrupts, 8 data bits */
        UCSRA=0x00;
        UCSRB=0xD8;
        UCSRC=0x86;

        strcpy(strb,stra);         // copy stra to strb
        strcpyf(strc,"World");     // copy "World" to strc

        puts(strb);          // transmit strb
        puts(strc);          // transmit strc

        while (1)
        {
        }
}
```

Results: The USART transmits, at 9600 baud,

```
Hello
      World
```

The line feed between the lines is from calling the *puts* function. *puts* appends a line feed at the end of the string as it sends it to the USART.

strcspn, strcspnf

#include <string.h>

unsigned char strcspn(char *str, char *set);

unsigned char strcspnf(char *str, char flash *set);

The *strcspn* and *strcspnf* functions return the index of the first character in the string *str* that matches a character in the string *set.* If none of the characters in the string *str* is in the string *set,* the length of *str* is returned. If the first character in the string *str* is in the string *set,* zero is returned. For the function *strspnf,* the string *set* must be located in FLASH.

Returns: Index of the first character in *str* that is in *set*

```
#include <string.h>

void main(void)
{
        char set[] = "1234567890-()";
        char stra[] = "1.800.555.1212";
        char index_1;
        char index_2;
```

```
        index_1 = strcspn(stra,set);
        index_2 = strcspnf(stra,".-()");

        while (1)
        {
        }
}
```

Results:

```
index_1 = 0

index_2 = 1
```

strlen, strlenf

#include <string.h>

For the TINY memory model:

```
unsigned char strlen(char *str);
```

For the SMALL memory model:

```
unsigned int strlen(char *str);
```

For either memory model:

```
unsigned int strlenf(char flash *str);
```

The *strlen* and *strlenf* functions return the length of the string *str*, not counting the null terminator. For the *strlen* function with the TINY memory model in use, the length can be from 0 to 255. If the SMALL memory model is in use and *strlen* is called, the length can be from 0 to 65,535. The function *strlenf* returns the length of a string located in FLASH. This length can be from 0 to 65,535 regardless of the memory model in use.

Returns: Length of the string *str*

```
#include <string.h>

void main(void)
{
        char stra[] = "1234567890";
        unsigned char len1;
        unsigned int len2;

        len1 = strlen(stra);
        len2 = strlenf("abcdefghijklmnopqrstuvwxyz");

        while (1)
        {
        }
}
```

Results:

```
len1 = 10

len2 = 26
```

*strncat, *strncatf

#include <string.h>

char *strncat(char *str1, char *str2, unsigned char n);

char *strncatf(char *str1, char flash *str2, unsigned char n);

The *strncat* and *strncatf* functions concatenate a maximum of *n* characters from string *str2* onto the end of string *str1*. The memory allocated for *str1* must be long enough to accommodate the new, longer string plus the null-terminating character or else unexpected results occur. For the function *strncatf*, *str2* must point to a string located in flash. A pointer to *str1* is returned.

Returns: **str1* (a pointer to the null-terminated concatenation of strings *str1* and *str2*)

```
#include <mega16.h>
#include <stdio.h>
#include <string.h>

/* quartz crystal frequency [Hz] */
#define xtal 7372000L

/* Baud rate */
#define baud 9600

void main(void)
{
        char stra[] = "abc";
        char strb[8];

        /* initialize the USART's baud rate */
        UBRRH=0x00;
        UBRRL=xtal/16/baud-1;
        /* initialize the USART control register
           RX & TX enabled, no interrupts, 8 data bits */
        UCSRA=0x00;
        UCSRB=0xD8;
        UCSRC=0x86;

        strcpyf(strb,"xyz");      // initialize strb!

        strncat(strb,stra,2);     // add stra to strb
        strncatf(strb,"123",1);   // add a flash string too!
```

```
        puts(strb);                 // put it out to the USART!
        while (1)
        {
        }
}
```

Results: The USART transmits, at 9600 baud,

```
xyzab1
```

strncmp, strncmpf

#include <string.h>

signed char strncmp(char *str1, char *str2, unsigned char n);

signed char strcnmpf(char *str1, char flash *str2, unsigned char n);

The *strncmp* and *strncmpf* functions compare at most *n* characters from string *str1* to string *str2*. For the function *strncmpf, str2* must point to a string located in FLASH. The functions start comparing with the first character in each string. When the character in *str1* fails to match the character in *str2*, the difference in the character values is used to determine the return value of the function. The return value of the function is the result of the subtraction of the ASCII code of the character from *str2* from the ASCII code of the character from *str1*. Any differences between the strings beyond the nth character are ignored.

Returns:

- negative value if *str1* < *str2*
- zero if *str1* = *str2*
- positive value if *str1* > *str2*

```
#include <string.h>

void main(void)
{
        char stra[] = "george";
        char strb[] = "georgie";
        signed char result;
        signed char resultf;

        result = strncmp(stra, strb,5);
        resultf = strncmpf(stra, "george",6);

        while (1)
        {
        }
}
```

Results:

```
result = 0

resultf = 0
```

*strncpy, *strncpyf

#include <string.h>

char *strncpy(char *dest, char *src, unsigned char n);

char *strncpyf(char *dest, char flash *src, unsigned char n);

The *strncpy* and *strncpyf* functions copy up to *n* characters from the string pointed to by *src* to the location pointed to by *dest.* If there are fewer than *n* characters in the *src* string, then the *src* string is copied to *dest* and null-terminating characters are appended until the total number of characters copied to *dest* is *n.* If *src* string is longer than or equal in length to *n,* then no terminating character is copied or appended to *dest.*

Returns: Pointer to *dest*

```
#include <mega16.h>
#include <stdio.h>
#include <string.h>

/* quartz crystal frequency [Hz] */
#define xtal 7372000L

/* Baud rate */
#define baud 9600

char stra[] = "Hello";
char strb[] = "HELLO";
char strc[6];

void main(void)
{

        /* initialize the USART's baud rate */
        UBRRH=0x00;
        UBRRL=xtal/16/baud-1;
        /* initialize the USART control register
           RX & TX enabled, no interrupts, 8 data bits */
        UCSRA=0x00;
        UCSRB=0xD8;
        UCSRC=0x86;

        strncpy(strb,stra,3);           // copy stra to strb
```

```
        strncpyf(strc,"World",10);      // copy "World" to strc
        printf("%s %s\n\r",strb,strc);

        while (1)
        {
        }
}
```

Results: Since "3" is less than the total length of "stra", the first three letters are copied over "strb", but no terminating null is copied. As a result, the original null termination is used to terminate "strb". So, the USART transmits, at 9600 baud,

```
HelLO World
```

*strpbrk, *strpbrkf

#include <string.h>

char *strpbrk(char *str, char *set);

char *strpbrkf(char *str, char flash *set);

The *strpbrk* and *strpbrkf* functions search the string *str* for the first occurrence of a character from the string *set*. If there is a match, the function returns a pointer to the character in the string *str*. If there is not a match, a null pointer is returned. For the function *strpbrkf*, the string *set* must be located in FLASH.

Returns: Pointer to the first character in *str* that matches a character in *set*

```
#include <mega16.h>
#include <stdio.h>
#include <string.h>

/* quartz crystal frequency [Hz] */
#define xtal 7372000L

/* Baud rate */
#define baud 9600

void main(void)
{
        char stra[] = "11/25/00";
        char set[] = "/.,!-";
        char strb[] = "November 25, 2000";
        char *pos;
        char *fpos;
```

```
        /* initialize the USART's baud rate */
        UBRRH=0x00;
        UBRRL=xtal/16/baud-1;
        /* initialize the USART control register
           RX & TX enabled, no interrupts, 8 data bits */
        UCSRA=0x00;
        UCSRB=0xD8;
        UCSRC=0x86;


        pos = strpbrk(stra,set); // find first
                                 // occurrence of something!
        fpos = strpbrkf(strb,",.-");
        printf("Initial Date: %s\n\r",stra);
        printf("String following match: %s\n\r",pos);
        printf("Just the year: %s\n\r",fpos+1);
        while (1)
        {
        }
}
```

Results: The USART transmits, at 9600 baud,

```
Initial Date: 11/25/00
String following match: /25/00
Just the year:  2000
```

strpos

#include <string.h>

char strpos(char *str, char c);

The *strpos* function locates the first occurrence of the character *c* in the string *str.* The index of the first occurrence of *c* is returned. If the character *c* is not found within the string, then –1 is returned.

Returns: Index of the first occurrence character *c* in string *str;* if *c* is not found in *str,* –1 is returned

```
#include <mega16.h>
#include <stdio.h>
#include <string.h>

/* quartz crystal frequency [Hz] */
#define xtal 7372000L

/* Baud rate */
#define baud 9600
```

```
void main(void)
{
   char stra[] = "11/25/2000";
   char month_day;
   char day_year;

   /* initialize the USART's baud rate */
   UBRRH=0x00;
   UBRRL=xtal/16/baud-1;
   /* initialize the USART control register
      RX & TX enabled, no interrupts, 8 data bits */
   UCSRA=0x00;
   UCSRB=0xD8;
   UCSRC=0x86;

   month_day = strpos(stra,'/');  // find first slash character
   day_year = strrpos(stra,'/');   // find last slash character

   printf("Starting String: %s\n\r",stra);

   // replace slash character with dashes
   stra[month_day] = '-';
   stra[day_year] = '-';

   printf("Modified string: %s\n\r",stra);

   while (1)
   {
   }
}
```

Results: The USART transmits, at 9600 baud,

```
Starting String: 11/25/2000
Modified String: 11-25-2000
```

*strrchr

#include <string.h>

char *strrchr(char *str, char c);

The *strrchr* function locates the last occurrence of the character *c* in the string *str.* If the character *c* is not found within the string, then a null pointer is returned.

Returns: Pointer to the last occurrence of character *c* in string *str;* if *c* is not found in *str,* a null pointer is returned

```
#include <mega16.h>
#include <stdio.h>
#include <string.h>
```

```
/* quartz crystal frequency [Hz] */
#define xtal 7372000L

/* Baud rate */
#define baud 9600

void main(void)
{
        char stra[] = "123.45.789";
        char *strb; // no need to allocate space, pointing
                    // into stra - already allocated!

        /* initialize the USART's baud rate */
        UBRRH=0x00;
        UBRRL=xtal/16/baud-1;
        /* initialize the USART control register
           RX & TX enabled, no interrupts, 8 data bits */
        UCSRA=0x00;
        UCSRB=0xD8;
        UCSRC=0x86;

        strb = strrchr(stra,'.');
        printf("Full String: %s\n\rNew String: %s\n\r",stra,strb);
        *strb = '6'; // replace the decimal point with a
        printf("Modified string: %s\n\r",stra);

        while (1)
        {
        }
}
```

Results: The USART transmits, at 9600 baud,

```
Full String: 123.45.789
New String: .789
Modified String: 123.456789
```

*strrpbrk, *strrpbrkf

#include <string.h>

char *strrpbrk(char *str, char *set);

char *strrpbrkf(char *str, char flash *set);

The *strrpbrk* and *strrpbrkf* functions search the string *str* for the last occurrence of a character from the string *set*. If there is a match, the function returns a pointer to the character in the string *str*. If there is not a match, a null pointer is returned. For the function *strrpbrkf*, the string *set* must be located in FLASH.

Returns: Pointer to the last character in *str* that matches a character in *set*

```
#include <mega16.h>
#include <stdio.h>
#include <string.h>

/* quartz crystal frequency [Hz] */
#define xtal 7372000L

/* Baud rate */
#define baud 9600

void main(void)
{
        char stra[] = "11/25/00";
        char set[] = "/.,!-";
        char strb[] = "November 25, 2000";
        char *pos;
        char *fpos;

        /* initialize the USART's baud rate */
        UBRRH=0x00;
        UBRRL=xtal/16/baud-1;
        /* initialize the USART control register
           RX & TX enabled, no interrupts, 8 data bits */
        UCSRA=0x00;
        UCSRB=0xD8;
        UCSRC=0x86;

        pos = strrpbrk(stra,set);      // find last
                                       // occurrence of something!
        fpos = strrpbrkf(strb,",.-@$");
        printf("Year in 11/25/00: %s\n\r",pos);
        printf("String following match: %s\n\r",fpos);

        while (1)
        {
        }
}
```

Results: The USART transmits, at 9600 baud,

```
Year in 11/25/00: /00
String following match: , 2000
```

strrpos

#include <string.h>

char strrpos(char *str, char c);

The *strrpos* function locates the last occurrence of the character *c* in the string *str*. The index of the last occurrence of *c* is returned. If the character *c* is not found within the string, then −1 is returned.

Returns: Index of the last occurrence of character *c* in string *str*; if *c* is not found in *str*, −1 is returned

```
#include <mega16.h>
#include <stdio.h>
#include <string.h>

/* quartz crystal frequency [Hz] */
#define xtal 7372000L

/* Baud rate */
#define baud 9600

void main(void)
{
  char stra[] = "11/25/2000";
  char month_day;
  char day_year;

  /* initialize the USART's baud rate */
  UBRRH=0x00;
  UBRRL=xtal/16/baud-1;
  /* initialize the USART control register
     RX & TX enabled, no interrupts, 8 data bits */
  UCSRA=0x00;
  UCSRB=0xD8;
  UCSRC=0x86;

  month_day = strpos(stra,'/');    // find first slash character
  day_year = strrpos(stra,'/');    // find last slash character

  printf("Starting String: %s\n\r",stra);

  // replace slash character with dashes
  stra[month_day] = '-';
  stra[day_year] = '-';

  printf("Modified string: %s\n\r",stra);

  while (1)
  {
  }
}
```

Results: The USART transmits, at 9600 baud,

```
Starting String: 11/25/2000
Modified String: 11-25-2000
```

strspn, strspnf

#include <string.h>

unsigned char strspn(char *str, char *set);

unsigned char strspnf(char *str, char flash *set);

The *strspn* and *strspnf* functions return the index of the first character in the string *str* that does not match a character in the string *set.* If all characters in the string *str* are in the string *set,* the length of *str* is returned. If no characters in the string *str* are in the string *set,* zero is returned. For the function *strspnf,* the string *set* must be located in FLASH.

Returns: Index of the first character in *str* that is not in *set*

```
#include <string.h>

void main(void)
{
        char set[] = "1234567890-()";
        char stra[] = "1.800.555.1212";
        char index_1;
        char index_2;

        index_1 = strspn(stra,set);
        index_2 = strspnf(stra,".1234567890-()");

        while (1)
        {
        }
}
```

Results:

```
index_1 = 1

index_2 = 14
```

*strstr, *strstrf

#include <string.h>

char *strstr(char *str1, char *str2);

char *strstr(char *str1, char flash *str2);

The *strstr* and *strstrf* functions search string *str1* for the first occurrence of string *str2.* If *str2* is found within *str1,* then a pointer to the first character of *str2* in *str1* is returned. If *str2* is not found in *str1*, then a null is returned. For the function *strstrf*, the string *str2* must be located in FLASH.

Returns: Pointer to the first character of *str2* in *str1* or null if *str2* is not in *str1*

```
#include <mega16.h>
#include <stdio.h>
#include <string.h>

/* quartz crystal frequency [Hz] */
#define xtal 7372000L

/* Baud rate */
#define baud 9600

void main(void)
{
      char stra[] = "Red Green Blue";
      char strb[] = "Green";
      char *ptr;
      char *ptrf;

      /* initialize the USART's baud rate */
      UBRRH=0x00;
      UBRRL=xtal/16/baud-1;
      /* initialize the USART control register
         RX & TX enabled, no interrupts, 8 data bits */
      UCSRA=0x00;
      UCSRB=0xD8;
      UCSRC=0x86;

      // grab a pointer to where Green is
      ptr = strstr(stra,strb);

      // grab a pointer to where the first B is
      ptrf = strstrf(stra, "B");

      printf("Starting String: %s\n\r",stra);
      printf("Search String: %s\n\r",strb);
      printf("Results String 1: %s\n\r",ptr);
      printf("Results String 2: %s\n\r",ptrf);

      while (1)
      {
      }
}
```

Results: The USART transmits, at 9600 baud,

```
Starting String: Red Green Blue
Search String: Green
Results String 1: Green Blue
Results String 2: Blue
```

*strtok

#include <string.h>

char *strtok(char *str1, char flash *str2);

The function *strtok* scans string *str1* for the first token not contained in the string *str2* located in FLASH. The function expects *str1* to consist of a sequence of text tokens, separated by one or more characters from the string *str2* (token separators). This function may be called repetitively to parse through a string (*str1*) and retrieve tokens that are separated by known characters (*str2*).

The first call to *strtok* with a non-null pointer for *str1* returns a pointer to the first character of the first token in *str1*. The function searches for the end of the token and places a null-termination character at the first token separator (character from *str2*) following the token. Subsequent calls to *strtok* with a null passed for *str1* return the next token from *str1* in sequence until no more tokens exist in *str1*. When no more tokens are found, a null is returned.

Note that *strtok* modifies *str1* by placing the null-termination characters after each token. To preserve *str1,* make a copy of it before calling *strtok.*

Returns: Pointer to the next token in *str1* or null if no more tokens exist in *str1*

```
#include <mega16.h>
#include <stdio.h>
#include <string.h>

/* quartz crystal frequency [Hz] */
#define xtal 7372000L

/* Baud rate */
#define baud 9600

char mytext[] = "(888)777-2222";
char flash separators[] = "()-";
void main(void)
{
        char area_code[4];
        char *prefix;
        char *postfix;
        char backup_copy[14];
```

```
        /* initialize the USART's baud rate */
        UBRRH=0x00;
        UBRRL=xtal/16/baud-1;
        /* initialize the USART control register
           RX & TX enabled, no interrupts, 8 data bits */
        UCSRA=0x00;
        UCSRB=0xD8;
        UCSRC=0x86;

        // we want to keep the original around too!
        strcpy(backup_copy,mytext);

        // grab a pointer to the area code
        strcpy(area_code,strtok(mytext, separators));

        // grab a pointer to the prefix
        prefix = strtok(0, separators);

        // grab a pointer to the postfix
        postfix = strtok(0, separators);


        printf("Starting String: %s\n\r",backup_copy);
        printf("Area Code: %s\n\r",area_code);
        printf("Prefix: %s\n\r",prefix);
        printf("Postfix: %s\n\r",postfix);

        while (1)
        {
        }
}
```

Results: The USART transmits, at 9600 baud,

```
Starting String: (888)777-2222
Area Code: 888
Prefix: 777
Postfix: 2222
```

tan

#include <math.h>

float tan(float x);

The *tan* function calculates the tangent of the floating point number *x*. The angle *x* is expressed in radians.

Returns: tan(x)

```
#include <math.h>

void main()
{
        float new_val;

        new_val = tan(5.121);

        while(1)
        {
        }
}
```

Results: new_val = –2.309

tanh

#include <math.h>

float tanh(float x);

The *tanh* function calculates the hyperbolic tangent of the floating point number *x*. The angle *x* is expressed in radians.

Returns: tanh(x)

```
#include <math.h>

void main()
{
        float new_val;

        new_val = tanh(5.121);

        while(1)
        {
        }
}
```

Results: new_val = 0.999

toascii

#include <ctype.h>

unsigned char toascii(char c);

The *toascii* function converts character *c* to a 7-bit ASCII format. This is performed by the following definition:

```
#define toascii(c) (c) & 0x7f
```

Returns: the ASCII value of *c* (values between 0 and 127)

```
#include <ctype.h>
void main()
{
        char ascii_value;
        ascii_value = toascii(0xB1);
        while (1)
        {
        }
}
```

Results: ascii_value = 0x31 = '1'

toint

#include <ctype.h>

unsigned char toint(char c);

The *toint* function interprets the ASCII character *c* as a hexadecimal digit and returns an unsigned char from 0 to 15. If the character is not a valid hexadecimal digit, the function returns 0.

Returns: Unsigned character from 0 to 15 representing ASCII character *c*

```
#include <ctype.h>
void main()
{
        unsigned char hex_value;
        hex_value = toint('a');
        while (1)
        {
        }
}
```

Results: hex_value = 0x0a

tolower

#include <ctype.h>

char tolower(char c);

The *tolower* function converts the ASCII character *c* from an uppercase character to a lowercase character. If *c* is a lowercase character, *c* is returned unchanged.

Returns: ASCII character *c* as a lowercase character

```
#include <ctype.h>
void main()
{
```

```
            char lower_case_c;
            lower_case_c = tolower('U');
            while(1)
            {
            }
      }
```

Results: lower_case_c = 'u'

toupper

#include <ctype.h>

char toupper(char c);

The *toupper* function converts the ASCII character *c* from a lowercase character to an uppercase character. If *c* is an uppercase character, *c* is returned unchanged.

Returns: ASCII character *c* as an uppercase character

```
      #include <ctype.h>
      void main()
      {
            char upper_case_c;
            upper_case_c = toupper('u');
            while(1)
            {
            }
      }
```

Results: upper_case_c = 'U'

vprintf

#include <stdio.h>

void vprintf(char flash *fmtstr, va_list argptr);

The *vprintf* function is identical to the function *printf* except that *argptr* is a pointer to a variable-length list of arguments. This pointer is of the type *va_list*. The *va_list* type is defined in the header file stdarg.h.

Returns: None

```
#include <mega16.h>
/* include the standard input/output library */
#include <stdio.h>
/* include the variable length argument lists macros */
#include <stdarg.h>

#define xtal 6000000L
#define baud 9600
```

```
/* create custom print routine with a variable length argument list */
void custom_print(char flash *cp_fmtstr, int data_len, ...)
{
  int i;
  int total;          /* total of the integer arguments */
  va_list cp_argptr; /* pointer to a variable length argument list */
  total = 0;          /* clear the tally */

  /* initialize the variable length argument list pointer */
  va_start(cp_argptr, data_len);

  printf("Start of Custom Message\n\r");

  /* use vprintf to print out the values passed to our function */
  vprintf(cp_fmtstr, cp_argptr);

  /* now use the va_arg macro to return the values as integers */
  for (i=0;i<data_len;i++)
      total += va_arg(cp_argptr,int);

  /* print the custom information about the data sent! */
  printf("data_len: %d\n\r",data_len);
  printf("total value: %d\n\r",total);
  printf("End of Custom Message\n\r");

  /* cleanup by calling the va_end macro to terminate
     the use of the pointer */
  va_end(cp_argptr);
}

void main(void)
{
  // USART initialization
  // Communication Parameters: 8 Data, 1 Stop, No Parity
  // USART Receiver: On
  // USART Transmitter: On
  // USART Mode: Asynchronous
  // USART Baud rate: 9600
  UCSRA=0x00;
  UCSRB=0xD8;
  UCSRC=0x86;
  UBRRH=0x00;
  UBRRL=xtal/16/baud-1;

  custom_print("Message data: %d %d %d %d %d %d\n\r",6,1,2,3,7,9,8);
```

```
    while (1)
        ;
  }
```

Results: The USART transmits, at 9600 baud,

```
Start of Custom Message
Message data: 1 2 3 7 9 8
Data_len: 6
total value: 30
End of Custom Message
```

vsprintf

#include <stdio.h>

void vsprintf(char *str, char flash *fmtstr, va_list argptr);

The *vsprintf* function performs almost exactly as *vprintf* performs, except that *vsprintf* copies the formatted text according to the format specifiers in the *fmtstr* string to the string *str* instead of to the USART. A null termination is appended to the end of *str* after the formatted text is copied to it. The memory for *str* should be large enough to accommodate the copied text and the null termination.

See *printf* for more information on format specifiers and *vprintf* for a code example.

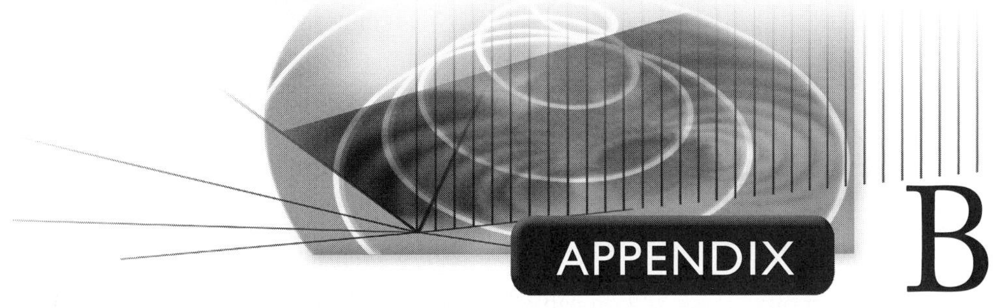

B

Getting Started with CodeVisionAVR and the STK500

FEATURES

- Installing and configuring CodeVisionAVR to work with the Atmel STK500 starter kit and AVR Studio debugger.
- Creating a new project using the CodeWizardAVR Automatic Program Generator
- Editing and compiling the C code
- Loading the executable code into the target microcontroller on the STK500 starter kit.

INTRODUCTION

The purpose of this document is to guide the user through the preparation of an example C program using the CodeVisionAVR C compiler. The example, which is the subject of this application note, is a simple program for the Atmel AT90S8515 microcontroller on the STK500 starter kit.

PREPARATION

Install the CodeVisionAVR C Compiler by executing the file **setup.exe**.

It is assumed that the program was installed in the default directory: **C:\cvavr**.

Install the Atmel AVR Studio debugger by executing the file **setup.exe**.

It is assumed that AVR Studio was installed in the default directory: **C:\Program Files\Atmel\AVR Studio**.

The demonstration program to be developed in the next few pages requires an Atmel AT90S8515 microcontroller and the STK500 starter kit.

Set up the starter kit according to the instructions in the STK500 User Guide.

(Reprinted courtesy of Pavel Haiduc and HP InfoTech S.R.L.)

Make sure the power is off and insert the AT90S8515 chip into the appropriate socket marked **SCKT3000D3**.

Set the **XTAL1** jumper. Also set the **OSCSEL** jumper between pins 1 and 2.

Connect one 10 pin ribbon cable between the **PORTB** and **LEDS** headers.

This will allow displaying the state of AT90S8515's PORTB outputs.

Connect one 6 pin ribbon cable between the **ISP6PIN** and **SPROG3** headers.

This will allow CodeVisionAVR to automatically program the AVR chip after a successful compilation.

In order to use this feature, one supplementary setting must be done:

Open the CodeVisionAVR IDE and select the **Settings|Programmer** menu option.

The following dialog window will open:

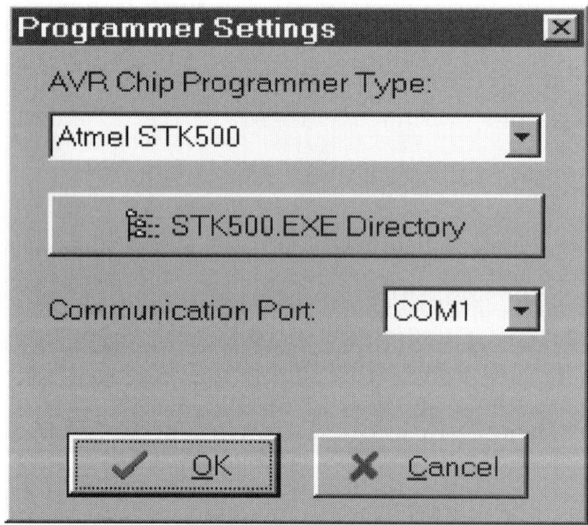

Make sure to select the **Atmel STK500 AVR Chip Programmer Type** and the corresponding **Communication Port,** which is used with the STK500 starter kit.

Then press the **STK500.EXE Directory** button in order to specify the location of the **stk500.exe** command line utility supplied with AVR Studio.

(Reprinted courtesy of Pavel Haiduc and HP InfoTech S.R.L.)

The following dialog window will open:

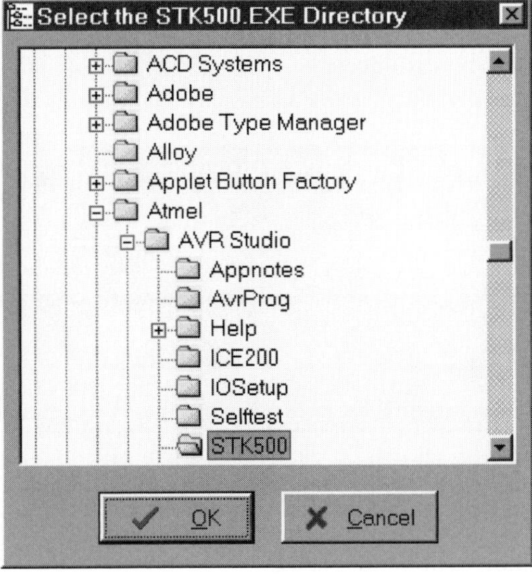

Select the **C:\Program Files\Atmel\AVR Studio\STK500** directory and press the **OK** button.

Then press the **OK** button once again in order to save the **Programmer Settings**.

In order to be able to invoke the AVR Studio debugger/simulator from within the CodeVisionAVR IDE, one final setting must be done.

Select the **Settings|Debugger** menu option. The following dialog window will open:

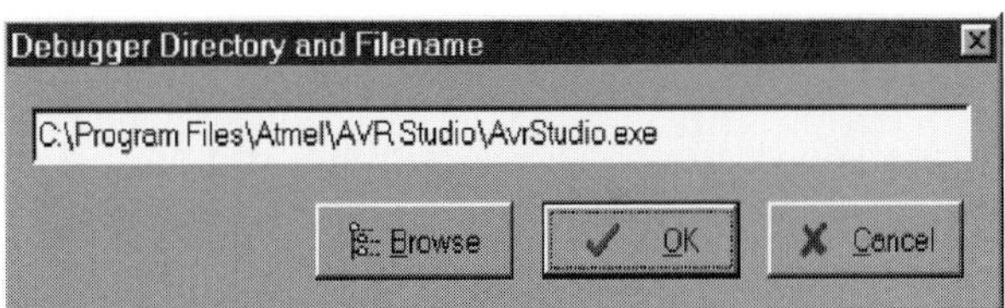

Enter **C:\Program Files\Atmel\AVR Studio\AvrStudio.exe** and press the **OK** button.

(Reprinted courtesy of Pavel Haiduc and HP InfoTech S.R.L.)

CREATING A NEW PROJECT

In order to create a new project, select the **File|New** menu option or press the toolbar button.

The following window will be displayed:

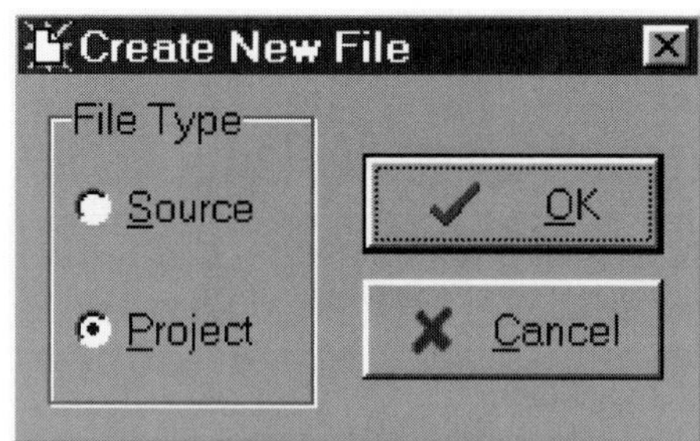

Select **Project** and press **OK**.

Then the following window will be displayed

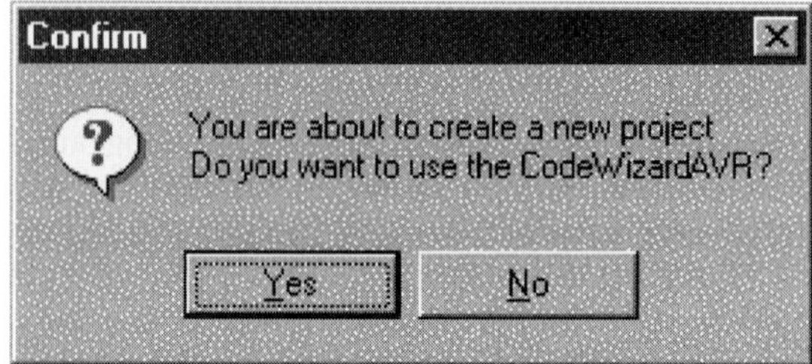

Press **Yes** to use the CodeWizardAVR Automatic Program Generator.

(Reprinted courtesy of Pavel Haiduc and HP InfoTech S.R.L.)

USING THE CODEWIZARDAVR AUTOMATIC PROGRAM GENERATOR

The CodeWizardAVR simplifies the task of writing start-up code for different AVR microcontrollers.

The above window opens and, for this example project, we shall select the AT90S8515 microcontroller and set the clock rate to 3.68 MHz since that is the clock on the STK500 starter kit.

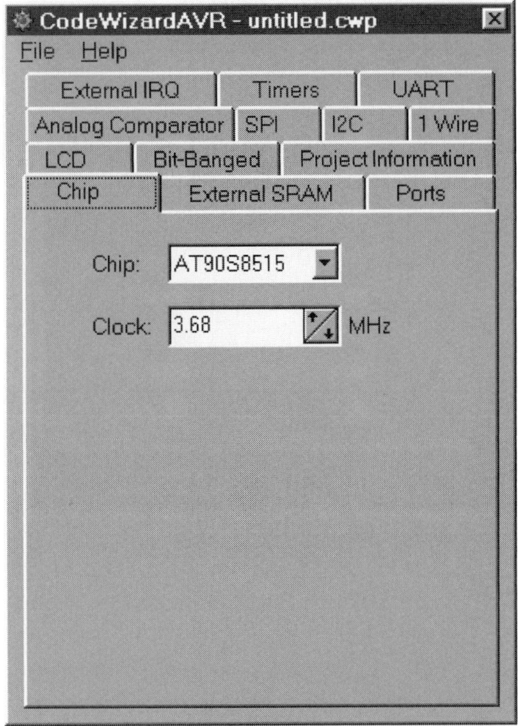

(Reprinted courtesy of Pavel Haiduc and HP InfoTech S.R.L.)

CONFIGURING THE INPUT/OUTPUT PORTS

Select the **Ports** tab to determine how the I/O ports are to be initialized for the target system.

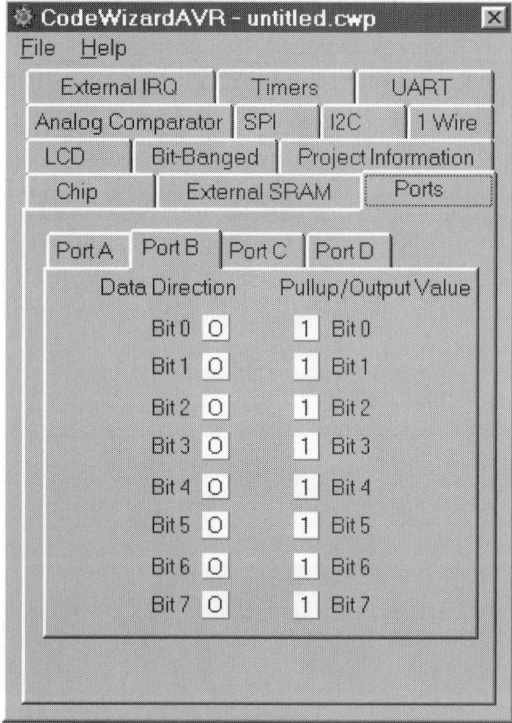

The default setting is to have the ports for all the target systems to be inputs (**Data Direction** bits to be all I's) in their Tri-state mode.

For this exercise, we want to set Port B (by selecting the **Port B** tab) to be all outputs and we do this by setting all the **Data Direction** bits to **O** (by clicking on them). We also set the **Output Values** to be all **1**'s since this corresponds to the LED's on the STK500 being off.

CONFIGURING TIMER 1

For this project, we want to configure Timer 1 to generate overflow interrupts.

(Reprinted courtesy of Pavel Haiduc and HP InfoTech S.R.L.)

We select the **Timers** tab and then select the **Timer 1** tab resulting in the window shown below.

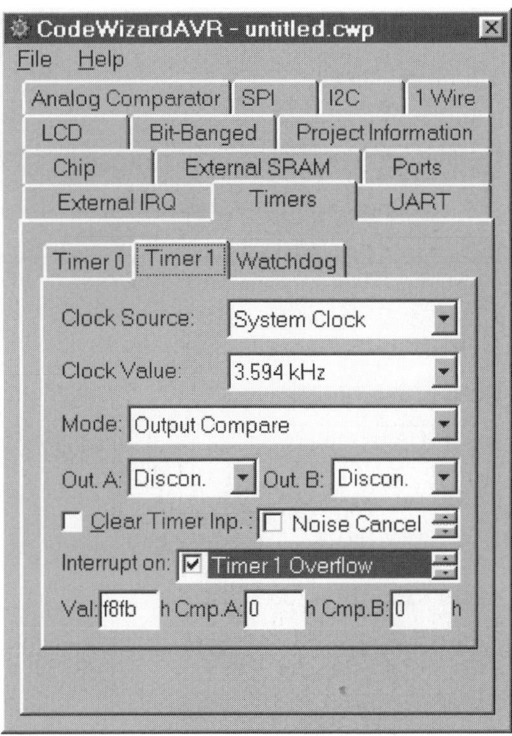

Set the options as shown in the above window. We have selected a clock rate of 3.594 kHz (the system clock of 3.68 MHz divided by 1024).

The timer is set to operate in the default "Output Compare" mode and to generate interrupts on overflow.

To obtain the frequency of LED movement of 2 per second we need to reinitialize the Timer 1 value to 0x10000-(3594/2)=0xF8FB on every overflow.

(Reprinted courtesy of Pavel Haiduc and HP InfoTech S.R.L.)

COMPLETING THE PROJECT

By selecting the **File|Generate, Save and Exit** menu option the Code Wizard will generate a skeleton C program with, in this case, the Port B and Timer 1 overflow interrupt set up correctly.

The following dialog window will appear:

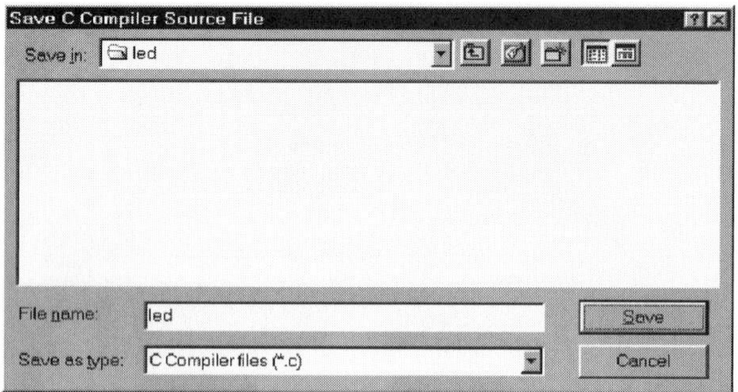

By pressing the button, a new directory **C:\cvavr\led** will be created.

It will hold all the files of our sample project.

Then we must specify the **File name** of the C source file: **led.c** and press the **Save** button.

A new dialog window will open:

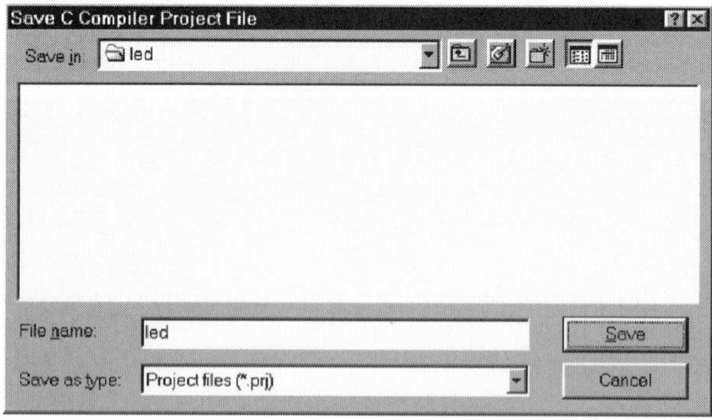

(Reprinted courtesy of Pavel Haiduc and HP InfoTech S.R.L.)

Here, we must specify the **File name led.prj** as the project name and put it in the same folder: **C:\cvavr\led.**

Finally, we will be prompted to save the CodeWizard project file:

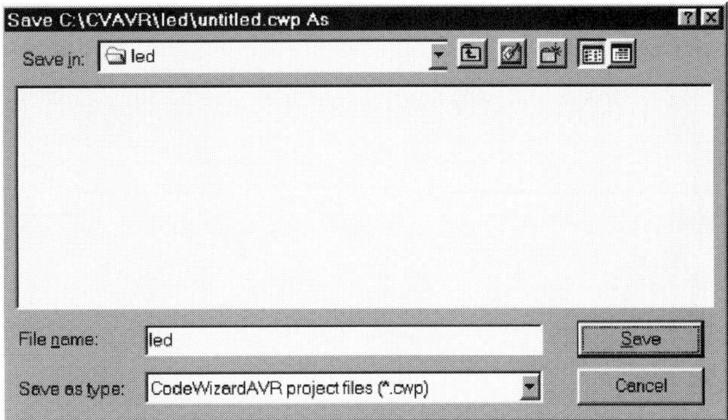

We must specify the **File name** as: **led.cwp** and press the **Save** button.

Saving all the CodeWizardAVR peripheral configuration in the **led.cwp** project file will allow us to reuse some of our initialization code in future projects.

The **led.c** source file is now automatically opened and available. One can then start editing the code produced by the CodeWizardAVR. The source listing is given on Appendix A of this application note.

In this example, only the interrupt handler code needs to be amended to manage the LED display.

The small bit of code which was added is shown highlighted; the remainder was supplied by the CodeWizardAVR.

VIEWING OR MODIFYING THE PROJECT CONFIGURATION

At any time, a project configuration may be changed using the **Project|Configure** menu option or by pressing the toolbar button.

(Reprinted courtesy of Pavel Haiduc and HP InfoTech S.R.L.)

The following dialog window will open:

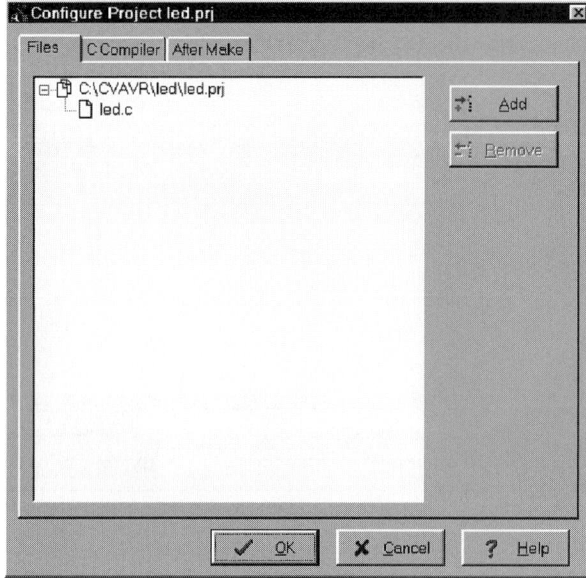

To add or remove files from the project, select the **Files** tab and use the **Add** or **Remove** buttons.

To change the target microcontroller, the clock rate, or the various compiler options select the **C Compiler** tab.

The following window opens and the configuration may be altered:

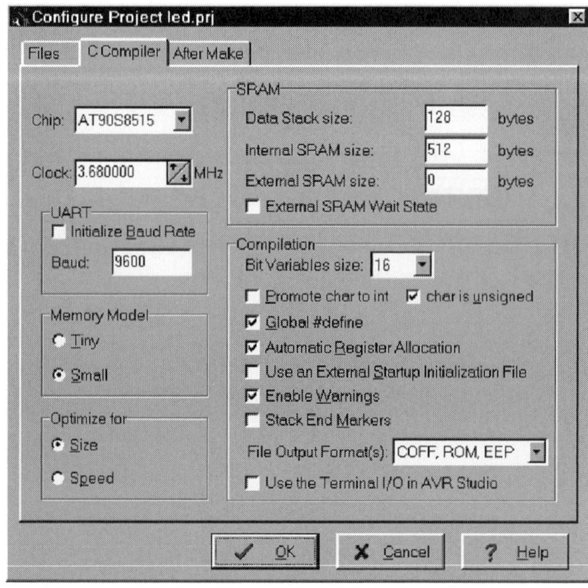

(Reprinted courtesy of Pavel Haiduc and HP InfoTech S.R.L.)

We may also select whether we wish to automatically program the target microprocessor after the Make or not.

This is chosen by selecting the **After Make** tab, which gives us the next window.

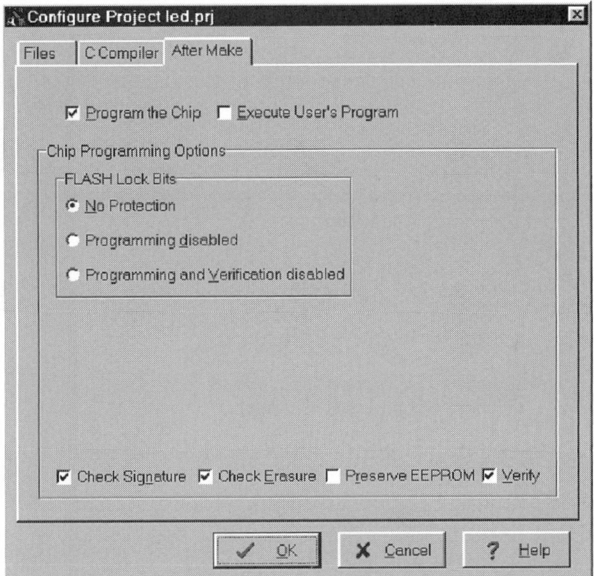

For the purposes of this example, the **Program the Chip** option must be checked.

This will enable automatic programming of the AVR chip after the Make is complete.

MAKING THE PROJECT

The **Project** pull-down menu gives the **Make** option. Click on it or on the button on the toolbar.

After a successful compile and assembly, the **Information** window will be displayed as shown below.

(Reprinted courtesy of Pavel Haiduc and HP InfoTech S.R.L.)

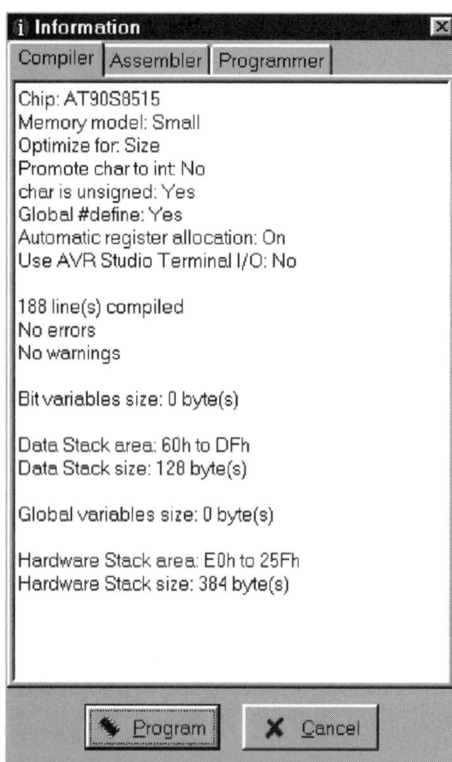

This window shows how the RAM memory was used by the compiler.

(Reprinted courtesy of Pavel Haiduc and HP InfoTech S.R.L.)

If the **Assembler** tab is clicked, the Assembler window shows the size of the assembled code, as shown below.

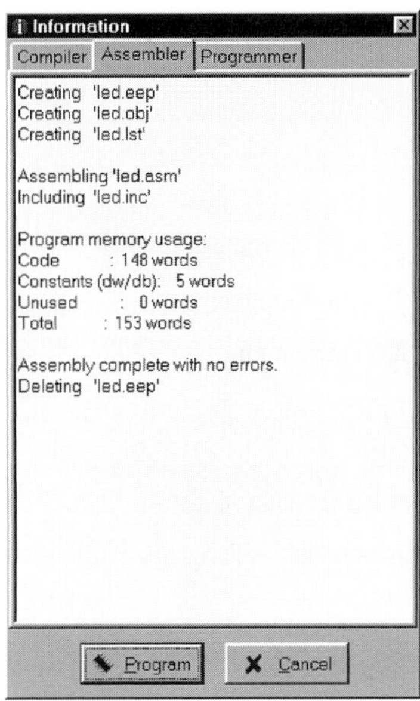

Selecting the **Programmer** tab displays the value of the **Chip Programming Counter.**

This counter can be initialized by pressing the **Set Counter** button.

(Reprinted courtesy of Pavel Haiduc and HP InfoTech S.R.L.)

If the Make process was successful, then power up the STK500 starter kit and press the **Program** button to start the automatic chip programming.

After the programming process is complete, the code will start to execute in the target microcontroller on the STK500 starter kit.

SHORT REFERENCE

Preparations:

1. Install the CodeVisionAVR C compiler
2. Install the Atmel AVR Studio debugger
3. Install the Atmel STK500 starter kit
4. Configure the STK500 programmer support in the CodeVisionAVR IDE by selecting: Settings->Programmer->

   AVR Chip Programmer Type: STK500->

   Specify STK500.EXE Directory: C:\Program Files\Atmel\AVR Studio\STK500->

   Communication Port
5. Configure the AVR Studio support in the CodeVisionAVR IDE by selecting:

   Settings->Debugger->

   Enter: C:\Program Files\Atmel\AVR Studio.

Getting Started:

1. Create a new project by selecting:

   File->New->Select Project
2. Specify that the CodeWizardAVR will be used for producing the C source and project files: Use the CodeWizard?->Yes
3. In the CodeWizardAVR window specify the chip type and clock frequency:

   Chip->Chip: AT90S8515->Clock: 3.86MHz
4. Configure the I/O ports: Ports->Port B->Data Direction: all Outputs->Output Value: all 1's
5. Configure Timer 1: Timers->Timer1->Clock Value: 3.594kHz->Interrupt on: Timer1 Overflow->Val: 0xF8FB
6. Generate the C source, C project, and CodeWizardAVR project files by selecting:

   File|Generate, Save and Exit->

   Create new directory: C:\cvavr\led->

   Save: led.c->Save: led.prj->Save: led.cwp
7. Edit the C source code

(Reprinted courtesy of Pavel Haiduc and HP InfoTech S.R.L.)

8. View or modify the project configuration by selecting Project->Configure->

   After Make->Program the Chip
9. Compile the program by selecting:

   Project->Make
10. Automatically program the AT90S8515 chip on the STK500 starter kit:

    Apply power->Information->Program.

THE SOURCE CODE

```
/*********************************************
This program was produced by the
CodeWizardAVR V1.0.1.8c Standard
Automatic Program Generator
© Copyright 1998-2001
Pavel Haiduc, HP InfoTech S.R.L.
http://infotech.ir.ro
e-mail: hpinfotech@xnet.ro, hpinfotech@xmail.ro

Project :
Version :
Date    :
Author  :
Company :
Comments:


Chip type           : AT90S8515
Clock frequency     : 3.680000 MHz
Memory model        : Small
Internal SRAM size  : 512
External SRAM size  : 0
Data Stack size     : 128
*********************************************/

#include <90s8515.h>

// the LED 0 on PORTB will be on
unsigned char led_status=0xFE;

// Timer 1 overflow interrupt service routine
interrupt [TIM1_OVF] void timer1_ovf_isr(void)
{
// Reinitialize Timer's 1 value
TCNT1H=0xF8;
TCNT1L=0xFB;
// Place your code here
```

(Reprinted courtesy of Pavel Haiduc and HP InfoTech S.R.L.)

```
// move the LED
led_status<<=1;
led_status|=1;
if (led_status==0xFF) led_status=0xFE;
// turn on the LED
PORTB=led_status;
}

void main(void)
{
// Input/Output Ports initialization
// Port A
PORTA=0x00;
DDRA=0x00;

// Port B
PORTB=0xFF;
DDRB=0xFF;


// Port C
PORTC=0x00;
DDRC=0x00;

// Port D
PORTD=0x00;
DDRD=0x00;

// Timer/Counter 0 initialization
// Clock source: System Clock
// Clock value: Timer 0 Stopped
// Mode: Output Compare
// OC0 output: Disconnected
TCCR0=0x00;
TCNT0=0x00;

// Timer/Counter 1 initialization
// Clock source: System Clock
// Clock value: 3.594 kHz
// Mode: Output Compare
// OC1A output: Discon.
// OC1B output: Discon.
// Noise Canceler: Off
// Input Capture on Falling Edge
```

(Reprinted courtesy of Pavel Haiduc and HP InfoTech S.R.L.)

```
TCCR1A=0x00;
TCCR1B=0x05;
TCNT1H=0xF8;
TCNT1L=0xFB;
OCR1AH=0x00;
OCR1AL=0x00;
OCR1BH=0x00;
OCR1BL=0x00;

// External Interrupt(s) initialization
// INT0: Off
// INT1: Off
GIMSK=0x00;
MCUCR=0x00;

// Timer(s)/Counter(s) Interrupt(s) initialization
TIMSK=0x80;

// Analog Comparator initialization
// Analog Comparator: Off
// Analog Comparator Input Capture by Timer/Counter 1: Off
ACSR=0x80;

// Global enable interrupts
#asm("sei")

// the rest is done by TIMER1 overflow interrupts
while (1);
}
```

(Reprinted courtesy of Pavel Haiduc and HP InfoTech S.R.L.)

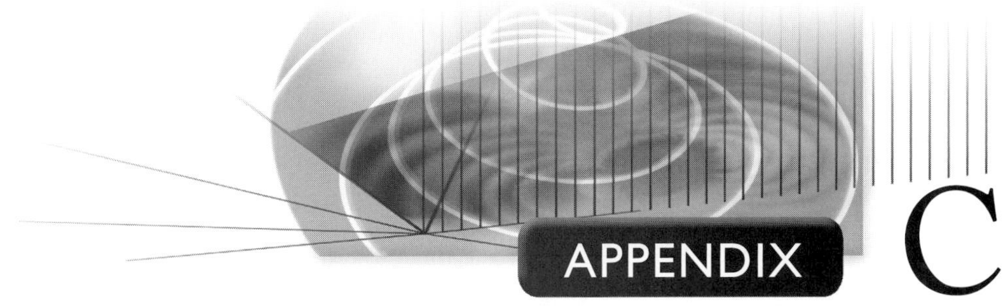

Programming the AVR Microcontrollers

The purpose of this appendix is to introduce the various programming methods available and to provide enough information to troubleshoot programming problems. The information given is not intended to be sufficient to enable the reader to create a programming device; see the microcontroller specification if you wish to create your own programmer.

All microcontrollers (and microcomputers of all types) require that all or part of the operating program be resident in the computer when it starts up. Microcontrollers and computers cannot do "nothing" (i.e., they must be executing code at all times when they are running). If they are in a "do-nothing" loop, they may be accomplishing nothing, but they are still executing the code that makes up the idle loop.

This means that the operating code must be permanently stored in a non-volatile section of memory. In the case of the Atmel AVR microcontrollers, this memory is composed of flash memory. And, assuming that you are not using an in-circuit emulator or a simulator, the code must be programmed into the flash memory in order for the code to be run.

Atmel provides on-board programming via the SPI port and via a parallel programming scheme. The SPI method is actually used almost exclusively to program the devices. The program bytes are applied to the SPI port in a specific sequence and are subsequently stored into the flash memory. This code is then executed the next time the microcontroller is reset. The huge advantage to SPI programming is that it almost always can be done in-circuit; the devices do not need to be removed from their application and programmed in a special programming device. This allows for easy field upgrades in the commercial world.

SPI PORT PROGRAMMING

As an example, consider the method of programming an AT90S8535, which is as follows:

1. After the microcontroller is powered up and running, hold RESET and SCLK low for at least 20 ms.
2. Send the *Programming Enable* command word into the microcontroller via the SPI port. During the transmission of the 3rd byte, 0x53 should be echoed back to the programming device. Finish the command by sending a dummy 4th byte.
3. Transmit the *Write Program Memory* command containing the address of the byte to be written and the byte itself (see Figure C.1 below).
4. Use the Read Program Memory command to poll the byte just sent. Read back the byte until the byte is correct (0xff will be sent back until the byte is programmed into the flash memory).
5. Repeat steps 3 and 4 until the entire program code is loaded into the flash memory.
6. Set RESET high to resume normal operation and to execute the code now loaded into the microcontroller.

Note: All of the programming command bytes require the transmission of 4 bytes to the microcontroller being programmed. All four must always be transmitted for synchronization purposes even though some may be dummy bytes.

Other operations may be executed on an AT90S8535 in a manner similar to the above by the selection of the appropriate command word from Figure C.1.

The actual SPI programming commands vary somewhat among the various Atmel AVR microcontrollers. Check the specification for your particular device for details.

COMMERCIAL PROGRAMMERS

There are a number of commercial programmers available to allow programming the flash code memory of the microcontrollers. Some, such as TheCableAVR made by Progressive Resources, LLC, use serial or USB communication to the PC. These typically use an Intel HEX file as their source for the code. These units have sophisticated PC software to control the programming process and include features such as automatic erase/program/verify sequencing. They also have additional electronics to convert the serial information into SPI-compatible communication.

A number of other devices also use PC software to program the microcontrollers, but they do so via the PC parallel port. These mostly manipulate the bits of the parallel port to simulate the action of the SPI bus in communicating with the microcontrollers. Devices made by Kanda Systems (STK 200+/300), Vogel Electronik (VTEK-ISP), and the programmer built into the Atmel STK500 development board are of this type. These latter programmers are also supported directly by CodeVisionAVR so that the microcontrollers may be programmed from within the CAVR IDE. These devices typically are somewhat slower and less sophisticated in their approach and flexibility than the serial device described above.

Command		**Command Bytes**				**Function**
		Byte 1	**Byte 2**	**Byte 3**	**Byte 4**	
Programming Enable	Send	10101100	01010011	xxxxxxxx	xxxxxxxx	Enable Programming Mode
	Rcv.	xxxxxxxx	xxxxxxxx	01010011	xxxxxxxx	
Chip Erase	Send	10101100	100xxxxx	xxxxxxxx	xxxxxxxx	Erase Flash and EEPROM
	Rcv.	xxxxxxxx	xxxxxxxx	xxxxxxxx	xxxxxxxx	
Read Program Memory—High Byte	Send	00101000	xxxxaaaa	aaaaaaaa	xxxxxxxx	Read High Byte from Program Memory word at aaaaaaaaaaaa
	Rcv.	xxxxxxxx	xxxxxxxx	xxxxxxx	bbbbbbbb	
Read Program Memory—Low Byte	Send	00100000	xxxxaaaa	aaaaaaaa	xxxxxxxx	Read Low Byte from Program Memory word at aaaaaaaaaaaa
	Rcv.	xxxxxxxx	xxxxxxxx	xxxxxxx	bbbbbbbb	
Write Program Memory—High Byte	Send	01001000	xxxxaaaa	aaaaaaaa	bbbbbbbb	Write High Byte to Program Memory word at aaaaaaaaaaaa
	Rcv.	xxxxxxxx	xxxxxxxx	xxxxxxx	xxxxxxxx	
Write Program Memory—Low Byte	Send	01000000	xxxxaaaa	aaaaaaaa	bbbbbbbb	Write Low Byte to Program Memory word at aaaaaaaaaaaa
	Rcv.	xxxxxxxx	xxxxxxxx	xxxxxxx	xxxxxxxx	
Read EEPROM Memory Byte	Send	10100000	xxxxxxxa	aaaaaaaa	xxxxxxxx	Read EEPROM Memory Byte at aaaaaaaaa
	Rcv.	xxxxxxxx	xxxxxxxx	xxxxxxxx	bbbbbbbb	
Write EEPROM Memory Byte	Send	11000000	xxxxxxxa	aaaaaaaa	bbbbbbbb	Write EEPROM Memory Byte at aaaaaaaaa
	Rcv.	xxxxxxxx	xxxxxxxx	xxxxxxxx	xxxxxxxx	

Notes:
1. a = address bit
   b = data bit
   x = don't care bit
2. There are also commands to poll and write the lock bits. See part specification for details.
3. "Chip Erase" requires a RESET and new "Programming Enable" after the erase is successful.

Figure C–1 *AT90S8535 SPI Programming Command Words*

BOOT LOADER PROGRAMMING

Some of the Atmel devices, such as the ATMega16, also allow for self programming via a *boot loader* program. These devices are capable of writing data to the flash code memory under the control of an on-board program called a *boot loader*.

The boot loader program must be written and stored in the boot loader section of flash code memory. This section of memory is protected from overwriting by setting the appropriate lock bits.

The advantage to a boot loader program is that it can be tailored to the application; i.e., if you want to be able to update the flash code memory from the serial port, then the boot loader can be written to accept data serially (formatted in some standard form such as an Intel HEX formatted file, for instance) and the data placed into flash memory.

To use a *boot loader*, the loader program must be programmed into the upper section of flash memory, called the *boot loader section*. Control bits are then set so that a RESET causes the *boot loader* to start watching the chosen input device (i.e., the serial port). A certain string of characters, followed by the program code, keys the boot loader to start re-programming the flash section of the device. Receipt of the wrong string of characters or the passage of a set amount of time without receipt of the control string will cause the boot loader to transfer control to the beginning of flash code memory to execute the stored program.

The advantage to a boot loader program is that it allows field reprogramming from almost any device wanted by tailoring the input device to suit the needs. For instance, some devices use the serial port and may be entirely reprogrammed by a PDA such as a Palm Pilot.

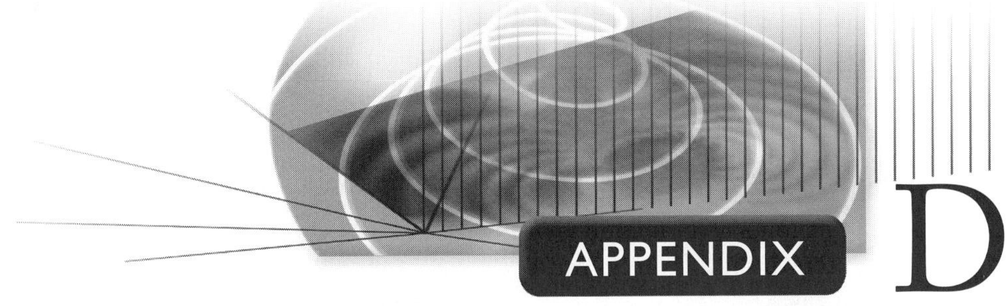

APPENDIX D

Installing and Using TheCableAVR

SYSTEM OVERVIEW

TheCableAVR is a tool designed to serially program the Atmel In-System Programmable microcontrollers. It consists of a Windows® application and a special downloading cable. By utilizing the specialized hardware in the downloading cable, the PC application allows the user to read, program, and verify many of the Atmel microprocessors that are in-system programmable. Figures D–1 and D–2 show TheCableAVR hardware and the user interface software respectively.

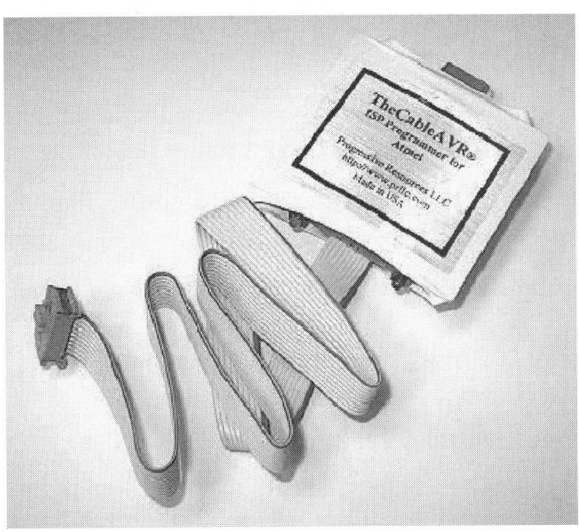

Figure D–1 *TheCableAVR Hardware*

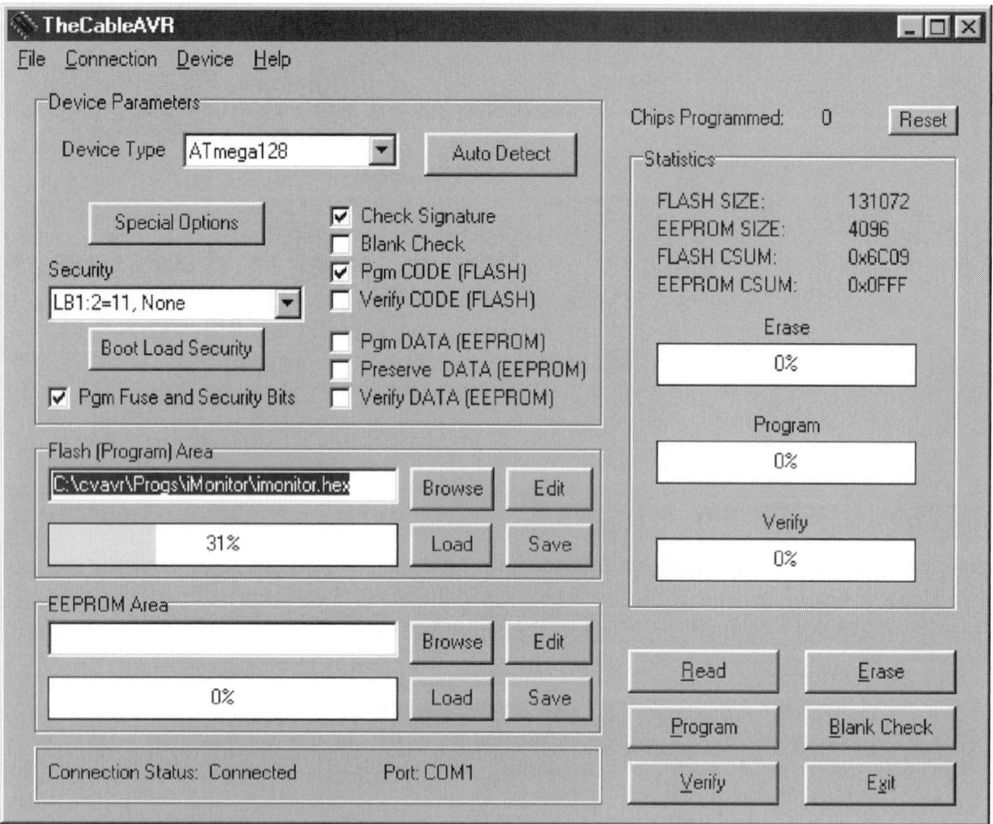

Figure D–2 *TheCableAVR Software*

SOFTWARE SYSTEM REQUIREMENTS

The minimum hardware and software requirements to ensure that the programmer PC application operates correctly are as follows:

- 100% IBM compatible 386DX or higher processor (minimum of 486 recommended)
- Windows 95® or higher
- Minimum 4MB RAM
- Minimum 1MB free hard disk space
- Spare PC serial port

GETTING STARTED

SOFTWARE INSTALLATION

1. Place the diskette labeled "Installation Diskette 1 of 2" in floppy drive (A: \ B:)
2. Select **Run** from the **Windows Start** menu.
3. Click **Browse**.
4. Navigate to the floppy drive (A: / B:) and click **Open**.
5. Click **OK** on the Run dialog box.
6. An introductory screen appears.
7. Follow the on-screen prompts to complete installation of the software.

On completion, the installation program installs the "TheCableAVR"' icon within a new program group called "TheCable."

To launch the software, double-click on the TheCableAVR icon.

HARDWARE INSTALLATION

- Plug TheCableAVR programmer into a serial port on the PC.
- Connect the IDC plug to the 10-pin IDC connector on the target board.
- Apply power to the target board. (TheCableAVR draws power from the target board.)
- Select **Connection** from the **File** pull-down menu in TheCableAVR software. In the pop-up window, select the appropriate COM port. Click **OK** to save the COM port selection and close the window.

To check the connection between the PC and the programming cable, open up the Connection pop-up window and click **Check**. If everything is connected correctly, a message pops up, "TheCableAVR programming cable is correctly connected."

If communication errors are encountered, make sure the target board is powered on (the power LED on the programmer should be lit). Also, ensure that no other software or hardware is conflicting with or using the selected COM port.

SOFTWARE OVERVIEW

TheCableAVR software allows the user to select the type of processor to be programmed, the areas(s) of the processor to program, and the security level to program. The software also allows the user to open and edit hex or binary files for the flash and/or data area of the processor to be programmed.

TheCableAVR software can also be called from the command line. This allows other applications to open and run TheCableAVR directly.

MAIN SCREEN

TheCableAVR software main screen can be divided into six sections: pull-down menus, Device Parameters frame, Flash (program) Area frame, EEPROM Area frame, Statistics frame, and control buttons group. Figure D–2 shows the main screen of TheCableAVR software.

PULL-DOWN MENUS

There are four main pull-down menus available: File, Connection, Device, and Help

File Menu:

- Open Project—Prompts the user to select a previously saved project to open and load.
- Save Project—Prompts the user to select a project filename to save the current settings under for future reference.
- Load Buffers—If a file is selected for the Flash and/or EEPROM buffers, load the file into the buffer.
- Export Project for Palm Pilot—Opens a window for downloading the project to TheCableAVR Companion software running on a Palm Pilot or similar device.
- Exit—Closes the application.

Connection Menu:

- Select Port—Opens the Connections pop-up window to allow the user to select a COM port.
- Check Connection—Attempts to communicate with TheCableAVR hardware. If communications are established, the message "TheCableAVR programming cable is correctly connected." appears to the user.
- Disconnect—Disconnect from the COM port. Upon the next command which requires communication with the hardware, the connection is re-established.

Device Menu:

- Read—Read the selected areas (flash and/or EEPROM) into the appropriate buffer and update the check sum field for all read buffers. Update the buffer filename to Flash_Buff or EEPROM_Buff to indicate that the contents of the buffer are from the processor and not a loaded file.
- Program—Program the processor according the options selected.
- Verify—Read and verify the processor for the areas selected (flash and/or EEPROM). If discrepancies are detected, the user is informed that differences exist and those differences are listed.
- Erase—Perform a Chip Erase operation on the processor.

- Blank Check—Read the flash and EEPROM areas of the chip. If all memory locations contain 0xFF, then inform the user that the chip is blank. If not, inform the user that the chip is not blank.
- Reset—Perform a Reset operation on the processor.
- Read Security—Read the current security level of the processor and display the level to the user.
- Write Security—Write the selected security level to the processor.

Help Menu:

- About TheCableAVR—Display software version and company information.
- Get Cable Version—Interrogate TheCableAVR programming cable hardware for the version number and display the number to the user. It may be necessary to call this twice to get the version number. Occasionally, the communications are out of sequence and the version reads as 0.0.

DEVICE PARAMETERS

The Device Parameters section allows the user to select the type of processor to perform operations on and which areas of the processor to include. The user can also select how the processor is handled during those operations, including whether or not to check the signature bytes and whether or not to blank check the processor during a program operation. There are seven fields (Device Type, Security, Check Signature, Blank Check, Pgm CODE, Pgm DATA, and Preserve DATA) and two command buttons (Auto Detect and Special Options) in the Device Parameters section.

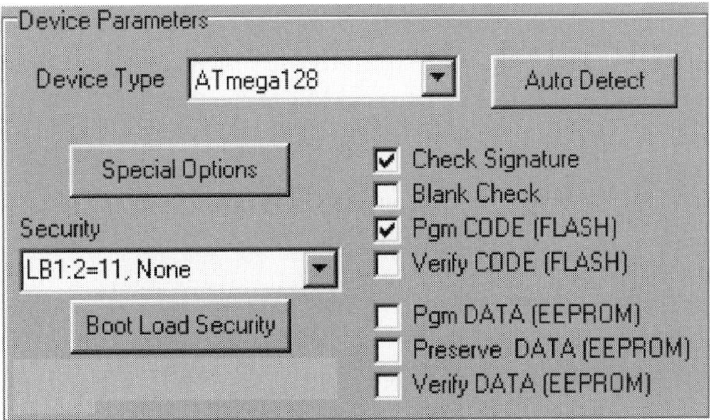

Figure D–3 *Device Parameters Fields*

DEVICE TYPE

Use this drop-down list to select the type of target processor. Selecting a processor automatically clears the flash and EEPROM buffers and reset the Statistics fields. If there are Special Options available for the processor, this command button is enabled. The Auto Detect command button automatically selects the processor type for the user if the processor is not protected.

AUTO DETECT

The Auto Detect command button automatically selects the processor type. When the Auto Detect command button is clicked, TheCableAVR reads the processor's signature bytes and determines the type of processor. The appropriate Device Type list item is then selected and the Statistics fields are updated accordingly.

Note: If the processor is read protected (security fuse bits enabled), then invalid signature bytes are returned and the type of processor cannot be determined. If this happens, a message is returned, "Device is Locked. Unable to determine type of device," and the Device Type field is not changed.

SECURITY

The Security drop-down list allows the user to select from three levels of security. These correspond to the lock bits on the processors.

- Level 1—No lock bits programmed; the device can be read and written to.
- Level 2—LB1 programmed, LB2 not programmed; further programming of the Flash and EEPROM is disabled.
- Level 3—LB1 and LB2 programmed; further programming and reading are both disabled for Flash and EEPROM.

The Security level can be read from or written to the processor at any time by using the Device pull-down menu items Read Security and Write Security.

SPECIAL OPTIONS

The Special Options command button is enabled for processors that have fuse bits that can be programmed. When it is clicked, a Special Options window is displayed to allow the user to set up the fuse bits for that particular processor.

CHECK SIGNATURE

When the Check Signature option is checked, TheCableAVR checks the signature bytes before performing a program, read, or verify on the processor. If the signature bytes do not match the selected processor, the user is given a warning message "Device Signature Error" and the selected operation is aborted.

BLANK CHECK

When the Blank Check option is checked, TheCableAVR automatically performs a blank check after the chip erase during the programming operation. If the processor did not completely erase, the user is issued a warning, "Flash (or EEPROM) Area is Not Blank," and the programming operation is aborted.

PGM CODE (FLASH)

The Pgm CODE (FLASH) option selects/deselects the flash area of the processor for programming, reading, and blank checking operations. Each of these operations is performed on the flash area only if the Pgm CODE (FLASH) option is selected. (If this option is not checked, the operation skips the flash area.)

PGM DATA (EEPROM)

The Pgm DATA (EEPROM) option selects/deselects the EEPROM area of the processor for programming, reading, and blank checking operations. Each of these operations is performed on the EEPROM area only if Pgm DATA (EEPROM) option is selected. (If this option is not checked, the operation skips the EEPROM area.)

Note: If the Preserve DATA (EEPROM) option is selected, the Pgm DATA (EEPROM) option is selected and disabled. Part of the Preserve Data operation is to program the EEPROM area, so this option has to be selected.

PRESERVE DATA (EEPROM)

In order to program the flash area of the processor, a chip erase must first be performed. This erases the contents of both the flash and the EEPROM areas. In some cases, it is desirable to preserve the current EEPROM data in the processor. When the Preserve DATA (EEPROM) option is selected, the EEPROM area of the processor is read and stored in the EEPROM buffer (of the PC application) before the chip erase is performed. After the flash area is programmed and verified, the EEPROM area is programmed from the EEPROM buffer.

Note: If the Pgm CODE (FLASH) option is not selected, this option is not available (because a chip erase is not performed during the program cycle).

FLASH (PROGRAM) AREA

The Flash (Program) Area section has two main parts—the buffer information boxes and the command buttons (see Figure D–4). The buffer information boxes tell what file (if any)

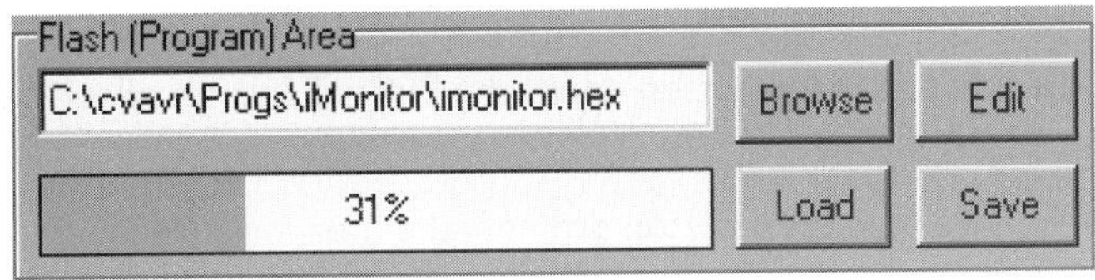

Figure D–4 *Flash (Program) Area*

has been selected for the buffer and whether or not that file has been loaded into the buffer. The command buttons control the file selection, loading the buffer from the file, editing the buffer, and saving the buffer to a file.

- Browse—The Browse command button prompts the user to select a file to load to the buffer. Once a file is selected, the filename is displayed in the edit field to the left of the Browse button.
- Load—The Load command button prompts the user to load the selected file into the buffer. The progress bar below the filename edit field shows how much of the available memory is used by the buffer.
- Edit—The Edit command button opens the Buffer Editor to allow the user to edit the contents of the buffer.
- Save—The Save command button prompts the user for a filename to save the current buffer contents under.

EEPROM AREA

The EEPROM Area section has two main parts—the buffer information boxes and the command buttons (see Figure D–5). The buffer information boxes tell what file (if any) has been selected for the buffer and whether or not that file has been loaded into the buffer. The command buttons control the file selection, loading the buffer from the file, editing the buffer, and saving the buffer to a file.

Figure D–5 *EEPROM Area*

- Browse—The Browse command button prompts the user to select a file to load to the buffer. Once a file is selected, the filename is displayed in the edit field to the left of the Browse button.
- Load—The Load command button prompts the user to load the selected file into the buffer. The progress bar below the filename edit field shows how much of the available memory is used by the buffer.
- Edit—The Edit command button opens the Buffer Editor to allow the user to edit the contents of the buffer.
- Save—The Save command button prompts the user for a filename to save the current buffer contents under.

STATISTICS

The Statistics section displays information about the processor selected, the flash and EEPROM buffers, and the progress of the operation being performed. See Figure D–6.

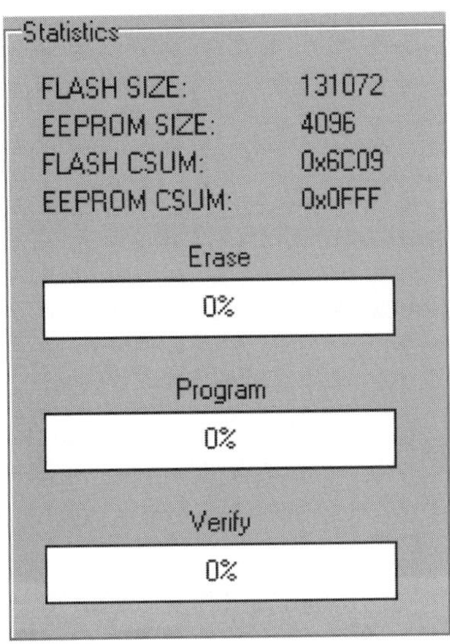

Figure D–6 *Statistics Fields*

```
FLASH SIZE:   Size of the flash area of the processor selected
EEPROM SIZE:  Size of the EEPROM area of the processor selected
FLASH CSUM:   Checksum of the flash buffer contents
EEPROM CSUM:  Checksum of the EEPROM buffer contents
```

The progress bars represent the percent completion of the operation being performed. The operation being performed is indicated by a caption above the progress bar.

COMMAND BUTTONS

There are six command buttons available from the main window: Read, Program, Verify, Erase, Blank Check, and Exit.

- Read—Read the selected areas (flash and/or EEPROM) into the appropriate buffer and update the check sum field for all read buffers. Update the buffer filename to Flash_Buff or EEPROM_Buff to indicate that the contents of the buffer are from the processor and not a loaded file.
- Program—Program the processor according the options selected.
- Verify—Read and verify the processor for the areas selected (flash and/or EEPROM). If discrepancies are detected, the user is informed that differences exist and those differences are listed.

- Erase—Perform a Chip Erase operation on the processor.
- Blank Check—Read the flash and EEPROM areas of the chip. If all memory locations contain 0xFF, then inform the user that the chip is blank. If not, inform the user that the chip is not blank.
- Exit—Close TheCableAVR application.

EDIT BUFFER SCREEN

The Edit Buffer screen is opened when the Edit command button is clicked for either the Flash or the EEPROM Area. The contents of the appropriate buffer (indicated in the window title bar) are displayed in this window for viewing and editing.

The starting address of each line is displayed in the left frame. The data of the buffer is displayed in the center frame in hexadecimal format. The right frame displays the data as ASCII characters. The selected value is highlighted in blue or a dashed outline. The dashed outline indicates that the value will be edited. To select a value, click the value with the mouse and type the new value in the field. The values can be edited as either hexadecimal values (by editing in the center frame) or ASCII characters (by editing in the right frame). The scroll bar at the right scrolls through the contents of the entire buffer.

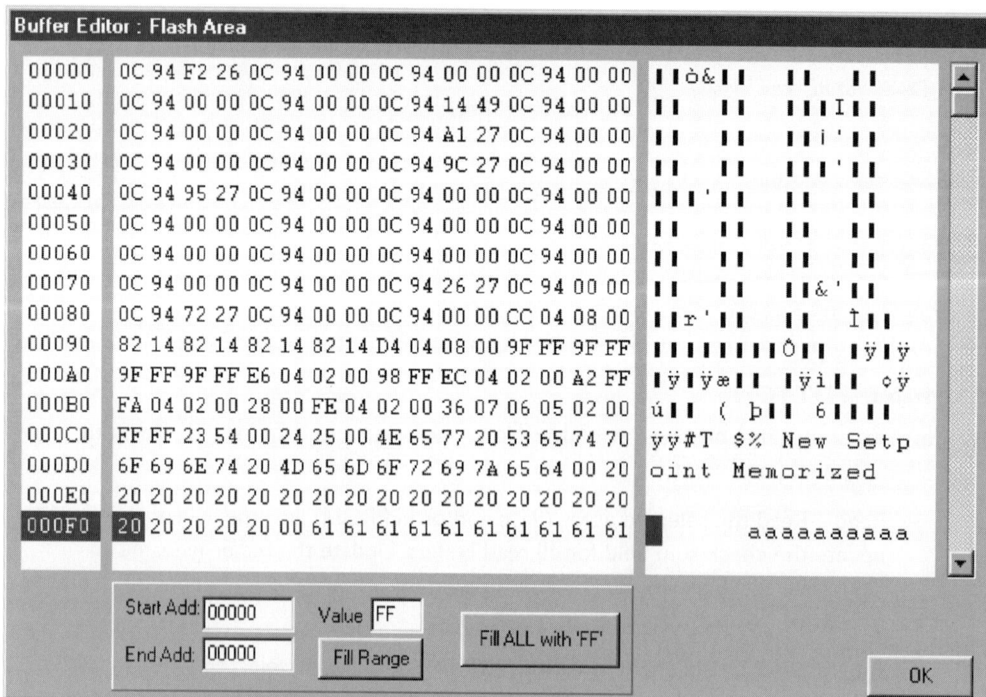

Figure D–7 *Buffer Editor Screen*

To fill a range with a particular value, edit the Start Add, End Add, and Value fields at the bottom of the window and click **Fill Range.** This fills the range of values from Start Add to End Add with Value. To fill the entire buffer with 0xFF, click **Fill ALL with 'FF'**. All values in this section should be entered as hexadecimal numbers.

Click **OK** to close the Edit Buffer window.

Note: Contents of the selected buffer are modified as the changes are made. If the user is unsure of the changes he is making, he should back up his file BEFORE modifying the buffer.

COM PORT SETUP SCREEN

The COM Port Setup Screen allows the user to select the COM port for connection to TheCableAVR hardware. Once the port is selected, the hardware is connected, and power has been applied to the target board, the user may click **Check** to check the setup. If everything is correctly set up, the message "TheCableAVR programming cable is correctly connected" appears to the user. Click **OK** to close the COM Port Setup screen and save the COM port setup.

ABOUT SCREEN

The About Screen simply displays software version information and contact information for Progressive Resources LLC.

COMMAND LINE SUPPORT

The command line support expects particular parameters to be passed when the application is called. These are:

- Project Filename (if not in the same directory as the executable, include the path).
- User Present Flag =

  1 User is present and message box will appear if an error occurs.

  2 User is not present and return value will have to be used to determine if the programming was successful.

An example with a user present would be

```
"C:\Program Files\Progressive Resources\TheCableAVR\TheCableAVR.exe"
"C:\PROJ FILES\PROJ.isp" "1"
```

An example without a user present would be

```
"C:\Program Files\Progressive Resources\TheCableAVR\TheCableAVR.exe"
"C:\PROJ FILES\PROJ.isp" "0"
```

The following values are returned by the application upon closing:

11	Unable to preserve EEPROM data due to locked device
10	Unable to program just the EEPROM due to locked device
9	Unable to program the security bits
8	Unable to program the fuse bits
5	EEPROM not blank after chip erase
4	Flash not blank after chip erase
3	Signature does not match selected device
2	Bad verify of flash or EEPROM after program
1	Unable to successfully communicate with TheCableAVR hardware
0	Program of device successfully completed
–1	Unable to successfully communicate with TheCableAVR hardware
–2	User cancelled programming of device
–3	Unable to open selected project file
–4	Unable to open either the Flash or the EEPROM file
–5	Flash file too large for selected device
–6	EEPROM file too large for the selected device
–7	Bad checksum (hex or eep files only) in Flash or EEPROM file

HARDWARE OVERVIEW

TheCableAVR hardware consists of a programming cable that has a 9-pin serial port connection on one end and a 10-pin IDC plug on the other. There are two indicator LEDs on the cable, Power and Busy. The green Power LED is lit any time the cable has power (applied through the target system). The red Busy LED is lit whenever the cable is holding the target system in reset to perform some programming operation (Read, Program, Verify, etc.).

PIN ASSIGNMENTS

The 10-pin IDC header on the target system should be pinned out as follows:

Pin	Name	Description	I/O
1	Vcc	Programmer Power (+Vcc)	-
2	NC	No Connect	-
3	NC	No Connect	-
4	MOSI	SPI—Master Out Slave In	O
5	NC	No Connect	-
6	MISO	SPI—Master In Slave Out	I
7	GND	Programmer GND	-
8	SCK1	SPI—Serial Clock	O
9	GND	Programmer GND	-
10	RESET	Target RESET Control Pin	O

Table D–1 *TheCableAVR Software*

Figure D–8 Shows a top view of the IDC header. Connector shown is the male 10-pin IDC connector as viewed from above the component side of the target board.

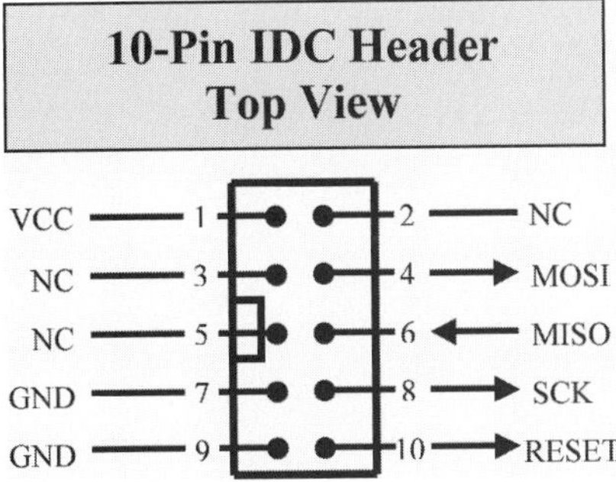

Figure D–8 *TheCableAVR 10-Pin IDC Header*

TARGET SYSTEM SETUP

For TheCableAVR programmer to work correctly, certain conditions must be met within the target system.

SPI ENABLE FUSE

The SPI Enable Fuse (SPIEN) must be ENABLED on the target processor in order for TheCableAVR programmer to work correctly. The SPIEN fuse can only be read/programmed in parallel programming mode. The default for the processors is SPIEN enabled.

RESET CIRCUIT

The serial programming mode of the Atmel AVR devices is initiated by setting the RESET pin to the appropriate RESET state. So, in order to place the target processor in serial programming mode, TheCableAVR hardware must be able to externally control the state of the RESET pin.

POWER

TheCableAVR hardware is powered from the target application. Pins 1 and 9 are VCC (+5 VDC) and GND, respectively.

TARGET SYSTEM OSCILLATOR

The target system oscillator must be running for TheCableAVR to access the target processor. The frequency must be an acceptable frequency for the target processor. The acceptable frequencies can be found in the microcontroller data sheets.

Note: The oscillator could be an external crystal/resonator or an internal RC oscillator.

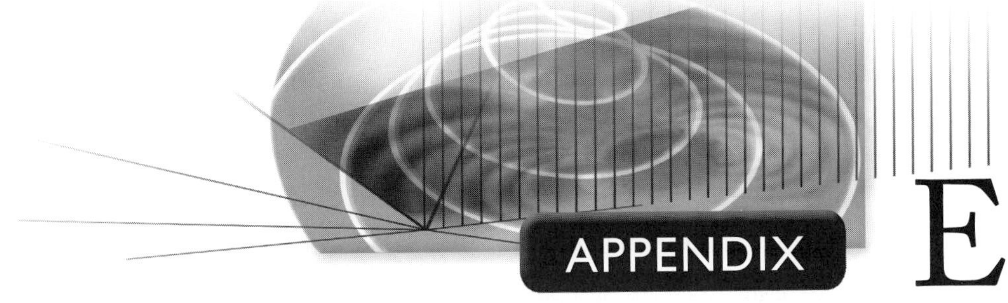

APPENDIX E

The MegaAVR-DEV Development Board

The **MegaAVR-Dev** development board by Progressive Resources LLC, Indianapolis, Indiana, is designed for prototyping and laboratory use. The board supports the AT90S8535, Mega16, Mega163, Mega32, or the Mega323 in a 40-pin DIP package.

Figure E–1 *MegaAVR-Dev Development Board*

The features of the development board include the following:

- Comes standard with an ATMega16-8PC processor.
- Comes preprogrammed with boot loading firmware. No other programming hardware is required. Just connect the MegaAVR-Dev to your PC with an RS232 cable and run the "AVRBL Boot Loader" Windows application (available free from: http://www.prllc.com/products/BootApp.zip).

- RS-232 through a 9-Pin D shell as well as screw terminals and a jumper header.
- Up to 32K of in-system programmable FLASH memory with up to 1K of EEPROM and up to 1K of Internal RAM (depending on processor selection).
- Eight 10-bit analog inputs, using either internal or user-supplied reference.
- Nine I/O-controlled LEDs, eight of which are jumper selectable.
- 32 kHz "watch" crystal for on-board real-time operations.
- A universal clock socket allows for "canned oscillators," as well as a variety of crystals, ceramic resonators, and passive terminations.
- 0.1" centered headers provide for simple connection to the processor special function pins and I/O.
- 10-pin, polarized, ISP and JTAG connections are provided for in-system programming and debugging. MegaAVR-Devs can also be programmed through RS-232 using an appropriate boot loader application.
- On-board regulation allows for power inputs from 8 to 38V DC with an LED power indicator.
- Termination is provided for 5V DC output at 250 mA.

For more information refer to the Progressive Resources web site http://www.prllc.com.

SPECIFICATIONS

Voltage range	8 V to 38 V DC
Power consumption	250 mW (nominal)
Dimensions	W 3.7 inches × H 3.7 inches
Mounting	Rubber feet, 4 places
Weight	~3 oz.
Operating temperature	0°C to +60°C
Storage temperature	0°C to +85°C
Humidity	0% to 95% at +50°C (non-condensing)

Table E–1 *MegaAVR-Dev Specification*

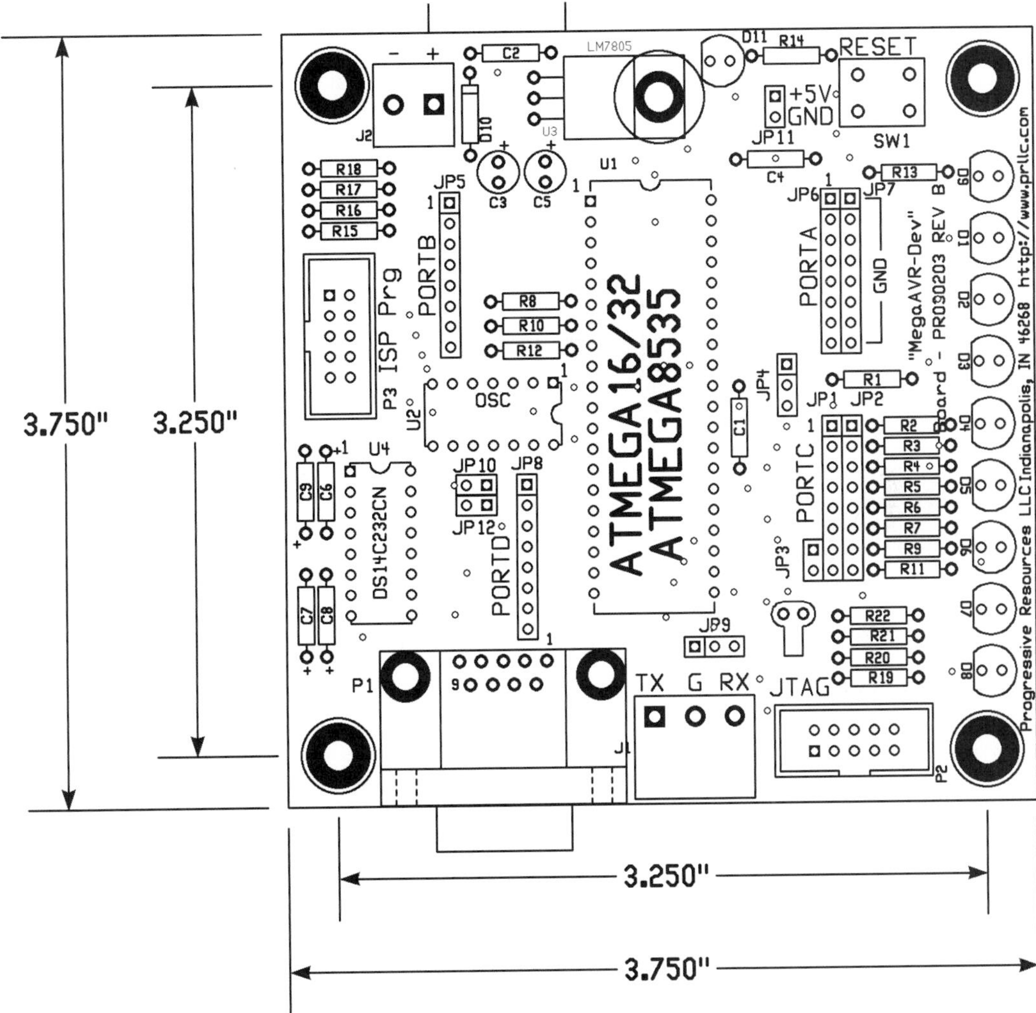

Figure E–2 *Connections, Headers, and Jumpers*

SCHEMATIC

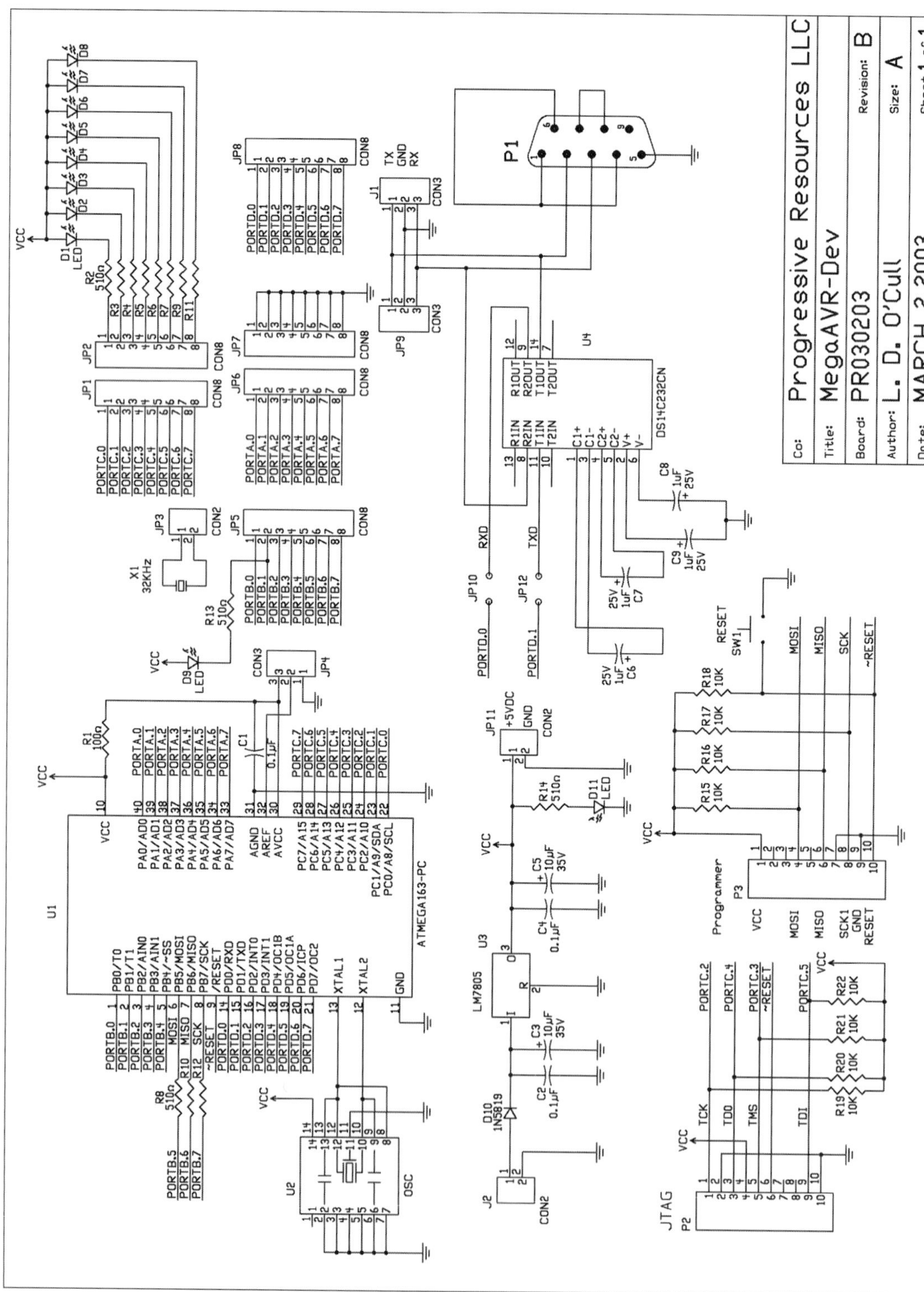

Figure E-3 *MegaAVR-Dev Schematic*

F

ASCII Table

American Standard Code for Information Interchange

Decimal	Octal	Hex	Binary	Value
000	000	000	00000000	NUL
001	001	001	00000001	SOH
002	002	002	00000010	STX
003	003	003	00000011	ETX
004	004	004	00000100	EOT
005	005	005	00000101	ENQ
006	006	006	00000110	ACK
007	007	007	00000111	BEL
008	010	008	00001000	BS
009	011	009	00001001	HT
010	012	00A	00001010	LF
011	013	00B	00001011	VT
012	014	00C	00001100	FF
013	015	00D	00001101	CR
014	016	00E	00001110	SO
015	017	00F	00001111	SI
016	020	010	00010000	DLE
017	021	011	00010001	DC1
018	022	012	00010010	DC2
019	023	013	00010011	DC3
020	024	014	00010100	DC4
021	025	015	00010101	NAK
022	026	016	00010110	SYN
023	027	017	00010111	ETB

Decimal	Octal	Hex	Binary	Value
024	030	018	00011000	CAN
025	031	019	00011001	EM
026	032	01A	00011010	SUB
027	033	01B	00011011	ESC
028	034	01C	00011100	FS
029	035	01D	00011101	GS
030	036	01E	00011110	RS
031	037	01F	00011111	US
032	040	020	00100000	SP(Space)
033	041	021	00100001	!
034	042	022	00100010	"
035	043	023	00100011	#
036	044	024	00100100	$
037	045	025	00100101	%
038	046	026	00100110	&
039	047	027	00100111	'
040	050	028	00101000	(
041	051	029	00101001	)
042	052	02A	00101010	*
043	053	02B	00101011	+
044	054	02C	00101100	,
045	055	02D	00101101	-
046	056	02E	00101110	.
047	057	02F	00101111	/
048	060	030	00110000	0
049	061	031	00110001	1
050	062	032	00110010	2
051	063	033	00110011	3
052	064	034	00110100	4
053	065	035	00110101	5
054	066	036	00110110	6
055	067	037	00110111	7
056	070	038	00111000	8
057	071	039	00111001	9
058	072	03A	00111010	:
059	073	03B	00111011	;
060	074	03C	00111100	<

Decimal	Octal	Hex	Binary	Value
061	075	03D	00111101	=
062	076	03E	00111110	>
063	077	03F	00111111	?
064	100	040	01000000	@
065	101	041	01000001	A
066	102	042	01000010	B
067	103	043	01000011	C
068	104	044	01000100	D
069	105	045	01000101	E
070	106	046	01000110	F
071	107	047	01000111	G
072	110	048	01001000	H
073	111	049	01001001	I
074	112	04A	01001010	J
075	113	04B	01001011	K
076	114	04C	01001100	L
077	115	04D	01001101	M
078	116	04E	01001110	N
079	117	04F	01001111	O
080	120	050	01010000	P
081	121	051	01010001	Q
082	122	052	01010010	R
083	123	053	01010011	S
084	124	054	01010100	T
085	125	055	01010101	U
086	126	056	01010110	V
087	127	057	01010111	W
088	130	058	01011000	X
089	131	059	01011001	Y
090	132	05A	01011010	Z
091	133	05B	01011011	[
092	134	05C	01011100	\
093	135	05D	01011101	]
094	136	05E	01011110	^
095	137	05F	01011111	_
096	140	060	01100000	`
097	141	061	01100001	a

Decimal	Octal	Hex	Binary	Value
098	142	062	01100010	b
099	143	063	01100011	c
100	144	064	01100100	d
101	145	065	01100101	e
102	146	066	01100110	f
103	147	067	01100111	g
104	150	068	01101000	h
105	151	069	01101001	i
106	152	06A	01101010	j
107	153	06B	01101011	k
108	154	06C	01101100	l
109	155	06D	01101101	m
110	156	06E	01101110	n
111	157	06F	01101111	o
112	160	070	01110000	p
113	161	071	01110001	q
114	162	072	01110010	r
115	163	073	01110011	s
116	164	074	01110100	t
117	165	075	01110101	u
118	166	076	01110110	v
119	167	077	01110111	w
120	170	078	01111000	x
121	171	079	01111001	y
122	172	07A	01111010	z
123	173	07B	01111011	{
124	174	07C	01111100	|
125	175	07D	01111101	}
126	176	07E	01111110	~
127	177	07F	01111111	DEL

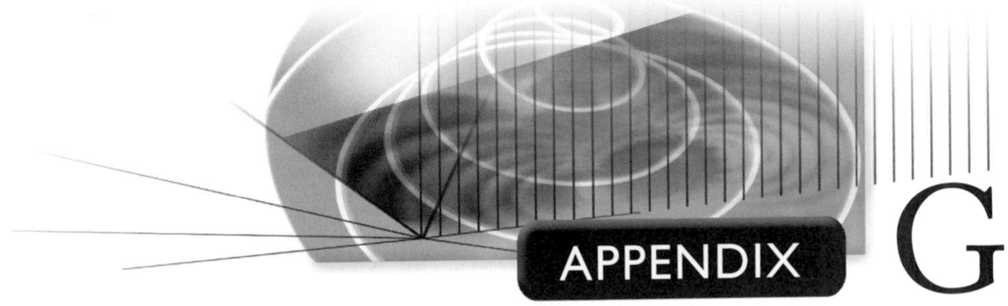

AVR Instruction Set Summary

Appendix G

AVR Instruction Set Summary

The complete data on the AVR processors can be found at Atmel's website: http://www.atmel.com

AVR Instruction Set Standard Nomenclature:

Status Register (SREG)	Symbols	Registers and Operands:	I/O Registers
C: Carry Flag	+ Add	Rd: Destination (and Source) Register	RAMPX,RAMPY,RAMPZ: Registers concatenated with X,Y, and Z to form indirect addresses > 64K
Z: Zero Flag	& AND	Rr: Source Register	
N: Negative Flag	<- Assignment	R: Result from instruction execution	
V: Twos complement overflow indicator	= Equal	K: Constant literal or data byte (8 bit)	
S: N ^ V Flags for signed tests	^ Exclusive OR	k: Constant address data for program counter	**Stack**
H: Half Carry Flag	* Multiply	b: Bit in register file or I/O register (3 bit)	
T: Transfer bit used by BLD and BST instructions	| OR	s: Bit in status register (SREG) (3 bit)	STACK: Stack for return address and pushed registers
I: Global interrupt enable/disable flag	- Subtract	X,Y,X: Indirect address register	
		P: I/O port address	
		q: Displacement for direct addressing (6 bit)	SP: Stack Pointer to the STACK

Mnemonics	Operands	Description	Operation	SREG Z	C	N	V	H	S	T	I	#Clocks
ARITHMETIC AND LOGIC INSTRUCTIONS												
ADC	Rd,Rr	Add wlth Carry two Registers	Rd <- Rd + Rr + C	•	•	•	•	•				1
ADD	Rd,Rr	Add two Registers	Rd <- Rd + Rr	•	•	•	•	•				1
ADIW	Rdl, K	Add Immediate to Word	Rdh:Rdl <- Rdh:Rdl + K	•	•	•	•		•			2
AND	Rd, Rr	Logical AND Registers	Rd.<- Rd & Rr	•		•	•					1
ANDI	Rd, K	Logical AND Register and Constant	Rd.<- Rd & K	•		•	•					1
CBR	Rd, K	Clear Bit(s) in Register	Rd <- Rd & ($FF - K)	•		•	•					1
CLR	Rd	Clear Register	Rd <- Rd ^ Rd	•		•	•					1
COM	Rd	One's Complement	Rd <- $FF - Rd	•	•	•	•					1
DEC	Rd	Decrement	Rd <- Rd - 1	•		•	•					1
EOR	Rd, Rr	Exclusive OR Registers	Rd <- Rd ^ Rr	•		•	•					1
FMUL	Rd, Rr	Fractional Multiply Unsigned	R1:R0 <- (Rd X Rr) << 1	•	•							2
FMULS	Rd, Rr	Fractional Multiply Signed	R1:R0 <- (Rd X Rr) << 1	•	•							2
FMULSU	Rd, Rr	Fractional Multiply Signed with Unsigned	R1:R0 <- (Rd X Rr) << 1	•	•							2
INC	Rd	Increment	Rd <- Rd + 1	•		•	•					1
MUL	Rd, Rr	Multiply Unsigned	R1:R0 <- Rd X Rr	•	•							2
MULS	Rd, Rr	Multiply Signed	R1:R0 <- Rd X Rr	•	•							2
MULSU	Rd, Rr	Multiply Signed with Unsigned	R1:R0 <- Rd X Rr	•	•							2
NEG	Rd	Two's Complement	Rd <- $00 - Rd	•	•	•	•	•				1
OR	Rd, Rr	Logical OR Registers	Rd.<- Rd | Rr	•		•	•					1
ORI	Rd, K	Logical OR (Immediate) Register and Constant	Rd.<- Rd | K	•		•	•					1
SBC	Rd, Rr	Subtract with Carry two Registers	Rd <- Rd - Rr - C	•	•	•	•	•				1
SBCI	Rd, K	Subtract with Carry Constant from Register	Rd <- Rd - K - C	•	•	•	•	•				1
SBIW	Rdl, K	Subtract Immediate to Word	Rdh:Rdl <- Rdh:Rdl - K	•	•	•	•		•			2
SBR	Rd, K	Set Bit(s) in Register	Rd <- Rd | K	•		•	•					1
SER	Rd	Set Register	Rd <- $FF									1
SUB	Rd, Rr	Subtract two Registers	Rd <- Rd - Rr	•	•	•	•	•				1
SUBI	Rd, K	Subtract (Immediate) Constant from Register	Rd <- Rd - K	•	•	•	•	•				1
TST	Rd	Test for Zero or Minus	Rd <- Rd & Rd	•		•	•					1

Appendix G

AVR Instruction Set Summary

The complete data on the AVR processors can be found at Atmel's website: http://www.atmel.com

AVR Instruction Set Standard Nomenclature:

Status Register (SREG)	Symbols	Registers and Operands:	I/O Registers
C: Carry Flag	+ Add	Rd: Destination (and Source) Register	RAMPX,RAMPY,RAMPZ: Registers concatenated with X,Y, and Z to form indirect addresses > 64K
Z: Zero Flag	& AND	Rr: Source Register	
N: Negative Flag	<- Assignment	R: Result from instruction execution	
V: Twos complement overflow indicator	= Equal	K: Constant literal or data byte (8 bit)	
S: N ^ V Flags for signed tests	^ Exclusive OR	k: Constant address data for program counter	**Stack**
H: Half Carry Flag	* Multiply	b: Bit in register file or I/O register (3 bit)	
T: Transfer bit used by BLD and BST instructions	| OR	s: Bit in status register (SREG) (3 bit)	STACK: Stack for return address and pushed registers
I: Global interrupt enable/disable flag	- Subtract	X,Y,X: Indirect address register	
		P: I/O port address	
		q: Displacement for direct addressing (6 bit)	SP: Stack Pointer to the STACK

Mnemonics	Operands	Description	Operation	SREG Z	C	N	V	H	S	T	I	#Clocks
BRANCH INSTRUCTIONS												
BRBC	s, k	Branch if Status Flag Cleared	if (SREG(s) =0) then PC <- PC + k + 1									1 / 2
BRBS	s, k	Branch if Status Flag Set	if (SREG(s) =1) then PC <- PC + k + 1									1 / 2
BRCC	k	Branch if Carry Cleared	If (C = 0) then PC <- PC + k + 1									1 / 2
BRCS	k	Branch if Carry Set	If (C = 1) then PC <- PC + k + 1									1 / 2
BREQ	k	Branch if Equal	If (Z = 1) then PC <- PC + k + 1									1 / 2
BRGE	k	Branch if Greater or Equal, Signed	If (N ^ V = 0) then PC <- PC + k + 1									1 / 2
BRHC	k	Branch If Half Carry Flag Cleared	If (H = 0) then PC <- PC + k + 1									1 / 2
BRHS	k	Branch If Half Carry Flag Set	If (H = 1) then PC <- PC + k + 1									1 / 2
BRID	k	Branch if Interrupt Disabled	If (I = 0) then PC <- PC + k + 1									1 / 2
BRIE	k	Branch if Interrupt Enabled	If (I = 1) then PC <- PC + k + 1									1 / 2
BRLO	k	Branch if Lower	If (C = 1) then PC <- PC + k + 1									1 / 2
BRLT	k	Branch if Less Than Zero, Signed	If (N ^ V = 1) then PC <- PC + k + 1									1 / 2
BRMI	k	Branch if Minus	If (N = 1) then PC <- PC + k + 1									1 / 2
BRNE	k	Branch if Not Equal	If (Z = 0) then PC <- PC + k + 1									1 / 2
BRPL	k	Branch if Plus	If (N = 0) then PC <- PC + k + 1									1 / 2
BRSH	k	Branch if Same or Higher	If (C = 0) then PC <- PC + k + 1									1 / 2
BRTC	k	Branch if Flag Cleared	If (T = 0) then PC <- PC + k + 1									1 / 2
BRTS	k	Branch if Flag Set	If (T = 1) then PC <- PC + k + 1									1 / 2
BRVC	k	Branch if Overflow Flag is Cleared	If (V = 0) then PC <- PC + k + 1									1 / 2
BRVS	k	Branch if Overflow Flag is Set	If (V = 1) then PC <- PC + k + 1									1 / 2
CALL	k	Direct Subroutine Call	PC <- k									4
CP	Rd, Rr	Compare	Rd - Rr	•	•	•	•	•				1
CPC	Rd, Rr	Compare with Carry	Rd - Rr - C	•	•	•	•	•				1
CPI	Rd, K	Cornpare Register with Immediate	Rd - K	•	•	•	•	•				1
CPSE	Rd, Rr	Compare, Skip if Equal	if (Rd = Rr) PC <- PC + 2 or 3									1 / 2 / 3
ICALL		Indirect Call to (Z)	PC <- Z									3
IJMP		Indirect Jump to (Z)	PC <- Z									2
JMP	k	Direct Jump	PC <- k									3
RCALL	k	Relative Subroutine Call	PC <- PC + k + 1									3
RET		Subroutine Return	PC <- STACK									4
RETI		Interrupt Return	PC <- STACK								•	4
RJMP	k	Relative Jump	PC <- PC + k + 1									2
SBIC	P, b	Skip if Bit in I/O Register Cleared	If (P(b)=0) PC <- PC + 2 or 3									1 / 2 / 3
SBIS	P, b	Skip if Bit in I/0 Register Is Set	If (P(b)=1) PC <- PC + 2 or 3									1 / 2 / 3
SBRC	Rr, b	Skip if Bit in Register Cleared	If (Rr(b)=0) PC <- PC + 2 or 3									1 / 2 / 3
SBRS	Rr, b	Skip if Bit In Register is Set	If (Rr(b)=1) PC <- PC + 2 or 3									1 / 2 / 3

Appendix G

AVR Instruction Set Summary

The complete data on the AVR processors can be found at Atmel's website: http://www.atmel.com

AVR Instruction Set Standard Nomenclature:

Status Register (SREG)	Symbols	Registers and Operands:	I/O Registers
C: Carry Flag	+ Add	Rd: Destination (and Source) Register	RAMPX,RAMPY,RAMPZ: Registers concatenated with X,Y, and Z to form indirect addresses > 64K
Z: Zero Flag	& AND	Rr: Source Register	
N: Negative Flag	<- Assignment	R: Result from instruction execution	
V: Twos complement overflow indicator	= Equal	K: Constant literal or data byte (8 bit)	
S: N ^ V Flags for signed tests	^ Exclusive OR	k: Constant address data for program counter	**Stack**
H: Half Carry Flag	* Multiply	b: Bit in register file or I/O register (3 bit)	
T: Transfer bit used by BLD and BST instructions	| OR	s: Bit in status register (SREG) (3 bit)	STACK: Stack for return address and pushed registers
I: Global interrupt enable/disable flag	- Subtract	X,Y,X: Indirect address register	
		P: I/O port address	
		q: Displacement for direct addressing (6 bit)	SP: Stack Pointer to the STACK

Mnemonics	Operands	Description	Operation	SREG Z	C	N	V	H	S	T	I	#Clocks
DATA TRANSFER INSTRUCTIONS												
ELPM		Extended Load Program Memory	R0 <- (RAMPZ:Z)									3
ELPM	Rd, Z	Extended Load Program Memory	Rd <- (RAMPZ:Z)									3
ELPM	Rd, Z+	Extended Load Program Memory and Post-Inc	Rd <- (RAMPZ:Z), RAMPZ:Z <- RAMPZ:Z + 1									3
IN	Rd, P	In Port	Rd <- P									1
LD	Rd, X	Load Indirect	Rd <- (X)									2
LD	Rd, X+	Load Indirect and Post-Inc.	Rd <- (X), X<- X + 1									2
LD	Rd, -X	Load Indirect and Pre-Dec.	X<- X - 1, Rd <- (X)									2
LD	Rd, Y	Load Indirect	Rd <- (Y)									2
LD	Rd, Y+	Load Indirect and Post-Inc.	Rd <- (Y), Y<- Y + 1									2
LD	Rd, - Y	Load Indirect and Pre-Dec.	Y<- Y - 1, Rd <- (Y)									2
LD	Rd, Z	Load Indirect	Rd <- (Z)									2
LD	Rd, Z+	Load Indirect and Post-Inc.	Rd <- (Z), Z<- Z + 1									2
LD	Rd, -Z	Load Indirect and Pre-Dec.	Z<- Z - 1, Rd <- (Z)									2
LDD	Rd, Y + q	Load Indirect with Displacement	Rd <- (Y + q)									2
LDD	Rd, Z + q	Load Indirect with Displacement	Rd <- (Z + q)									2
LDI	Rd, K	Load Immediate	Rd <- K									1
LDS	Rd, k	Load Direct from SRAM	Rd <- (k)									2
LPM		Load Program Memory	R0 <- (Z)									3
LPM	Rd, Z	Load Program Memory	Rd <- (Z)									3
LPM	Rd, Z+	Load Program Memory and Post-Inc.	Rd <- (Z), Z<- Z + 1									3
MOV	Rd, Rr	Move Between Registers	Rd <- Rr									1
MOVW	Rd, Rr	Copy Register Word	Rd+1:Rd <- Rr+1:Rr									1
OUT	P, Rr	Out Port	P <- Rr									1
POP	Rd	Pop Register from Stack	Rd <- STACK									2
PUSH	Rr	Push Register onto Stack	STACK <- Rr									2
SPM		Store Program Memory	(Z) <- R1:R0									-
ST	X, Rr	Store Indirect	(X) <- Rr									2
ST	X+, Rr	Store Indirect and Post-Inc.	(X) <- Rr, X<- X + 1									2
ST	-X, Rr	Store Indirect and Pre-Dec.	X<- X - 1, (X) <- Rr									2
ST	Y, Rr	Store Indirect	(Y) <- Rr									2
ST	Y+, Rr	Store Indirect and Post-Inc.	(Y) <- Rr, Y<- Y + 1									2
ST	-Y, Rr	Store Indirect and Pre-Dec.	Y<- Y - 1, (Y) <- Rr									2
ST	Z, Rr	Store Indirect	(Z) <- Rr									2
ST	Z+, Rr	Store Indirect and Post-Inc.	(Z) <- Rr, Z<- Z + 1									2
ST	-Z, Rr	Store Indirect and Pre-Dec.	Z<- Z - 1, (Z) <- Rr									2
STD	Y + q, Rr	Store Indirect with Displacement	(Y + q) <- Rr									2
STD	Z + q, Rr	Store Indirect with Displacement	(Z + q) <- Rr									2
STS	k, Rr	Store Direct to SRAM	(k) <- Rr									2

Appendix G

AVR Instruction Set Summary

The complete data on the AVR processors can be found at Atmel's website: http://www.atmel.com

AVR Instruction Set Standard Nomenclature:

Status Register (SREG)	Symbols	Registers and Operands:	I/O Registers
C: Carry Flag	+ Add	Rd: Destination (and Source) Register	RAMPX,RAMPY,RAMPZ: Registers concatenated with X,Y, and Z to form indirect addresses > 64K
Z: Zero Flag	& AND	Rr: Source Register	
N: Negative Flag	<- Assignment	R: Result from instruction execution	
V: Twos complement overflow indicator	= Equal	K: Constant literal or data byte (8 bit)	
S: N ^ V Flags for signed tests	^ Exclusive OR	k: Constant address data for program counter	**Stack**
H: Half Carry Flag	* Multiply	b: Bit in register file or I/O register (3 bit)	
T: Transfer bit used by BLD and BST instructions	| OR	s: Bit in status register (SREG) (3 bit)	STACK: Stack for return address and pushed registers
I: Global interrupt enable/disable flag	- Subtract	X,Y,X: Indirect address register	
		P: I/O port address	
		q: Displacement for direct addressing (6 bit)	SP: Stack Pointer to the STACK

Mnemonics	Operands	Description	Operation	SREG Z	C	N	V	H	S	T	I	#Clocks
BIT AND BIT-TEST INSTRUCTIONS												
ASR	Rd	Arithmetic Shift Right	Rd (n) <- Rd(n + 1), n =0..6	•	•	•	•					1
BCLR	s	Flag Clear	SREG(s) <- 0		SREG							1
BLD	Rd, b	Bit load from T to Register	Rd(b) <- T									1
BSET	s	Flag Set	SREG(s) <- 1		SREG							1
BST	Rr, b	Bit Store from Register to T	T <- Rr(b)							•		1
CBI	P, b	Clear Bit in I/O Register	I/O (P,b) <- 0									2
CLC		Clear Carry	C <- 0		•							1
CLH		Clear Half Carry Flag in SREG	H <- 0					•				1
CLI		Global Interrupt Disable	I <- 0								•	1
CLN		Clear Negative Flag	N <- 0			•						1
CLS		Clear Signed Test Flag	S <- 0						•			1
CLT		Clear T in SREG	T <- 0							•		1
CLV		Clear Twos Complement Overflow	V <- 0				•					1
CLZ		Clear Zero Flag	Z <- 0	•								1
LSL	Rd	Logical Shift Left	Rd (n + 1) <- Rd(n), Rd (0) <- 0	•	•	•	•					1
LSR	Rd	Logical Shift Right	Rd (n) <- Rd(n + 1), Rd (7) <- 0	•	•	•	•					1
ROL	Rd	Rotate Left Through Carry	Rd (0) <- C,Rd(n + 1) <- Rd (n), C <- Rd(7)	•	•	•	•					1
ROR	Rd	Rotate Right Through Carry	Rd (7) <- C,Rd(n) <- Rd (n + 1), C <- Rd(0)	•	•	•	•					1
SBI	P, b	Set Bit In I/O Register	I/O (P,b) <- 1									2
SEC		Set Carry	C <- 1		•							1
SEH		Set Half Carry Flag in SREG	H <- 1					•				1
SEI		Global Interrupt Enable	I <- 1								•	1
SEN		Set Negative Flag	N <- 1			•						1
SES		Set Signed Test Flag	S <- 1						•			1
SET		Set T in SREG	T <- 1							•		1
SEV		Set Twos Complement Overflow	V <- 1				•					1
SEZ		Set Zero Flag	Z <- 1	•								1
SWAP	Rd	Swap Nibbles	Rd (3..0) <--Rd(7..4) <- Rd (7..4) <- Rd(3..0)									1
MCU CONTROL INSTRUCTIONS												
NOP		No Operation										1
SLEEP		Sleep	Refer to manufacturer data for Sleep function description									1
WDR		Watchdog Reset	Refer to manufacturer data for WDR/timer function description									1
BREAK		Break	For "On-Chip-Debug" only, refer to manufacturer data for description									N/A

Answers to Selected Exercises

CHAPTER 1

*2. Create an appropriate declaration for the following:

A. A constant called "x" that will be set to 789.

Answer:

```
flash int x = 789;
    or
const int x = 789;
    or
const int flash x = 789;
```

Notes: *flash* and *const* are considered the same type of memory in non-volatile space. An integer (int) value is declared in this case because it is a 16-bit number (+/– 32767 or 0–65535) and is the next larger size variable type from a character (char or unsigned char), which is only 8 bits (+/–128 or 0–255).

B. A variable called "fred" that will hold numbers from 3 to 456.

Answer:

```
unsigned int fred;
        or
int fred;
```

Note: In this case, the numbers are always positive so int or unsigned int will work, but the range of the number is such that it will not fit into a char or unsigned char.

C. A variable called "sensor_out" that will contain numbers from –10 to +45.

Answer:

```
signed char sensor_out;
```

Note: In this case the numbers will fit into a character (+/– 127).

D. A variable array that will have 10 elements each holding numbers from –23 to 345.

Answer:

```
int array[10];
```

Note: In this case, the numbers are such that they will not fit into a char because they can exceed 128 (8 bits).

E. A character string constant that will contain the string "Press here to end".

Answer:

```
flash char press_string[] = "Press here to end";
                or
const char  flash press_string[] = "Press here to end";
                or
const char  press_string[] = "Press here to end";
```

Notes: *flash* and *const* are considered the same type of memory in non-volatile space. All three of these declarations will result in the same generated code.

F. A pointer called "array_ptr" that will point to an array of numbers ranging from 3 to 567.

Answer:

```
unsigned int *array_ptr;
```

Notes: When declaring the type of a pointer, it is not the size of the pointer but what it is pointing to that is of concern. In this case, the pointer is pointing to values that are larger than 8 bits but will easily fit into a 16-bit integer.

G. Use an enumeration to set "uno", "dos", "tres" to 21, 22, 23, respectively.

Answer:

```
enum { uno=21, dos, tres };
```

Notes: The default for an enumeration starting label value is 0. So the first must be assigned to the desired starting value and the subsequent values will follow.

*4. Evaluate as true or false as if used in a conditional statement:

For all problems: x = 0x45; y = 0xc6

A. (x == 0x45)

Answer:

```
TRUE
```

B. (x | y)

Answer:

```
TRUE  (since 0x45 | 0xC6 = 0xC7, and 0xC7 is not zero)
```

C. (x > y)

Answer:

```
FALSE
```

D. (y – 0x06 == 0xc)

Answer:

```
FALSE   (since 0xC6 - 0x06 = 0xC0, and that is not equal to 0x0C)
```

*6. Evaluate the value of the variables after the fragment of code runs:

```
unsigned char cntr = 10;
unsigned int value = 10;

do
{
        value++;
} while  (cntr < 10);
             // value = ??        cntr = ??
```

Answer:

```
value = 11       cntr = 10
```

Notes: The variable *value* gets incremented, then the value of *cntr* is tested but not modified.

*10. Write a fragment of C code to declare an appropriate array and then fill the array with the powers of 2 from 2^1 to 2^6.

One possible solution:

```
unsigned char twos[6];  /* size array to hold the 6 figures.. */
char x,y;               /* declare a couple of indexes */

y = 0;
for(x = 1; x!=0x80; x<<=1) /* shift a one left until it is 2^7 */
  twos[y++] = x;    /* store value into array and increment index */
```

CHAPTER 2

*2. Describe the following memory types and delineate their uses:

A. FLASH Code Memory

B. Data Memory

C. EEPROM Memory

Partial answer: EEPROM memory is non-volatile memory used to store data that must not be lost when the processor loses power. It has a limited number of possible write-cycles and so is usually not used for variable storage.

*6. Write a fragment of C language code to initialize External Interrupt 1 to activate on a falling edge applied to the external interrupt pin.

Answer:

```
GIMSK = 0b1x000000;   //enable INT 1
MCUCR = 0bxxxx10xx; //set INT1 to react to a falling edge
```

Note: *x* is not an acceptable digit, but is used to show that the *x* bits do not affect the problem. Only those bits set to *1* or *0* actually affect the answer.

*8. Write a fragment of C language code to initialize the Port C pins so that the upper nibble can be used for input and the lower nibble may be used for output.

Answer:

```
DDRC = 0x0f;        //initialize Port C
```

*11. Sketch the waveform appearing at the output of the USART when it transmits an "H" at 9600 baud. The sketch should show voltage levels and the bit durations in addition to the waveform.

Answer: From the ASCII table we find that the code for "H" is 0x48 and the time-per-bit is 1/9600 = 104 microseconds. Each column below is one bit and is 104 microseconds wide.

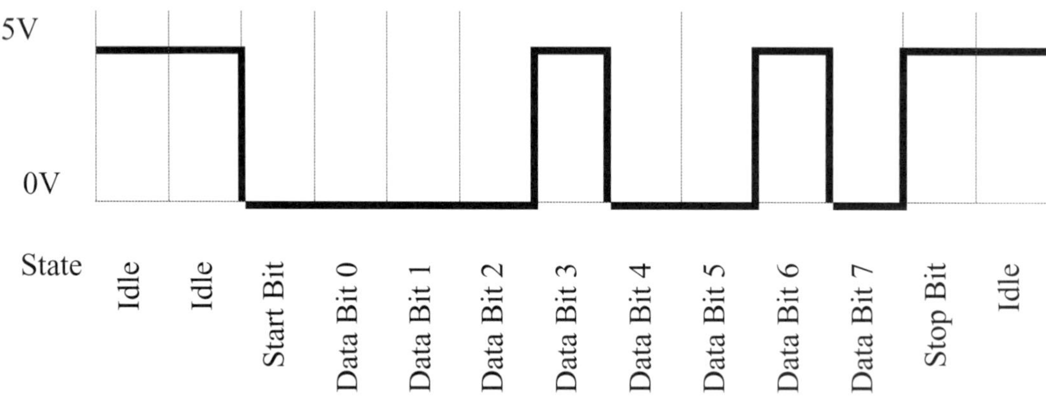

*13. Compute the missing values to complete the following table relating to analog to digital conversion:

V_{in}	$V_{fullscale}$	**Digital Out**	**# of bits**
4.2V	10V	107 or 0x6B	8
1.6V	5V		10
	5V	123	10
	10V	223	8

Answer to line 1 (the rest are up to you):

$$\frac{V_{in}}{V_{fullscale}} = \frac{X}{2^n - 1}$$

$$\frac{4.2V}{10V} = \frac{X}{2^8 - 1}$$

$$\frac{4.2(2^8 - 1)}{10V} = X = 107_{10} = 0x6B$$

Note: Because of integer math limits, any decimal portion is dropped from the answer.

CHAPTER 3

*1. Write a macro using the #define directive to alias the function *putchar()* with a function named send_the_char().

Answer:

```
#define send_the_char(a) putchar(a)
```

*4. Write a function that prints its compile date and time. Use the compiler's internal tag names to get the date and time values.

Answer:

```
putsf("\n\rCompile Date :\r");
putsf(__DATE__);
putsf("\rCompile Time : \r");
putsf(__TIME__);
```

Notes: The function *putsf()* prints a flash-based string (constant string) to the standard output.

*6. Write a function that inputs a 16-bit hexadecimal value and then prints the binary equivalent. Note: there is no standard output function to print binary—so it is all up to you!

One possible solution:

```
void put_bin16(unsigned int value)
{
     int k;                    // working counter
     for(k=0; k < 16; k++)     // do 16 digits
     {
          if(value & (1<<k))   // if 2^k then print a '1'
          putchar('1')
     else
          putchar('0');        // else print a '0'
     }
}
```

Notes: In this example, a 1 is shifted left, creating a mask to test each bit position, for all 16 positions of the integer.

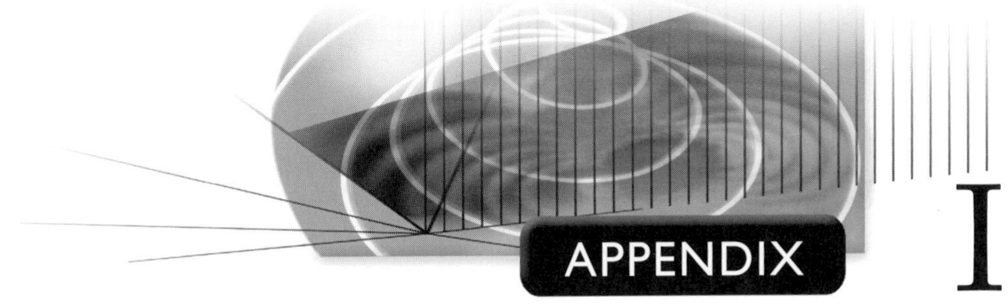

I

A "Fast Start" to Embedded C Programming and the AVR

This appendix is a tutorial example showing how to enter, compile, and download a program using CodeVisionAVR, the AVRBL Bootloader Windows® Application, and the MegaAVR-DEV board. The methods described here may be used to enter and test all of the code examples and exercises from this text.

1. INSTALLING CODEVISIONAVR

Locate the CodeVisionAVR C Compiler installation on the included CD in the "CODEVISION EVAL" directory. Run **setup.exe** to install. Install the CodeVisionAVR compiler into the default directory: "C:\cvavr". If there is an old version of CodeVisionAVR present, you may install the new one over it. Also on the first run, under Windows® NT4, 2000, or XP, you must use **Administrator** privileges. On subsequent runs, **Power User** privileges may be also used.

2. INSTALLING THE AVRBL WINDOWS APPLICATION

Locate the AVRBL application setup on the included CD in the "AVRBL APP" directory. Run **setup.exe** to install. Install the AVRBL application into the default directory: "C:\Program Files\AVR Bootloader\AVR Bootloader.exe". To properly install the AVRBL application under Windows® NT4, 2000, or XP, you must have **Administrator** privileges.

3. SETTING UP A CODEVISIONAVR PROJECT

First, create a folder in the "C:\cvavr\" directory called "test". Your resulting path will be "C:\cvavr\test".

Open the CodeVisionAVR application and create a new file. New projects are created with the **File|New** menu command or by clicking the **Create New File** button on the toolbar. These actions open the Create New File dialog box as in Figure I–1, requesting that you specify the type of file to open. Select the **Project** option and click **OK.** A second dialog box asks you to confirm use of the CodeWizardAVR to create the new project. This dialog box is depicted in Figure I–2.

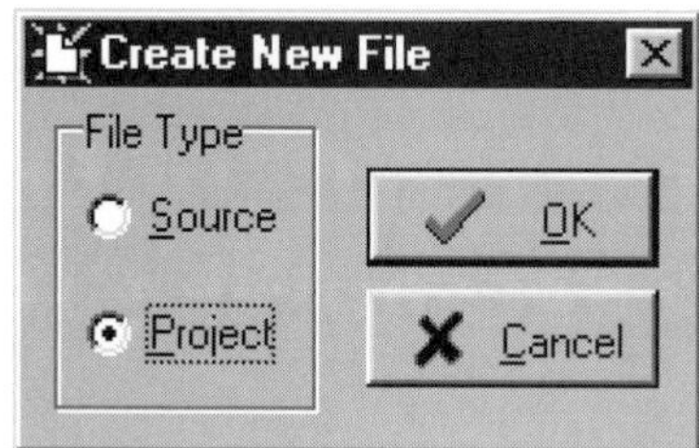

Figure I–1 *Create New File Dialog Box*

To create a project without using the code wizard, select **No** when prompted to use it.

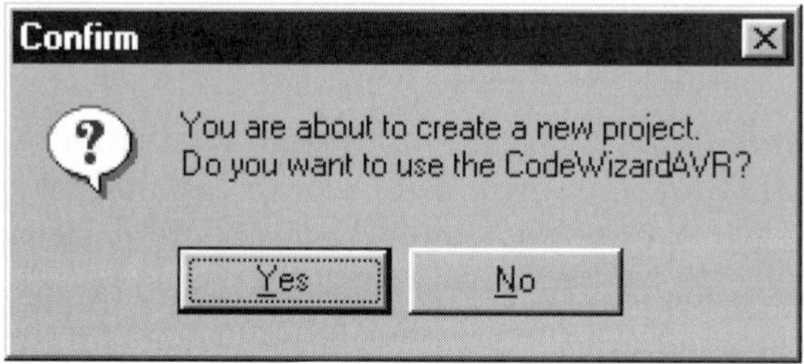

Figure I–2 *CodeWizardAVR Confirm Dialog Box*

A final dialog box prompts you to specify a project file name and location. Select the "C:\cvavr\test" folder that you created earlier, and place the project in it, named "test".

When these steps are completed, CodeVisionAVR creates a new project and opens the **Configure|Project** dialog box for the new project. This dialog box has several tabbed frames of options that allow you to customize the project for your specific application, as shown in Figure I–3.

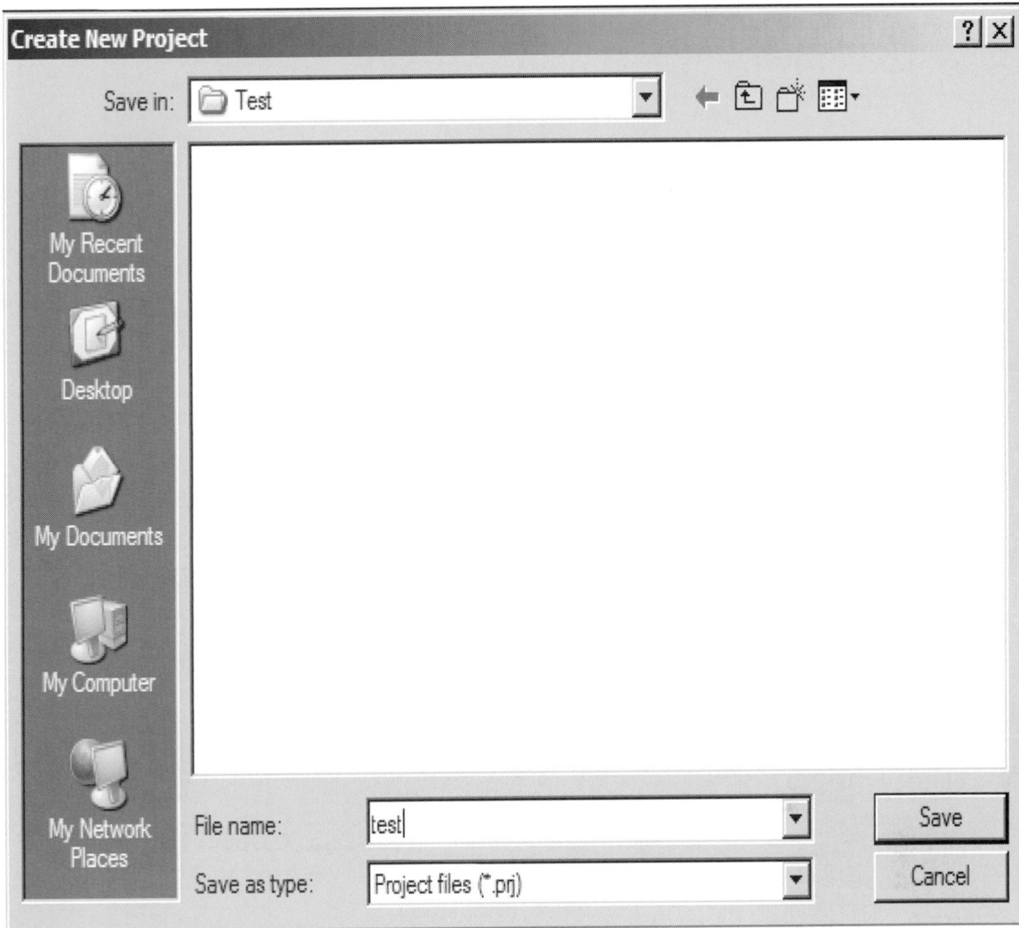

Figure I–3 *CodeVisionAVR after Creating a New Project*

The project is configured with the **Project|Configure** menu command or the **Project Configure** toolbar button. Selecting either of these opens the Configure Project dialog box. There are three tabbed frames in this dialog box: Files, C Compiler, and After Make. The C Compiler tab, shown in Figure I–4, allows you to set properties for the project pertaining to the target device and the executable program file.

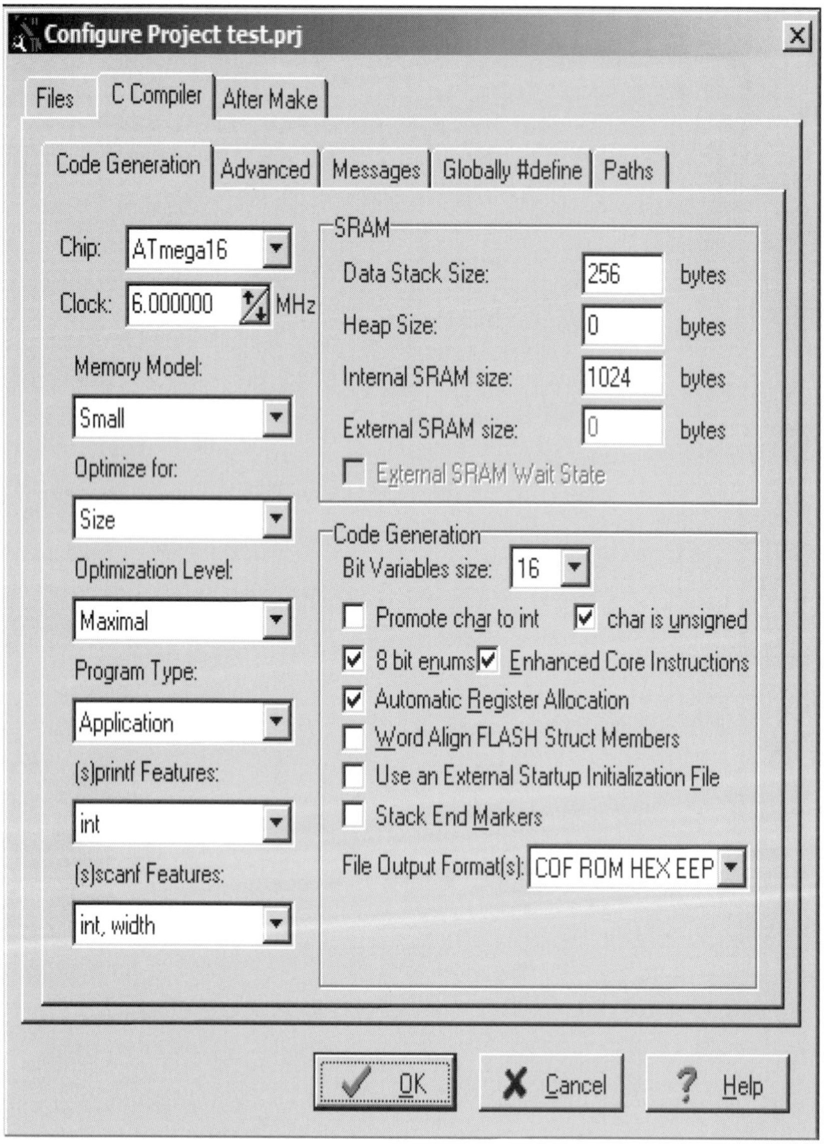

Figure I–4 *Configure Project Dialog Box*

In this dialog, make the following selections under the **Code Generation** tab:

```
Chip:    ATmega16
Clock:   6.000000 MHz
```

All the other settings should be left in the default condition. Click **OK.**

Next, select **File|New** from the menu and **Source** from the pop-up window, as shown in Figure I–5.

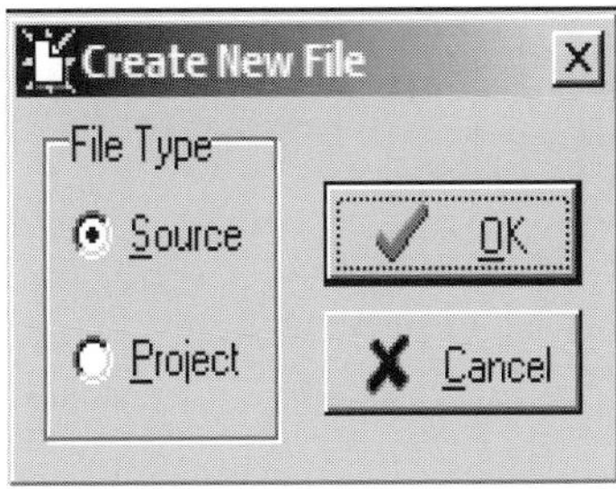

Figure I–5 *New File Type Pop-Up Window*

This will create a file "Untitiled.c" in the CodeVisionAVR workspace. Save this empty file as "test.c" in your "C:\cvavr\test" project directory.

Now, going back to the **Project|Configure** menu, bring up the Configuration dialog box. Under the Files tab, click the Add button and select the "test.c" file you added to your project directory. The resulting window should appear as it does if Figure I–6:

Figure I–6 *Project Configuration—Files Dialog*

Click OK.

4. WRITING SOME TEST CODE

Enter the code below. It will set up the USART at 9600 baud and send a message every ½ second out to the PC. You will notice that it really does not take much code to make things happen.

```
#include <MEGA16.h>
#include <stdio.h>
#include <delay.h>

main()
{
        // USART Baud rate: 9600
        UCSRA=0x00;
        UCSRB=0x08;
        UCSRC=0x86;
        UBRRH=0x00;
        UBRRL=0x26;

        while(1)
        {
                printf("I'm a C coding Genius!!\n\r");
                delay_ms(500);
        }
}
```

5. COMPILING

Select **Project|Make** to compile your code. You should get a resulting pop-up window titled "Information," as shown in Figure I–7:

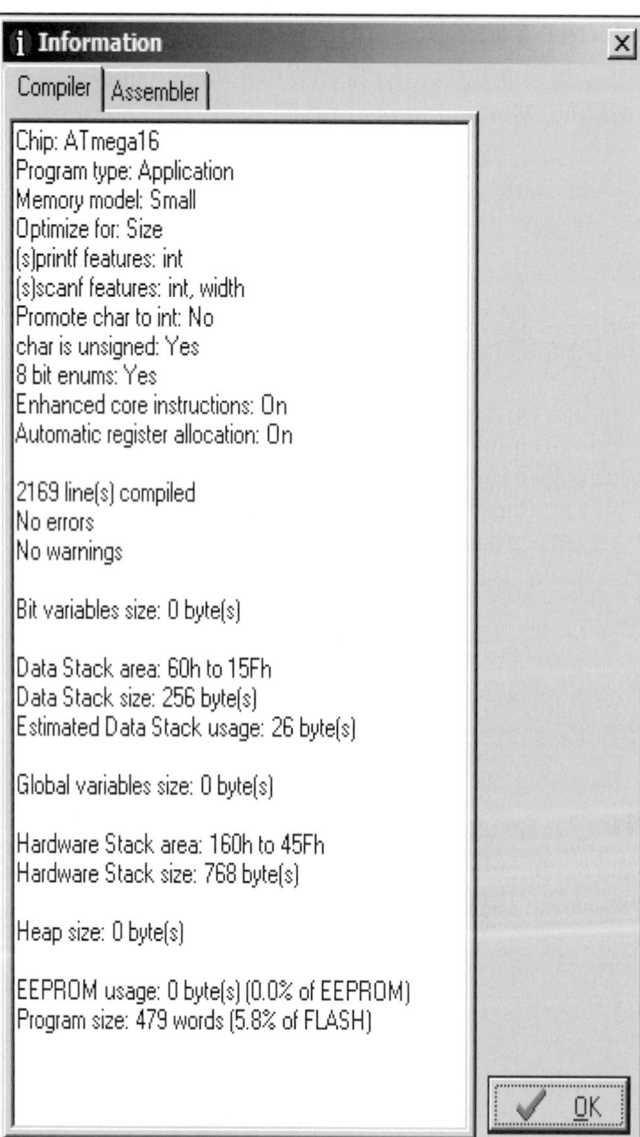

Figure I–7 *Compile Information Pop-Up Window*

If there are any errors or warnings, you should go back and review the code that you entered and make sure it matches the code in Section I–4. If it is correct and there are still errors, check the project settings and make sure that they match the settings called out previously.

6. DOWNLOADING TO THE MEGAAVR-DEV BOARD

The MegaAVR-DEV board has a serial connection that can be used to download code using the AVRBL Windows application. P1 is a standard DB-9 connector usually used to connect to a PC. JP10 and JP12 are jumpers which connect the processor serial signals (RXD and TXD) to/from the RS-232 driver chip. These jumpers must be in place for the RS-232 serial connection to work.

Open the AVRBL Windows application.

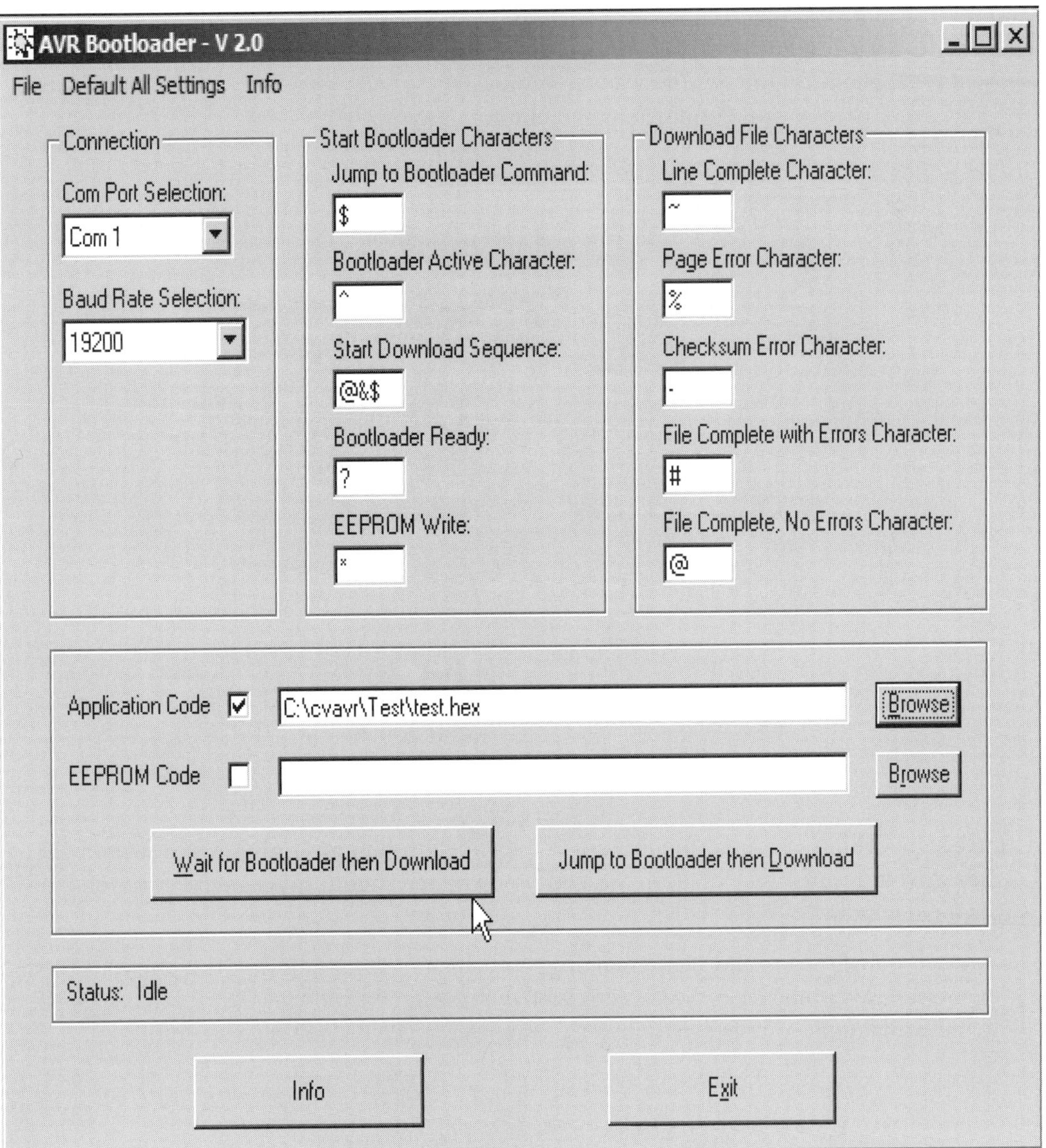

Figure I–8 *AVRBL Windows Application Main Screen*

Set the **Comm Port Selection,** shown in Figure I–8, for the appropriate port to your PC connection. Set the **Baud Rate Selection** to **19200.** Browse the **Application Code** and select your "C:\cvavr\test\test.hex", generated by the compiler. Apply power to your board according to the MegaAVR-DEV datasheet. Click on the Wait for Bootloader then Download button on the AVRBL Windows application, then press the reset button on your MegaAVR-DEV board. The status bar should indicate that the program is being downloaded to FLASH. When it is complete, click **OK.**

7. THE THRILL OF VICTORY

Return to the CodeVisionAVR IDE. The **Settings|Terminal** menu will bring up a dialog box, as shown in Figure I–9. Set up the terminal **Baud rate** for 9600 baud and the appropriate **Port** port to your connection, and click **OK.**

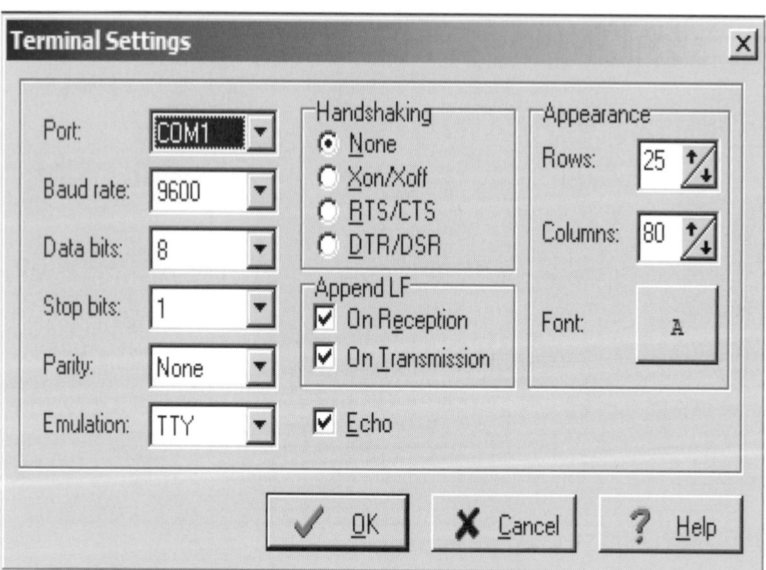

Figure I–9 *Terminal Settings Dialog Box*

Select the **Tools|Terminal** menu item. This will bring up a terminal window and should display your results. The line "I'm a C coding Genius!!" should appear every ½ second in the terminal window.

You can now go back and play, changing, embellishing, and tailoring the program to your own taste. Remember that the CodeVisionAVR terminal window will need to be closed before downloading your updated program using AVRBL to avoid a port conflict.

Good luck, and may the force be with you.

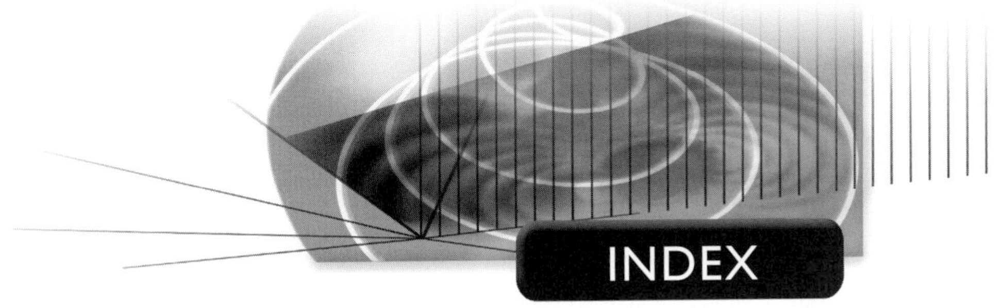

SYMBOLS

A

B

C

D

E

K

L

Q

R

V

W